Springer Series in Materials Science

Volume 329

The Springer Series in Materials Science covers the complete spectrum of materials research and technology, including fundamental principles, physical properties, materials theory and design. Recognizing the increasing importance of materials science in future device technologies, the book titles in this series reflect the state-of-the-art in understanding and controlling the structure and properties of all important classes of materials.

Walter Arnold · Klaus Goebbels · Anish Kumar

Non-destructive Materials Characterization and Evaluation

Springer

Walter Arnold
Department of Material Science
and Engineering
Saarland University
Saarbrücken, Germany

Klaus Goebbels
Saarbrücken, Germany

Anish Kumar
Metallurgy and Materials Group,
Indira Gandhi Centre for Atomic Research
A CI of Homi Bhaba National Institute
Kalpakkam, India

ISSN 0933-033X ISSN 2196-2812 (electronic)
Springer Series in Materials Science
ISBN 978-3-662-66487-2 ISBN 978-3-662-66489-6 (eBook)
https://doi.org/10.1007/978-3-662-66489-6

This Springer imprint is published by the registered company Springer-Verlag GmbH, DE, part of Springer Nature.
The registered company address is: Heidelberger Platz 3, 14197 Berlin, Germany

This book is dedicated to the memory of Dr. Paul Höller and Dr. Baldev Raj, both pioneers in the sciences of non-destructive testing (NDT), non-destructive materials characterization (NDMC), and their industrial applications. Professor Höller was founding director of Fraunhofer IZFP, Saarbrücken, Germany and Dr. Raj, group leader, department head, and eventually Director of the Indira Gandhi Centre for Atomic Research (IGCAR) in Kalpakkam. Both knew each other and initiated Indo-German collaborative projects financed by the International Bureau of the DLR in Cologne in the field of NDT and NDMC that supported many exchange visitors from BARC and IGCAR to Fraunhofer IZFP and vice versa, also two of us (WA and AK). These projects were extended to other institutions in India.

Preface

Parallel to Non-destructive Testing (NDT) for the detection and evaluation of defects in materials and components, Non-destructive Materials Characterization (NDMC) techniques have been developed to quantify micrometer-sized features in the materials microstructure. These developments include acoustic, micro-magnetic, optical, X-ray and other methods. In most NDMC-methods, one generates correlations between the destructively determined microstructure and mechanical properties and non-destructively determined parameters. Based on these correlations, calibrations are made to determine the materials properties by non-destructive tests in components, at specific points and—if of interest—repeatedly over the time of use. By these procedures, NDMC allows to enhance the productivity during fabrication and in-situ monitoring for reliable operation to detect any potential changes.

The book discusses methods and equipment for NDMC for metals, polymers, elastomers, glasses, composite materials and describes their physical and technical base. It is written for engineers in R&D, design, production, quality assurance and non-destructive evaluation. The relevance of NDMC is to achieve higher reliability, safety and productivity for both monitoring production processes and for in-service inspections.

Saarbrücken, Germany — Walter Arnold
Saarbrücken, Germany — Klaus Goebbels
Kalpakkam, India — Anish Kumar

Acknowledgements

We are very grateful to Birgit Conrad-Markschläger and to Oliver Sandmeyer, librarians at the Fraunhofer IZFP, for supporting us in literature searches and finding many references which were often difficult to unearth, in particular conference proceedings. Their help was invaluable. One of us (WA) thanks Dipl. Ing. Jörg Köhncke for his efficient advice when problems arose with software and computer hardware.

Furthermore, we thank Dipl. Ing. Michael Becker, Dr. Michael Maisl, both from Fraunhofer IZFP, Prof. Joseph A. Turner, Department of Mechanical Engineering, University of Nebraska at Lincoln, Prof. Rainer Birringer, Physics Department, Saarland University, Dr. Martin Spies, Baker Hughes, Stutensee, and Prof. David C. Jiles, Department of Material Science, Iowa State University, for a critical reading of the various chapters of the manuscript. AK thanks colleagues at NDE Division, IGCAR for providing simulations and experimental data for the book. Finally, we thank the colleagues from our institutions, Fraunhofer IZFP, Saarland University and IGCAR for their fruitful collaboration on the subjects discussed in this book. We thank also our Ph.D., master and exchange students for their enthusiasm when working on their theses and study projects. Last but not least, we thank the staffs of Springer Nature for their support and patience when writing this book.

Introduction

Materials are used for components in applications with specific boundary conditions: mechanical loads, thermal loads, corrosive environments, radiation attacks and others. In most cases, several types of loads occur simultaneously. To withstand these loads over the lifetime of the components, the materials out of which they are made of must have appropriate properties. In fact, during the centuries of technical, technological and industrial developments, a limited number of materials properties were found and defined, which are important for the material's fitness for purpose: e.g., chemical composition, elasticity, strength, toughness, hardness, fatigue and creep behavior. These properties to a large extent governed or are determined by the material's microstructure.

In general, the required materials properties are evaluated by destructive tests on coupon basis before the material is brought into use. They are carried out at test specimens taken from the production process whereas the product may not be analyzed with respect to the required properties. In the case that the product is examined by non-destructive testing (NDT) techniques, they are focused on the detection of defects like cracks, voids and inclusions. These procedures on components are in use either at the production stage or at recurring tests during operation. Examples of NDT techniques for volumetric evaluation are X-ray inspection, dating back to Röntgen's inspection of his hunting rifle, shortly after the discovery of X-rays in 1895 and ultrasonic defect testing, continuously developed and widely used since the 1950s. For more than 100 years, non-destructive testing has evolved as an important and effective quality control tool, using X-rays, ultrasound, magnetic and electromagnetic testing methods, penetrant testing and many others. They are used worldwide, and they grew out of research and technology developments, often in specialized R&D centers like the Fraunhofer IZFP (short for "Fraunhofer Institute for Non-destructive Testing") in Germany, the NDE Center in Ames, Iowa or the Indira Gandhi Center for Atomic Research (IGCAR) in Kalpakkam, India, just to name a few. The applications are controlled by regulations and norms. Out of these efforts, non-destructive materials characterization and evaluation has emerged as a separate technology. Because NDMC grew as a spin-off from NDT, the boundary between the two is not sharply defined.

The physical background and the understanding of the tools for NDMC have been continuously developed during the past in parallel with applications. At various conferences devoted to NDT (e.g., the series of World conference on NDT), it became obvious, that NDT-methods for defect detection may also be applied for the evaluation of material properties. For example, low-frequency ultrasonic signals of about 2 MHz with a corresponding wavelength of 3 mm in steel are reflected at flaws of comparable and larger dimensions. Ultrasonic signals at higher frequencies, approximately 20 MHz with a wavelength of 0.3 mm in the same material will be scattered considerably by the grain-structure, allowing one to determine its grain size, which correlates with the yield strength (Kumar et al., 1999). Although numerous books, handbooks and standards are devoted to describe in-depth NDT techniques for defect detection and characterization, one book appeared on NDMC, written 25 years ago by one of us (Goebbels, K., 1994. Materials Characterization for Process Control and Product Conformity; CRC Press.). The present book is partially based on this text extending it considerably. Furthermore, a number of conference proceedings deal with NDMC, in particular the series on Non-destructive Materials Characterization. The corresponding conferences were organized by R. E. Green and now by S. Kenderian and B. Djordjevic with varying co-organizers and editors, with the recent proceedings appearing with Springer.

A dedicated and comprehensive book on NDMC is very appropriate now as (i) in the last few decades NDMC has grown to a very high level of maturity through theoretical, simulation and experimental studies; (ii) however, the information is scattered in many research papers and a few review papers, and (iii) with the rapid development in electronics and computers, the experimental and characterization tools have changed dramatically in the past few decades in all the fields including NDMC.

Our book focuses on NDMC because the technology has developed to such an extent that for many materials, especially for metals, their properties can be determined during production and over their lifetime. Applying such methods will lead to a significant improvement of quality, safety and reliability of products and will help to reduce expensive destructive testing of specimens. Finally, we hope to give some impetus to push forward NDMC procedures like continuous process-control and structural health monitoring.

The emphasis of our book is on the non-destructive determination of a material's properties and state, whether it is fit for applications under specified conditions. These properties may comprise

- Chemical composition,
- Density and porosity,
- Microstructural parameters, such as grain size and volume fraction of constituent phases,
- Internal stresses and texture,
- Elastic parameters such as Young's modulus, shear modulus and Poisson's ratio,
- Mechanical properties such as yield strength, tensile strength, hardness, ductility and toughness.

Some of the NDMC procedures described already belong to the state of the art, e.g., hardness-testing with eddy-current probes and determination of the chemical composition by X-ray-fluorescence analysis. Some are less common, like the use of magnetic Barkhausen emission for determining mechanical strength parameters, or ultrasonic scattering to evaluate the microstructure of polycrystalline metals.

The acceptance of NDMC techniques instead of destructive tests of material properties depends on the maturity of the technology, but also on the level of understanding between company divisions responsible for design, purchasing, controlling, production and quality control including independent inspection authorities. The authors want to contribute to this effort. Therefore, the book is directed for all levels of management. If NDMC is applied properly, it may lead to huge cost reductions and enhanced reliability of components.

Some of the NDMC-methods lead directly to the wanted material property, e.g., the Poisson's ratio by measuring the longitudinal and shear-wave velocity. Others like Barkhausen emission measurements require calibration procedures to determine yield strength. The corresponding calibration techniques will be covered as well. The book's content is not restricted to metals. NDMC techniques for ceramic, polymer and composite materials will be described too. Finally, we discuss future perspectives of NDMC.

Gases, liquids or solids consist of atoms and molecules. Atoms in solids are arranged in crystalline structures like body centered cubic, face centered cubic, hexagonal or in many other crystalline lattice or in amorphous phases or in combination of both. In polymers, elastomers and composites, macromolecules built up the structure. In the gaseous and in liquid phases, atoms and molecules can move more or less freely, but collide with each other continuously. In the solid state, the atoms are bound in a lattice with binding energies determined by the interatomic potentials. The interatomic forces determine the elastic properties which govern the propagation of ultrasonic waves and hence at least the ultrasonic part of NDMC.

While in glasses, polymers and elastomers, the amorphous microstructure extends up to macroscopic dimensions, in polycrystalline materials the size of the grains are the limiting boundaries, playing an important role in particular for the mechanical properties. The grains as small single crystals with their spatial orientation are embedded into the ensemble of the other grains with different orientations. The grain boundaries, mismatched in respect to the orientation of different grains, result in free space for vacancies, and for foreign atoms/molecules, thus creating local compressive or tensile micro-stresses, which may act as sources of stationary or mobile dislocations under external load.

Electric, magnetic, thermal and optical properties of crystalline solids are determined by the so-called elementary excitations, i.e., electrons, magnons, phonons, excitons, polaritons, dislocations and many others. They influence elasticity, hence sound velocity as well as internal friction and ultrasonic absorption up to very high frequencies. Likewise, the real and imaginary parts of the optical conductivity are determined by elementary excitations. Understanding these properties of solids requires quantum mechanical principles. Some of these texts are classics despite that their first editions appeared decades ago at the time of the rapid developments in electronics and optics. Material science enters when these solid properties have to be incorporated in devices. Then, properties like strength, toughness, fatigue resistance and so on play a decisive role, so that these devices can be used reliably and reach their "fitness for purpose." Here, in this book we discuss the use of ultrasonic propagation in solids and engineering materials, of electromagnetic properties and X-ray radiation to monitor the fitness for purpose of technical components. The parameters to be monitored are strength, toughness, fatigue resistance, creep damage, mechanical stresses and many other parameters. This task of non-destructive material characterization (NDMC) renders its double role visible. It is a discipline of materials science based to a large extent on principles of solid-state physics.

In a generalized term, NDMC comprises of exposing the test material to some form of energy (elastic, magnetic, electromagnetic, radiation, etc.) and studying the interaction of the energy with the material microstructure or response of the material to the imposed energy through NDMC sensors without causing any damage or degradation to the material under investigation. Almost all forms of energies and all physical properties of materials have been exploited in NDMC. The underlying principles of NDMC can be understood from Fig. 1. Composition and crystal structure in combination with thermal/thermomechanical treatments and exposure to service conditions govern the material microstructure. Microstructure can be visualized directly by microscopy techniques such as optical microscopy, scanning electron microscopy (SEM), transmission electron microscopy (TEM), scanning acoustic microcopy (SAM), and atomic force microscopy (AFM). The microstructure affects both mechanical and physical properties. Mechanical properties are measured conventionally using destructive testing techniques, whereas measurement of physical properties using a non-destructive method is the basis of NDMC. The NDMC parameters are correlated with a microstructural or mechanical property parameter measured on a calibration sample or are based on known physical laws. The established correlation along with NDMC measurement on an unknown sample/component is used to unravel the microstructure and mechanical property of the component.

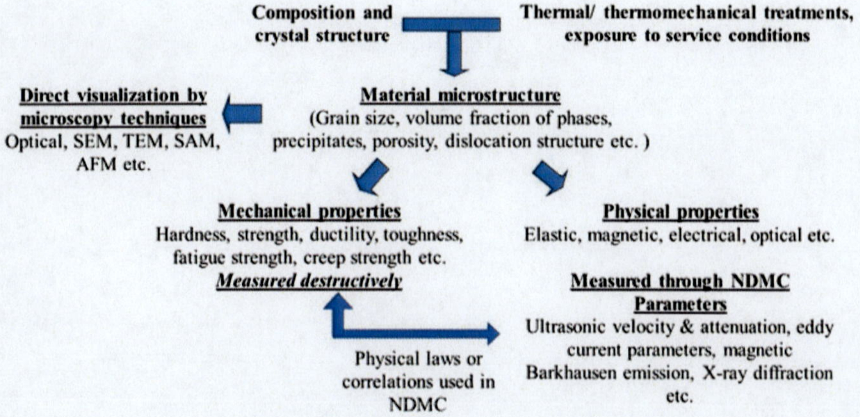

Fig. 1 Principles of non-destructive materials characterization (NDMC)

Contents

Chapter 1
Ultrasonic Non-destructive Materials Characterization

Abstract The propagation of ultrasonic waves through materials is essentially governed by its elastic and anelastic properties. Following the basics of interatomic forces and elasticity, ultrasonic velocities in single crystals and polycrystalline materials are discussed in chapter 1 in detail, providing means for materials characterization using ultrasonic velocity measurements. The third-order elastic constants in polycrystalline materials describe the acoustoelastic constants which form the basis for the evaluation of elastic stresses in materials. Stresses lead to anisotropic velocities and so do textures. It is discussed how these can be separated. The anisotropy of the constituents of the microstructure of a material leads to ultrasonic scattering and therefore to attenuation of the propagating waves, whereas the anelasticity in lattice structure leads to internal friction. Both effects are discussed with various examples. Fatigue and creep manifest themselves in the parameters describing ultrasonic propagation and hence can be exploited for non-destructive materials characterization (NDMC). Thus, in chapter 1, we emphasize the principles forming the basis for NDMC using ultrasonics and demonstrate their applications through suitable examples.

1.1 Interatomic Forces and Elasticity

In order to understand the basic elastic properties of a solid, one has to have a closer look to the binding forces between the atoms within a solid. There are the attractive forces and the repulsive forces. Attractive interactions are caused by Van der Waals forces, electro-static or ionic forces, overlap of wave-functions in the covalent bonding, and the metallic bonding due to the interaction between the atomic core and the electrons. Repulsive forces are caused by quantum mechanical effects. The details of these bonding types are discussed in many textbooks in material science and solid state physics, see for example [73, 129, 300].

W. Arnold et al., *Non-destructive Materials Characterization and Evaluation*, Springer Series in Materials Science 329,
https://doi.org/10.1007/978-3-662-66489-6_1

For the purpose of this book, it is sufficient to consider the Lennard–Jones potential as a representative of various potential forms [162]. It is given by

$$V(r) = 4E_0\left[\frac{1}{4}\left(\frac{r_0}{r}\right)^{12} - \frac{1}{2}\left(\frac{r_0}{r}\right)^{6}\right] \tag{1.1}$$

In (1.1), E_0 is the dissociation energy for a pair of atoms and r_0 is their equilibrium position, i.e. the interatomic distance. Magnitudes are $E_0 \approx 10^{-21}$ J and $r_0 \approx 0.5$ nm, respectively. The forms of the potential and its derivative which is the force between two atoms $F = dV \big/ dr$ are shown in Fig. 1.1a, b.

Often the interatomic forces are approximated by springs with the spring constant k_s (Fig. 1.1b). The corresponding potential is $V = k_s(\Delta a)^2/2$ where Δa is the extension of the spring. In order that such a potential reflects the potential of (1.1) as accurate as possible, it must be fitted to the interatomic potential at the equilibrium position r_0. This entails that not only the equilibrium position r_0 must be the same, but also their slopes and their curvatures. In Fig. 1.1c, a macroscopic force F_{mac} is applied on the solid at the area A. The force acting between the atoms is $F/n = F_{mac}{r_0}^2/A$ and the strain at the atoms $\delta r/r_0$ is equal to the macroscopic strain $\delta l/l = N\delta r/Nr_0$, because $\delta r = (-\ dr/d(F(r)))\delta F$, where n and N are the numbers of atoms in each row and column, respectively. Combining these equations and using that the stress $\sigma = F_{mac}/A$ yields for the Young's modulus $E = -\ 1/r_0\ (dF(r)/dr)$. Furthermore, the force is the derivative of the potential and hence we arrive at [300]:

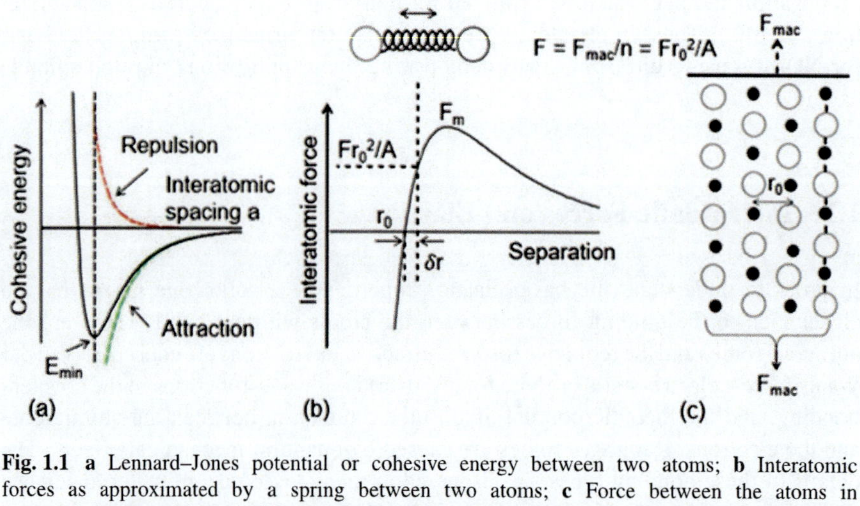

Fig. 1.1 **a** Lennard–Jones potential or cohesive energy between two atoms; **b** Interatomic forces as approximated by a spring between two atoms; **c** Force between the atoms in a crystal lattice with two different atoms. The force between the atoms is F_{mac}/n, where n is the number of rows across the area A and F_{mac} is the force applied on the solid. The interatomic potential in the equilibrium position r_0 can be approximated by a parabola, which results in a spring-like force as indicated in (**b**). The maximum force between the two atoms is F_m and the minimum in potential energy is $E_{\min}$ which corresponds to the dissociation energy (Fig. 1.1a, b in modified form from [300], with permission by Oxford Publishing Limited)

$$E = \frac{1}{r_0}\frac{\partial^2 V(r)}{\partial r^2} \tag{1.2}$$

This relation is of very general nature and can be applied for the shear modulus as well, provided it is correspondingly adapted [78].

As just discussed, the Lennard–Jones potential can be approximated by a spring potential. Deviations from this potential are taken into account by admitting higher-order elastic constants, as will be discussed later in more detail. The stored elastic energy in the solid can be described by [200]:

$$\Phi(\varepsilon) = \frac{1}{2}c_{ijkl}\varepsilon_{ij}\varepsilon_{kl} + \frac{1}{6}c_{ijklmn}\varepsilon_{ij}\varepsilon_{kl}\varepsilon_{mn} \tag{1.3}$$

Here, c_{ijkl} and c_{ijklmn} are tensors for the elastic properties of the material, also called second and third order elastic constants. For small displacements and hence strains, we obtain Hooke's law:

$$\sigma_{ij} = c_{ijkl} \cdot \varepsilon_{kl} \tag{1.4}$$

In the above equations, the summation convention holds. In one-dimensional form one obtains $\sigma = E\varepsilon$ from (1.4), in accordance with the derivation above. The SI units of the elastic constants are N/m^2 or Pa, whereas the strain ε is dimensionless. Typical strain values for ultrasonic waves are of the order of 10^{-6} leading to stress values of ~ 50–200 kPa in metals.

1.2 Elastic Constants

The elastic constants for solids extend from some GPa in polymers to $\approx$ 100 GPa in metals and to several 100 GPa in ceramics and covalently bonded crystals. For example, in diamond $c_{11} = 1075$ GPa. The large differences reflect the differences in the curvatures of the interatomic potentials and hence the bonding type, see above.

In response to small external forces F acting on a solid of area A or volume V, the solid deforms elastically and stresses develop. If there is a uniaxial strain, the stress is given by $\sigma = \varepsilon E$. If the external force induces simple shearing, the shear stress is $\tau = G\alpha$, where α is the shearing angle and G is the shear modulus. Finally for uniform compression ΔP, the bulk modulus B determines the relative volume change by $\Delta P = -\, B\, \Delta V/V$ (see Fig. 1.2a–c). In these cases, the solid returns to its original shape when the stresses are removed. When the stresses exceed a certain threshold, for example the yield stress, plastic deformation takes place and the solid changes its shape permanently. There are many possibilities to store permanent deformations in a solid, for example by dislocations in a lattice or disorder in the grain boundaries of the microstructure. Once there is too much plastic deformation stored, micro-voids/micro-cracks may occur, which will combine or grow into a

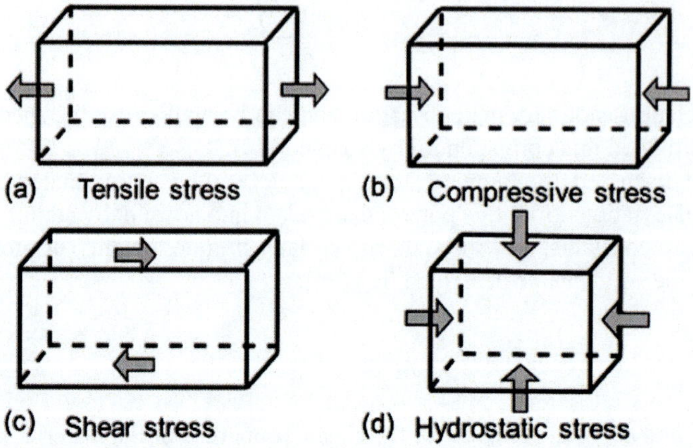

Fig. 1.2 Elastic deformation of a solid: **a**, **b** $\sigma(\varepsilon) = E\varepsilon$ holds for a one-dimensional stress state either in tension or in compression; **c** If a solid is simply sheared, the shear stress is $\tau(\alpha) = G\alpha$ where α is the shearing angle; **d** In case of a hydrostatic compression (equal stress from all directions) the relative volume change is proportional to the hydrostatic pressure divided by the bulk modulus $\Delta V/V = -\Delta P/B$

macro-crack upon further loading leading to component failure at a critical load. The precursors to the formation of macro-cracks must be monitored and is a foremost task of non-destructive materials characterization, in addition to ensuring the desired microstructure and mechanical properties of the components. One set of techniques to achieve this are ultrasonic methods which are discussed in this chapter.

1.3 Ultrasonic Waves

Acoustic waves in air with frequencies above the hearing limit of 20 kHz for humans at a young age, are called ultrasonic waves. This term is now generally used for all mechanical waves above 20 kHz. For specialists in the field however, the term acoustic wave is used dominantly for waves in gases and liquids. In solids different modes of elastic waves can propagate. The difference is that gases and liquids cannot sustain shear strains whereas solids sustain shear elastic deformations and hence transverse waves can propagate in solids.

For a homogeneous solid, there is no theoretical upper limit in frequency. A practical limit is given by attenuation. As solids consist of atoms, there is a lower limit in wavelength however, which is π/r_0 and called the Brillouin zone. In frequency, this limit is in the Terahertz range, i.e. $\approx 10^{13}$ Hz [129]. Although there were many studies of phenomena in solid state physics using GHz ultrasonics [291], the ultrasonic frequencies employed for engineering applications are much lower, except for surface acoustic wave devices deposited on single crystals used as filters in mobile telephone technology, where frequencies up to 5 GHz are common. For non-destructive testing applications for inspecting into the depth of components, the ultrasonic frequencies employed are at most 50 MHz. A frequency of 50 MHz allows inspections to a depth of a few centimeters. The penetration limit is caused by scattering and absorption in most engineering materials, as discussed later in Sect. 1.17. The corresponding ultrasonic imaging technology is often called C-SAM imaging, a combination of C-scan and Scanning Acoustic Microscopy (SAM) and is applied dominantly in the inspection of small scale components, for example in the electronic industry [7].

Scanning Acoustic Microscopy operated at frequencies up to 2 GHz allows one to image the microstructure of materials with a resolution comparable to optical imaging within a penetration depth of a few wavelengths, i.e. some 10 μm. In SAM, the contrast is determined by the elastic properties and their variations and the density differences of the surface constituents of the material examined, whereas the contrast in optical images is determined by changes in the material's refractive index [27], see Sect. 1.6.

Pulsed ultrasonic waves are used in pulse-echo techniques (also in pitch-catch or double-ended techniques) in order to localize defects by time-of-flight measurements. Material characterization can be carried out by observing backscattered signals, by measuring the overall attenuation, or by measuring the velocity of ultrasonic waves in the material. All types of ultrasonic waves are of interest for non-destructive materials characterization, especially for examining the microstructure and for determining physical and engineering material parameters.

In general, there are three types of ultrasonic waves:

- Longitudinal waves (L), with the particle vibration or the displacement vector parallel to the propagation direction.
- Shear or transverse waves (T), with the particle vibrations or the displacement vector perpendicular to the propagation direction.
- Guided Waves like surface waves, Lamb waves or shear horizontal waves. They are acoustic or elastic waves restricted in propagation by the geometry of the components, for example by the walls of ducts in acoustics and/or by plate boundaries in non-destructive testing. The restrictions entail boundary conditions when solving the wave equation for a certain geometry.

Surface acoustic waves (SAW) also called Rayleigh waves (RW) are one prominent type of guided waves. They are bound to a free surface. In SAW's, the displacement vector has components parallel and perpendicular to the propagation direction. The particle vibration undergoes a retrograde elliptical motion. The penetration depth of the SAW's displacement is approximately one wavelength. The transverse component vibrates perpendicular to the surface, i.e. out-of-plane or in shear-vertical (SV) mode, see Fig. 1.3. Lamb waves are guided waves in plates with the polarization out-of-plane. They belong to the family of shear-vertical polarized waves (SV-waves). For shear horizontal waves in plates the polarization is in-plane (SH-waves).

Rayleigh waves play an important role in both engineering and science. For examples, earthquakes release their energy to a large extent into surface waves [29]. Mobile phone technology is not possible without SAW filters [34]. In non-destructive testing, the interest in Rayleigh waves is based on using them to monitor surface properties

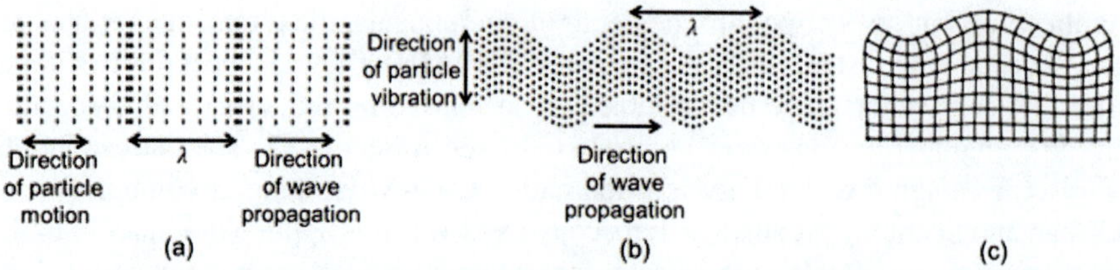

Fig. 1.3 Schematic display of the particle motion for **a** longitudinal, **b** shear, and **c** Rayleigh or surface waves, see for example [74, 295]. Particles mean that the volume elements are very small compared to the wavelength. Once the size of these elements becomes so small that the crystalline or atomic structure of the solid becomes noticeable, the local action of the forces between the atoms (see Fig. 1.1) has to be taken into account. When additionally the wavelength becomes comparable to the interatomic distance a, the sound wave velocity becomes strongly dispersive, and waves with k-vectors $k = \pi/a$ (Brillouin zone) do not propagate anymore, i.e. their group velocity becomes zero [129]

of layered materials. In a homogeneous and an isotropic material the Rayleigh wave velocity is dispersion-less. In contrast, Rayleigh waves become dispersive if elastic properties change with depth within a wavelength. This allows e.g. to control and to determine the homogeneity and thickness of hardened surfaces, based on their microstructure's influence on the surface wave velocity. Another advantage of surface waves is their ability to travel along curved surfaces and bends without much loss in amplitude provided that the radius of curvature is much larger than their wavelength.

Other types of guided waves displayed schematically in Fig. 1.4 are of interest for NDT and NDMC as well:

- Elastic waves in rods (Fig. 1.4a) which are used for defect detection for correspondingly shaped components.

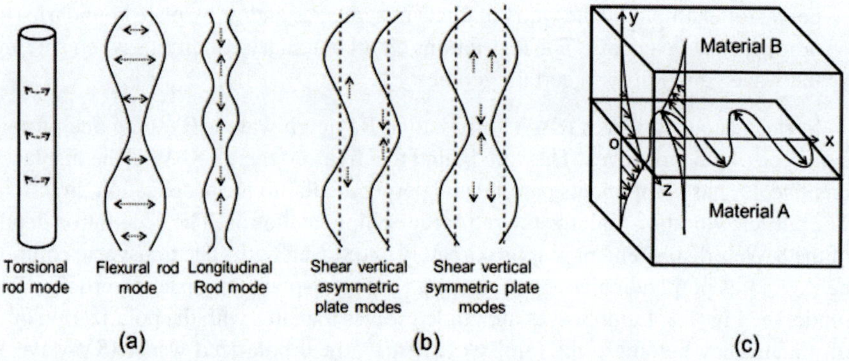

Fig. 1.4 Mode shapes and direction of particle vibrations (arrows) of different guided wave modes: **a** rod modes; **b** shear-vertical (SV) plate wave modes; and **c** waves along the boundary of two materials. These types of waves are in general dispersive, see for example [74]

- Plate waves (Fig. 1.4b): These waves belong to the class of shear vertical (SV) and shear horizontal (SH) waves because their displacement vector is either parallel or perpendicular to the surface vector of the plate, i.e. perpendicular to the surface plane or in-plane, and perpendicular to their propagation direction.
- Interface waves (Fig. 1.4c): They propagate along an interface between two (or several) elastic media. The wave is of maximum amplitude at the interface and decreases exponentially away from the interface into both the media. These waves are of interest to evaluate strength parameters or degradation in layered materials, e.g. laminated structures.

Before we discuss details and applications of the various wave modes in NDMC, let us consider the one-dimensional case for wave propagation. We write for the stress–strain relation (see 1.4):

$$\sigma_1 = c_{11}\varepsilon_1 \tag{1.5}$$

Here, ε_1 is the strain and σ_1 is the stress in 1-direction. Taking the derivatives and considering that the gradient of the stress $\partial\sigma_1/\partial x_1$ is equal to the mass element accelerated in the corresponding volume, one obtains from (1.5):

$$\frac{\partial\sigma_1}{\partial x} = c_{11}\frac{\partial\varepsilon_1}{\partial x} = \rho\frac{\partial^2 u}{\partial t^2} \tag{1.6}$$

Furthermore, since the gradient of the displacement is equal to the strain ($\varepsilon_1 = \partial u_1/\partial x_1$), we obtain the wave equation for the displacement u:

$$\frac{\partial^2 u}{\partial t^2} - \left(\frac{c_{11}}{\rho}\right)\frac{\partial^2 u}{\partial x^2} = 0 \tag{1.7}$$

The factor (c_{11}/ρ) is the sound velocity squared. The displacement u propagates with the sound velocity through the material. In a similar way the wave equation for shear waves can be formulated. Taking into account the boundary conditions for the various wave modes, one obtains the velocity and dispersion for these modes and their displacement fields. This is discussed in the appropriate sections of this book where needed. Otherwise, the reader is referred to the many textbooks on ultrasonic wave propagation; see for example [14, 40, 74, 128, 173, 243, 281, 290, 291].

1.4 Ultrasonic Velocity and Attenuation

As is common for all wave modes, ultrasonic wave propagation is described by several parameters. The most prominent are phase and group velocities, polarization or wave type, and attenuation. The material parameters which determine the velocity are the density and the elastic constants. The unit for the ultrasonic velocities is mostly given in m/s or in mm/μs as a subunit (1 mm/μs corresponds to 10^3 m/s or 1 km/s). Likewise, the wavelength λ is mostly given in mm in NDT and NDE and the attenuation coefficient α is given in dB/mm or Neper/mm, also written as Np/mm which is equal to mm^{-1}. If the amplitude (pressure) attenuation coefficient is α, the ultrasonic echo pattern decays like

$$A = A_0 e^{-\alpha x} \tag{1.8}$$

For example, if the amplitudes of the two consecutive backwall echoes obtained in a plate of thickness 10 mm are 80% and 40% of the full screen height, the attenuation coefficients are given as:

$$\alpha_{Np/mm} = \frac{ln(80/40)}{20} = 0.03466\, Np/mm\, and$$
$$\alpha_{db/mm} = \frac{20\, log\,(80/40)}{20} = 0.301\, dB/mm$$

The conversion factor for α in units of Np/mm to dB/mm is 8.686 (= 20 × log(e)). However, if it is defined that the intensity of an ultrasonic beam decays as $I = I_0 e^{-\alpha_{int} x}$, the conversion factor for α_{int} in units of Np/mm to dB/mm is 4.343 (= 10 × log(e)), since $I \propto A^2$. The use of the unit dB/length is preferable because it is unambiguous in respect to whether (1.8) is used or $I = I_0 e^{-\alpha_{int} x}$. In any case $2\alpha = \alpha_{int}$ holds.

The frequency f and the wavelength λ are related to each other by the material's sound velocity v:

$$v = \lambda f \quad \text{or} \quad \omega = vk \tag{1.9}$$

Without dispersion, the wave velocity v is frequency independent. If there is dispersion, i.e. $v = v(\omega)$ or $v = v(\lambda)$, more generally $\omega = \omega(k)$, the phase velocity ω/k and the group velocity $v_{gr} = \partial\omega/\partial k$ are no longer identical [40]. For example, if there is attenuation, the phase velocity becomes dispersive and in a formal way the wave vector becomes complex [290]. Due to the Kramers–Kronig relations the dispersive velocity $v(\omega)$ and the attenuation $\alpha(\omega)$ are related to each other, which is discussed in detail in Sect. 1.24. Finally, dispersion also arises because of the boundary conditions in wave propagation. For example for plate waves $v = v(d/\lambda)$ where d is the thickness of the plate [295].

The frequency-dependent attenuation coefficient consists of several parts:

$$\alpha_{tot} = \alpha_{geom} + \alpha_{mat} \tag{1.10}$$

with

$$\alpha_{mat} = \alpha_A + \alpha_S \tag{1.11}$$

where α_{tot} corresponds to the apparent total loss in the signal amplitude measured experimentally. The geometrical part α_{geom} takes into account the divergence of the ultrasonic beam due to its finite aperture and transmission/reflection losses at boundaries that are large compared to the wavelength. This geometrical spreading of the ultrasonic beam due to its finite size is of particular importance at low frequencies when the associated wavelengths become comparable to the transducer size [128, 214].

The material-dependent part α_{mat} contains an absorption part α_A and a scattering part α_S. The absorption coefficient α_A describes the inelastic part of the wave propagation, i.e. the energy losses transferred eventually into heat, and the scattering coefficient α_S stands for the elastic scattering losses. Scattering processes change the k-vector of the ultrasonic waves and hence lead to attenuation. They are accompanied by mode conversions from longitudinal waves into transverse wave and surface waves and vice versa. After some time, an equilibrium distribution develops. Due to its frequency dependence, scattering also leads to dispersion. This can be exploited as an additional tool for materials characterization. Details will be discussed in what follow.

1.5 Generation and Detection of Ultrasound

Different techniques exist for the excitation and the detection of ultrasound. The most important are piezoelectric transducers, electromagnetic acoustic transducers (EMAT) for metals, in particular ferromagnetic materials, capacitive transducers, and laser-generated ultrasound with optical interferometers as detectors. All ultrasonic transmitters exhibit a so-called near-field with a characteristic length N. Within this near-field, interference effects dominate the amplitude distribution which may render it difficult to unambiguously evaluate defects. The near-field pattern depends on the distribution of the elementary sources within the aperture of the transmitter. Beyond the near-field, the beam spreads out as a sphere of radius r originating from the center of the transducer with an amplitude decay proportional to $1/r$, see also Sect. 1.5.2.

1.5.1 Piezoelectric Transducers

In many cases ultrasound for NDMC applications is generated by piezoelectric transducers. Their working principle can easily be seen by a one-dimensional consideration, here in 1-direction with the displacement in parallel. The piezoelectric equations relating stress and strain under the action of an electric field are given by [128, 291]

$$\sigma_1 = c_{11}\varepsilon_1 - e_{11}E_1 \tag{1.12}$$

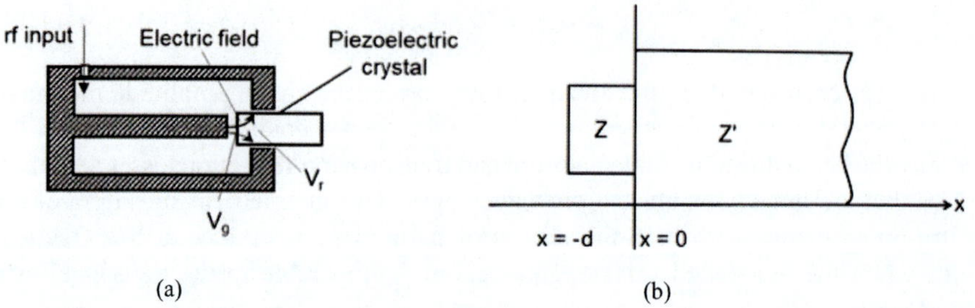

Fig. 1.5 **a** Generation of ultrasound across the surface of a piezoelectric material; **b** principle of generation of ultrasonic waves due to piezoelectricity in a transducer of thickness d. Equation (1.14) shows that the gradient of the electric field is the source of ultrasound, occurring here at the surface of the piezoelectric crystal (**a**) or at $x = -d$ and $x = 0$ (**b**)

$$P_1 = e_{11}\varepsilon_1 - \eta_{11}E_1 \tag{1.13}$$

Here, E_1 is the electric field, ε_1 is the strain, e_{11} is the piezoelectric constant, η_{11} is the dielectric susceptibility, σ_1 is the stress, and P_1 is the electric polarization.

Inserting (1.12) into (1.6) and applying an electrical field $E_1(t) = E_0 e^{i\omega t}$ across the surface of a piezoelectric crystal (Fig. 1.5a), yields for the wave equation [14]:

$$\frac{\partial^2 u}{\partial x^2} - \frac{1}{v_1^2}\frac{\partial^2 u}{\partial t^2} = d_{11}\frac{\partial E_1}{\partial x} \tag{1.14}$$

With $d_{11} = e_{11}/c_{11}$ and the boundary condition for a free transducer that the surface is stress free

$$\frac{\partial u}{\partial x} = \varepsilon_1 = d_{11}E_1 \tag{1.15}$$

one obtains

$$u = i(d_{11}E_0/k)e^{i\omega t}e^{-ik(x-x_0)} \tag{1.16}$$

for the surface displacement. Equation (1.14) shows that the *gradient* of the electric field is the source of ultrasound. This holds for all piezoelectric transducers. If the transducer is placed in a microwave resonator or more general in a capacitor (Fig. 1.5b), the radiated ultrasonic power is given by

$$S = P_{in}\left[c_{11}d_{11}^2 v A_s Q/\left(\omega\varepsilon_r\varepsilon_0\left(\varepsilon_r V_g + V_r\right)\right)\right] \tag{1.17}$$

A_s is the cross-section of the rod, Q is the quality factor of the electrical resonator which drives the transducer, P_{in} is the input electrical intensity. Here, V_g is the volume of the capacitor which is not filled with the transducer material and V_r is the volume filled with the material. In case the electrodes are deposited on the transducer, $V_g \approx 0$.

If the transducer is coupled to a sample of impedance Z', waves are generated at both surfaces of the piezoelectric crystal and radiate the acoustic power S into the component [291]:

$$S = \left(\frac{Z'}{Z}\right) \frac{A_s v_1 \varepsilon_1 K_{11}^2 E_0^2 (1 - \cos kd)^2}{2\left(\sin^2 kd + \left(\frac{Z'}{Z}\right)^2 \cos^2 kd\right)} \tag{1.18}$$

Here, d is the thickness of the piezoelectric element, Z' and Z are the impedances of the material and transducer, respectively, $K_{11}^2 = e_{11}^2/c_{11}\varepsilon_1$ is the piezoelectric coupling factor, here for longitudinal waves of velocity v_l, ε is the dielectric permittivity, and A_s is the area of the transducer, see Fig. 1.5b. The transducer is in resonance for $d = \lambda/2$, $d = 3\lambda/2$, $d = 5\lambda/2\ldots$ Even resonances are not permitted because they do not fulfill the electrical boundary conditions. For broadband transducers, the resonances are suppressed by corresponding layers which mechanically load the transducer. Further details can be found in a number of text-books on piezoelectric materials, see for example [183]. Analog considerations hold for shear wave transducers.

1.5.2 Near-Field and Beam Divergence of Transducers

Because of the limited geometrical extension of the active area of ultrasonic transducers, often called probes, the ultrasonic beam excited at their surface diverges into an increasing volume beyond a distance called near-field length N, weakening continuously the vibration or strain amplitude which locally interacts with the materials microstructure, see Fig. 1.6 [128]. The near-field effect is caused due to the constructive and destructive interferences of waves emanating from different points on the surface of the transducer. Continuous reduction in the beam amplitude beyond the near-field also reduces the sensitivity for detection of defects in the so-called far-field as defined below. The decrease of strain is equivalent to attenuation of the ultrasonic signals or wave-packets and is often called geometrical attenuation. It also causes a dispersive phase delay of the different frequencies or wavelengths in a wave packet. For accurate attenuation and velocity measurements, these effects must be taken into account [214].

The amplitude decay of a typical pulse-echo pattern as shown in Fig. 1.7, generated by a circular piston source (Fig. 1.6) in the pulse-echo or transmit-receive technique, can be described by the following equation:

$$A_{(x=vt)} = A_0 \frac{N}{x} e^{-\alpha x} \tag{1.19}$$

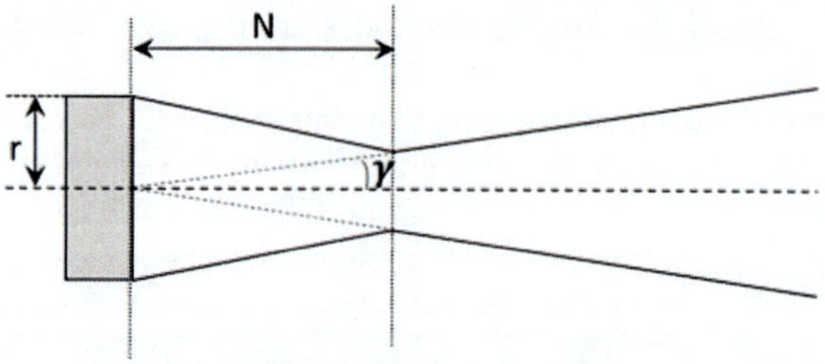

Fig. 1.6 Divergence of the wave-field emanating from a circular piston-source of radius r. There is a near-field length N and half-opening angle γ. In the far-field, i.e. for distances larger than N, the source behaves like a point-source, see for example [128]

Fig. 1.7 Ultrasonic back-wall-echo sequence obtained in aluminum. The path length was $d = 10$ mm. Here, the near-field of the piezoelectric transducer was about 30 mm ($D = 12$ mm, $\lambda \approx 1.2$ mm, hence $N = 30$ mm)

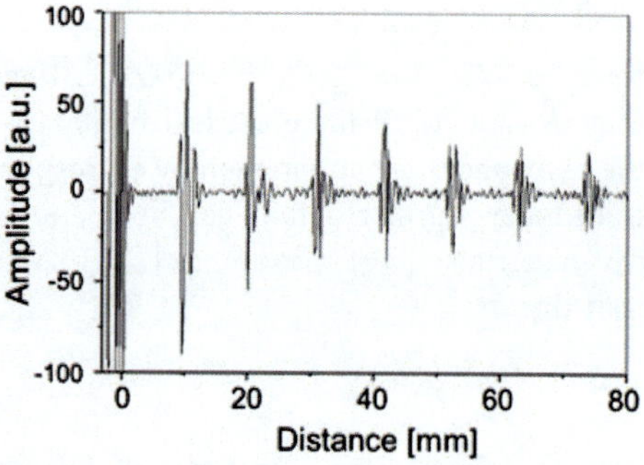

$A(x)$ is the received, reflected signal amplitude after complete round-trips (forward and back) of the sound along a path of length $x = 2d$ with the time-of-flight t. A_0 is the amplitude of the transmitted signal at $t = 0$, $x = 0$. Here, N is the near-field length. For example for a circular disc transducer with diameter $D = 2r$, $N = D^2/(4\lambda)$ [128, 133]. The beam half-opening angle γ is given by $\gamma = 1.22\lambda/D$. The variable N/x describes the spreading beyond the near-field length (Fig. 1.6).

1.5.3 Polarization of Piezoelectric Shear-Wave Transducers

Ultrasonic angle beam inspection of weld joints is most commonly performed using shear waves. Exploiting Snell's law, the shear vertical waves are produced by mode conversion of longitudinal waves generated by a piezoelectric transducer which is coupled to the surface by a wedge under an oblique angle. It is also possible to generate shear waves normal to the transducer surface by using a piezoelectric crystal cut along a specific plane (Y-cut quartz crystal) or by poling a ceramic piezoelectric element in a specific direction. Such transducers are very useful in NDMC for measurement of shear wave velocity for determining the elastic properties and also

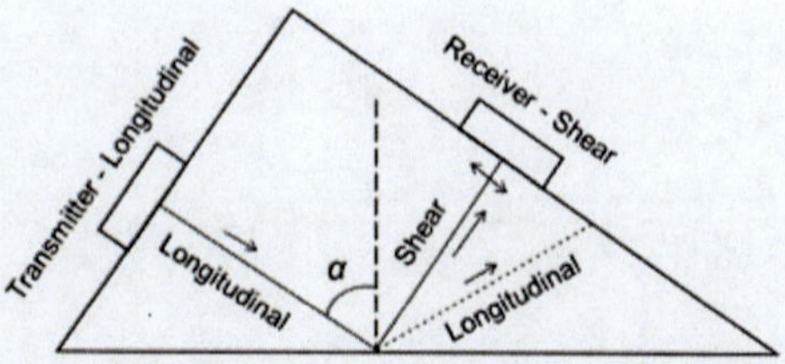

Fig. 1.8 A longitudinal wave is generated at the left surface of a wedge by an ultrasonic transducer. The reflected wave at the lower side mode converts to a shear wave with its polarization parallel to the sagittal plane and which is received by a shear wave transducer. For a certain input angle α, which depends on the ratio of the longitudinal to the shear wave velocity, and hence on Poisson's ratio, the mode conversion is maxium into shear waves

for studying the anisotropy in the test samples due to texture or stress as discussed in Sects. 1.11 and 1.16.

It is often required to know the polarization of the shear waves emanating from the transducer. The manufacturer of such transducers usually orients the piezoelectric plate such that its displacement direction (polarization) points to the feeder cable pin. If the polarization of the probe is not known however, one may exploit mode conversion in a wedge, typically made of Perspex, fused quartz, or a fine-grained metal, in order to determine the polarization. The reflected longitudinal wave at the angled side is partially mode-converted to a shear wave. This shear wave is polarized in the sagittal plane at a reflection angle governed by the Snell's law and the ratio of the longitudinal to the shear wave velocity. For a certain input angle α, which depends on the ratio of the longitudinal to the shear wave velocity, and hence on Poisson's ratio, the mode conversion is fully into shear waves. For fused quartz, this occurs at 42°. It can be received by a shear-wave transducer which shows maximum output amplitude if its polarization is turned parallel to the sagittal plane [128], and in this way the polarization can be determined (Fig. 1.8).

1.5.4 Electromagnetic Acoustic Transducers (EMATs)

Besides the piezoelectric probes, electromagnetic acoustic transducers (EMAT's) are of great interest for applications in metallic materials [97]. Their advantages for NDE are that no mechanical coupling medium of the EMAT-probe with the material is necessary, i.e. applications for high temperatures become possible. This is valid for the generation as well as for the detection of ultrasonic signals. The working principles of EMAT transducers can be summarized as:

(1) The excitation of the ultrasound occurs in the conducting material by Lorentz forces when induced oscillatory eddy currents j_{ec} and an applied static or quasi-static magnetic field B_0, are present in the material. Magnetic forces contribute by the action of the oscillating magnetic fields on the magnetic state. In addition,

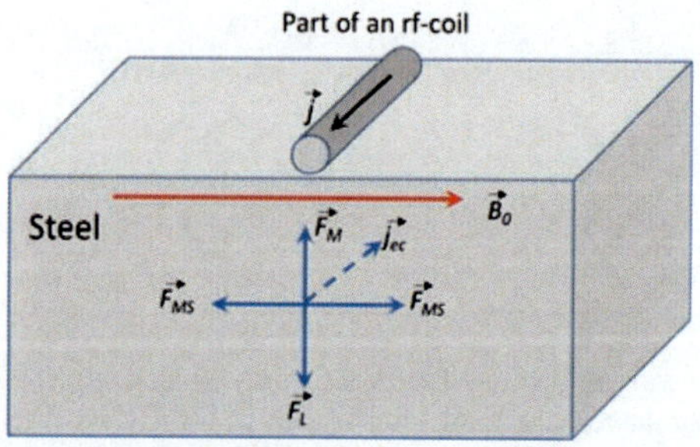

Fig. 1.9 Working principle of an electromagnetic acoustic transducer (EMAT). The line conductor represents a part of an rf-coil which generates the eddy currents j_{ec} in the metallic component to be tested. Together with the static field B_0, here tangential to the surface, the oscillatory eddy currents exert Lorentz forces F_L on the electrons and the ions of the metallic component at the same oscillatory frequency. In addition, magneto-static forces and magneto-strictive forces are generated, see [97]

when the material is ferromagnetic, magnetostrictive forces generate ultrasound. The wave modes depend on the direction of the eddy currents relative to the magnetic field, Fig. 1.9. The frequency is determined by the frequency of the ac current in the excitation coils.

(2) All types of wave modes, including longitudinal waves, shear waves, and guided waves (i.e. Rayleigh-, SV-, and SH modes), can be excited and detected.
(3) A certain mode and its wave vector can be enhanced by a suitable placement of the magnets and by matching the shape of the coil to the wavelength to be generated.

1.5.5 *Laser-Generated Ultrasound and Optical Interferometric Detection*

Laser-based ultrasound is an interesting method in non-destructive testing for materials characterization, and process control [193, 218, 219, 264]. An example is shown in Fig. 1.10 depicting the principle of in-line thickness measurement of a hot tube in a production line. The main advantage of laser ultrasonics is that it is contact-free, so that it can be employed for investigations at hot, fast moving or complicated shaped components. The narrow pulses of the laser-generated ultrasound allow precise measurements of time-of-flight and therefore of component thickness or sound velocity. The large bandwidth of the generated narrow pulses makes it possible to measure dispersive effects. Another advantage of laser-generated ultrasound is that the source of the ultrasound can be adapted to the testing problem by variation of the laser parameters such as pulse energy, pulse duration, size, and the shape of the point of generation. In order to optimize the laser parameters for the testing problem, it is important to adapt the generating mechanism. It is well-known

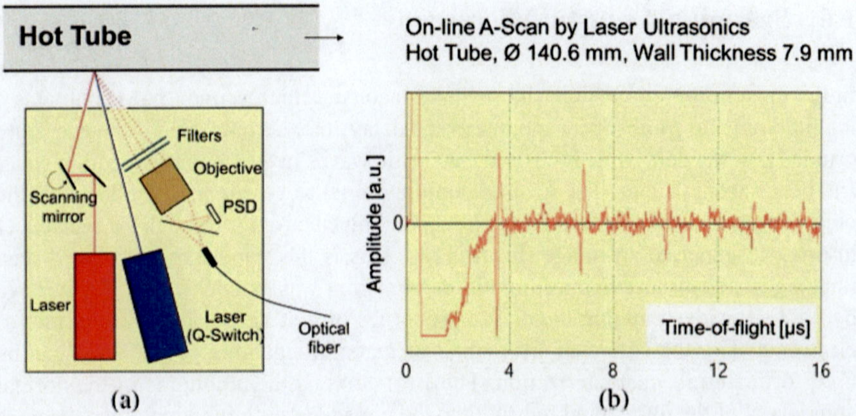

Fig. 1.10 **a** Principle of the measurement of the wall-thickness of a hot pipe-line tube in the production line. Ultrasonic signals are generated by the Q-switch laser and are detected by a Fabry–Perot interferometer. The light gathered for the interferometer from the optically seen rough tube is measured continuously by a photo-diode (PSD). Because the tube is moving, there are different surface spots illuminated. If there is sufficient light power received to obtain good SNR of the signals, the generating laser is fired and the received signal in the optical fiber passing through the interferometer is captured; **b** The thickness is evaluated from the time-of-flight of the echoes [219]. Figure reproduced from [217] with permission from Dr. M. Paul, Mahle GmbH, Stuttgart, Germany. A similar system was developed by [219]

that there are two different effects which play a role in generating ultrasound by optical irradiation:

(1) Thermoelastic effect: At intensities below a certain level, which depends on the material and optical wavelength, heating of the target and the ensuing surface expansion within the thermal skin effect results in stresses tangential to the surface [257],
(2) Ablation and plasma generation: At higher laser intensities, the ablation of material generates a large pressure normal to the target [110].
(3) Both excitation mechanisms generate distinct stress-profiles which propagate as ultrasonic signals through the component under inspection. They can only be differentiated if the detector has a large enough bandwidth.

The detection of ultrasound by contact-less optical interferometers has been reviewed several times [50, 192, 299]. It can be seen from these reviews that most interferometers are designed to detect out-of-plane signals. In order to detect in-plane displacements, a specially designed speckle interferometer has to be used [186]. Although the sensitivity of interferometers is impressive, at the shot noise limit of the photo-detectors about $\approx 10^{-6}$ nm/$\sqrt{}$Hz, one has to keep in mind that this number holds for a light power of 1 mW collected from the sample. That number is not so easy to attain for rough surfaces. There are a number of interferometer designs to maximize the collected light power, see the reviews mentioned above. Interferometers for the detection of ultrasound are now commercially available.

1.6 Scanning Acoustic Microscopy

In the most common version, the scanning acoustic microscope (SAM) contains a longitudinal wave transducer mounted on a delay-line, here sapphire. The curvature grinded into the delay-line focuses the acoustic waves in the coupling medium water and hence acts as a lens. The focused compressional waves are partly reflected at the object surface and are captured by a corresponding electronics through the transducer in order to generate an image (Fig. 1.11a). This is the standard set-up for C-scan imaging in non-destructive testing. C-scan imaging was extended to the GHz range by [227]. Besides imaging the elastic properties of materials and defects in micro-electronics [7], the so-called $V(z)$-curve or material signature of the SAM can be used for materials characterization. The $V(z)$-curve is the amplitude of the normal component of the ultrasound returning to the transducer as a function of the distance between the acoustic lens and the object surface, see Fig. 1.11b. The oscillating shape of the $V(z)$-curves results from the interference of the directly reflected central part of the ultrasonic beam with the ultrasound emitted from the Rayleigh waves, which are generated on the object surface if the opening angle of the lens includes the critical angle for surface wave generation of the material under investigation. Therefore, the Rayleigh wave velocity of the sample material and hence, some of its elastic properties, can be determined from the distance Δz between the maxima and minima of measured $V(z)$-curves. A comprehensive description of the theoretical foundations and the applications of acoustic microscopy and $V(z)$-curves can be found in [26]. It has been shown that the Rayleigh wave velocity can be obtained not only from the interference maxima in a $V(z)$-curve, but also from the distance between the maximum of specular reflection and the first interference maximum [105]. This is advantageous if rigid materials such as ceramics are investigated, because in these cases acoustic microscopes operating at GHz frequencies yield $V(z)$-curves with one maximum only, besides that from the specular reflection. Calibrating the specular reflection amplitudes in an SAM, it is possible to obtain images whose contrast is determined by the acoustic impedance distribution at the surface of the material [106].

1.7 Acoustic Impedance

During the propagation in a material, besides the amplitude decrease due to beam spreading [214], the exponential decay $e^{-\alpha x}$ in polycrystalline materials is to a large extent the consequence of local changes of sound velocity due to elastic anisotropy, or due to density variations, which causes scattering at the grain boundaries often described qualitatively by the acoustic impedance Z:

$$Z = \rho \cdot v \tag{1.20}$$

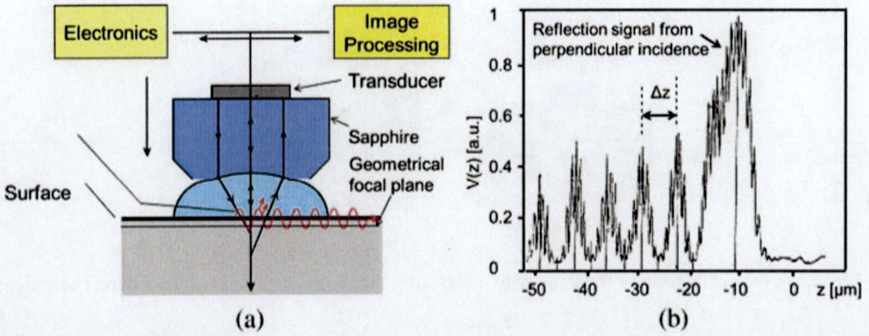

Fig. 1.11 **a** Principle of Scanning Acoustic Microscopy (SAM). An acoustic lens focuses ultrasound on a surface. Under the critical angle surface waves are generated which radiate back into the lens and interfere with the central ray. This generates a so-called $V(z)$-curve, **b** which can be exploited for non-destructive materials characterization (Fig. 1.11b in modified form from [104] with permission from Elsevier)

Here, ρ is the materials density. Equation (1.20) is valid for plane waves. For non-planar waves, the impedance gets more complicated [197]. If Z changes in the path of the beam, volume elements whose extent are large compared to the wavelength, lead to reflections, like at defects or at the back-wall of a component. The amplitude of the refleceted wave is proportional to the reflection coefficient $R_{0,x}$:

$$R_{0,x} = \frac{\Delta Z}{\Sigma Z} = \frac{\rho_x v_x - \rho_0 v_0}{\rho_x v_x + \rho_0 v_0} \tag{1.21}$$

where the indices 0 and x characterize the different media adjacent to each other, i.e. the media in which the wave is incident and refracted, respectively. A negative $R_{0,x}$ indicates a phase shift of 180° at the boundary between the two media when $\rho_x v_x < \rho_0 v_0$. Different angles of incidence and geometrical structures, including the surfaces, directly influence the reflection amplitude. Volume elements with dimensions comparable to the wavelength lead to resonance effects, while volume elements much smaller vibrate and hereby scatter the ultrasonic waves away in all directions.

The back-wall echo from a plane-parallel plate of density ρ_x, whose lateral dimension is large compared to the transducer diameter, yields a signal amplitude due to the differences between the velocities and the densities in the plate (e.g. steel) and air, i.e. due to the impedance mismatch. The density ρ_x of the plate can be determined by measuring the echo amplitude relative to the amplitude of a material of known density ρ_0 and velocity v_0 using (1.21), which can be rewritten as

$$\rho_x = \rho_0 \left(\frac{v_0}{v_x} \right) \left(\frac{1 - R_{0,x}}{1 + R_{0,x}} \right) \tag{1.22}$$

In order to obtain the density, the following steps are taken: (1) The reflection amplitude ($= -A_0$) of the back-wall echo in a solid reference material like perspex

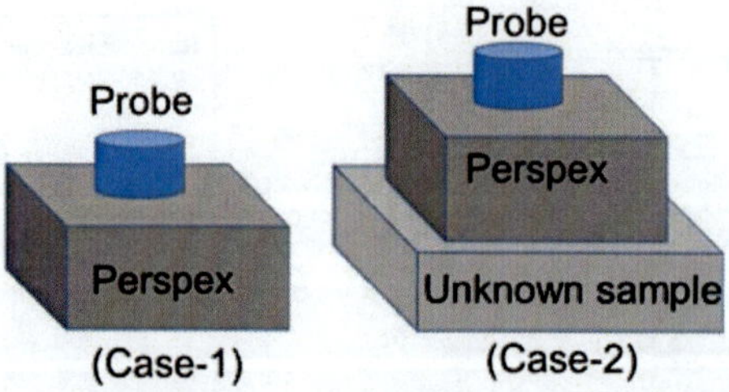

Fig. 1.12 Determination of the density by measuring the impedance through reflection coefficient

(case 1 in Fig. 1.12) is measured for which the coefficient $R_{0,air}$ is − 1, because the reflection of the perspex material against air is very close to 100% due to the low impedance value of air; (2) The amplitude A_I of the reflected echo from the interface between the reference material perspex and the unknown material x is also measured, which is smaller than 100% because of the finite impedance difference (case 2 in Fig. 1.12). For the two materials perspex-unknown material bonded together by a coupling medium, the experimentally measured normalized reflection coefficient ($R_{0,x} = -A_1/A_0$) allows to determine the density by removing the effects of other unknowns including transmission losses at the probe/Perspex interface. The sound velocity v_x in the unknown material must be known beforehand.

For an example with $R_{0,x} = -0.60$, $\rho_x = 4\rho_0 v_0/v_x$ using (1.22). The accuracy in the measurement improves by using the reference material having acoustic impedance close to that of the unknown material. An error of ~ 3% in amplitude measurement may lead to ~ 30% and ~ 10% errors, respectively in the density for steel and aluminum, using the above method with perspex as the reference material.

1.8 Ultrasonic Velocity in Single Crystals

The wavelength and pulse-length of ultrasonic beams in solid components depend on the frequency and the frequency bandwidth of the transducer employed, respectively. Pulse-lengths and wavelengths are typically some 100 μm to some mm. Thus, one could conclude that by ultrasonic waves one cannot probe the atomic structure of a material, except if the wavelength is comparable to the interatomic distance. However, this conclusion is not valid. Because the strain $\varepsilon = \Delta l/l$, i.e. the relative extension or dilatation of the material, acts with the same magnitude at all length scales, the effects on the microstructural or crystal lattice scale transfer to macroscopic scales, not only in the sound velocity but also in the absorption coefficient due to internal friction.

Let us assume that the central frequency of a transducer is 1 MHz. For a velocity of 6 mm/µs this means a wavelength of 6 mm. For half a wavelength the strain is either positive or negative and for a 1.5 cm beam diameter, the volume uni-directionally strained is about 0.5 cm³. In case of iron (steel), this volume contains about 4×10^{22} atoms.

The ultrasonic oscillations are transferred from atom to atom by the interatomic forces, see Fig. 1.1, with a phase delay. The phase delay eventually determines the ultrasonic phase velocity. The stress–strain relation is described by Hooke's law, where c_{ijkl} $(i, j, k, l = 1, 2, 3)$ is the elasticity tensor [129]:

$$\sigma_{ij} = c_{ijkl}\varepsilon_{kl} \tag{1.23}$$

In this equation Einstein's sum-convention is used, i.e. the sum is carried out over indices which are not present on the left side *and* on the right side of the equation. The first index describes the stress direction and the second stands for the plane, on which the stress acts, see Fig. 1.13. Because there are no preferred directions, the situation reduces to nine different stress values. Due to the equilibrium condition $\sigma_{ij} = \sigma_{ji}$, only six tensor components remain: σ_{11}; σ_{22}; σ_{33}; $\sigma_{12} = \sigma_{21}$; $\sigma_{23} = \sigma_{32}$; $\sigma_{13} = \sigma_{31}$. This reduces (1.23) to

$$\sigma_i = c_{ij}\varepsilon_j \tag{1.24}$$

The indices now have to be interpreted as follows (Fig. 1.14):

Tensor notation	11	22	33	23, 32	13, 31	12, 21
Matrix notation	1	2	3	4	5	6

The strain ε generated for a given ultrasonic amplitude is the same on an atomic and macroscopic scale, see Sect. 1.1, and they are of the order of 10^{-6} and less. Strains of the order of 10^{-3} and more already produce shock waves.

The c_{ij} are the second order elastic constants (SOEC). For the cubic crystals, three independent elastic constants exist: c_{11}, c_{12}, and c_{44}. They are directly related to the sound velocity and its directional dependence in a single crystal (Fig. 1.13) [290]. Seven different velocities exist, three for longitudinal waves and four for shear or transverse waves:

$$v_L[100] = \sqrt{c_{11}/\rho} \tag{1.25}$$

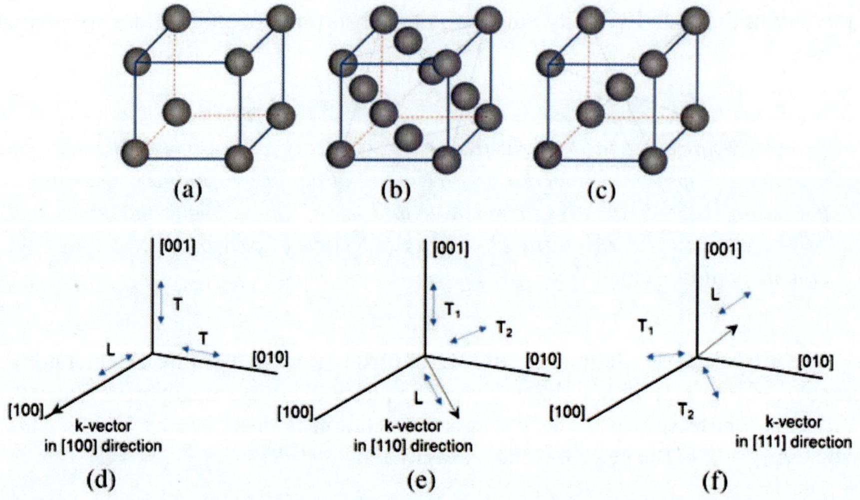

Fig. 1.13 **a** Basic cubic unit cell with a = b = c; **b** face centered cubic structure; and **c** body centered cubic structure. The cubic crystals are widely studied due to materials like steel, aluminum, nickel, copper, and others. There are three main propagation directions with different velocities for longitudinal (L) as well as for transverse waves (T): **d** along the cube's edge in [100] direction (longitudinal (1.25) and transverse polarization (1.28)); **e** across the cube's face in [110] direction [longitudinal (1.26)] and transverse waves with polarizations in [001] (1.29), respectively [110] direction (1.30); and **f** diagonal through the cube in [111] direction, [longitudinal (1.27)] and transverse (1.31). The double arrows indicate the polarization of the transverse waves. In an isotropic cubic lattice $A = 0$ (1.32)

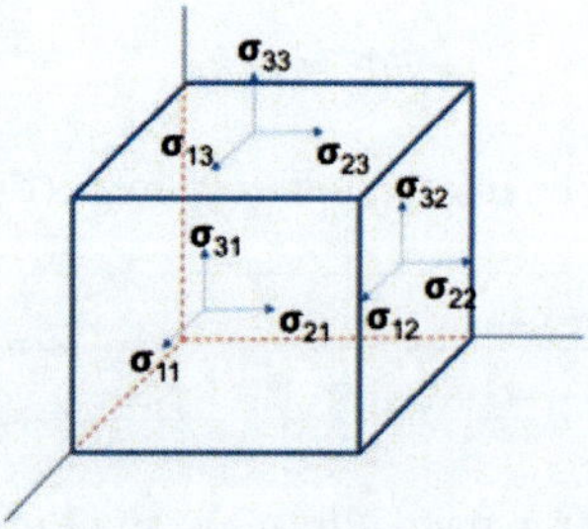

Fig. 1.14 Compressional and shear stresses in a small volume of a solid. When the linear size of the volume becomes comparable to the interatomic distances, the forces and their direction between the atoms have to be considered and the picture for macroscopic stresses is no longer valid

$$v_L[110] = \sqrt{1/2(c_{11} + c_{12} + 2c_{44})/\rho} = \sqrt{(c_{11} - (1/2)A)/\rho} \tag{1.26}$$

$$v_L[111] = \sqrt{1/3(c_{11} + 2c_{21} + 4c_{44})/\rho} = \sqrt{(c_{11} - (2/3)A)/\rho} \tag{1.27}$$

$$v_T[100] = \sqrt{\frac{c_{44}}{\rho}} \text{ (transverse isotropic)} \tag{1.28}$$

$$v_T[110] = \sqrt{\frac{c_{44}}{\rho}} \text{ (polarization parallel to [001])} \tag{1.29}$$

$$v_T[110] = \sqrt{(c_{44} + (1/2)A)/\rho} \text{ (polarization parallel to [110])} \tag{1.30}$$

$$v_T[111] = \sqrt{(c_{44} + (1/3)A)/\rho} \text{(independent of polarization)} \tag{1.31}$$

In the above equations A is a measure of the anisotropy of the crystalline structure:

$$A \equiv c_{11} - c_{12} - 2c_{44} \tag{1.32}$$

Some anisotropy factors are displayed in Fig. 1.15. Another quite common definition of the anisotropy factor A_f is:

$$A_f \equiv 2c_{44}/(c_{11} - c_{12}) \tag{1.33}$$

For an isotropic material $A_f = 1$ holds. For completeness, it is mentioned that for elastic wave propagation in cubic lattices, the quantity

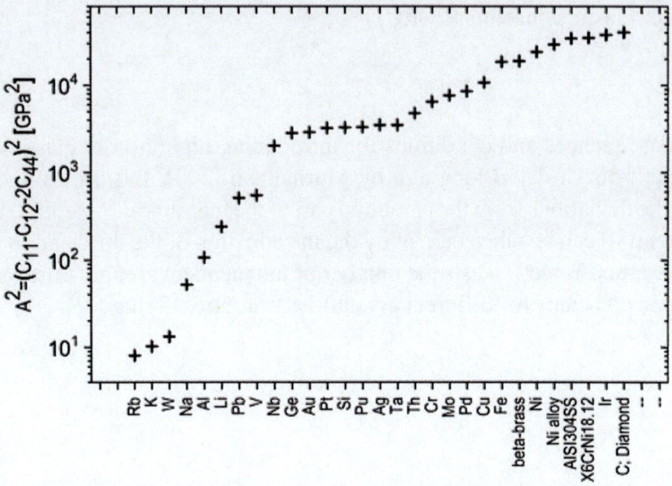

Fig. 1.15 Squared anisotropy factor $A^2 = (c_{11} - c_{12} - 2c_{44})^2$ (1.32) for single crystals of different cubic materials. The materials are listed on the x-axis in the order of the increasing anisotropy factor

$$A_E = 2(c_{11} - c_{12} - 2c_{44})/(c_{11} - c_{12}) \tag{1.34}$$

has also been used as a measure of elastic anisotropy [78].

The anisotropy factor A plays an important role in the ultrasonic characterization of single crystals as well as of polycrystalline materials as will be discussed in the following sections. Table 1.2 lists the single crystal elastic constants c_{ij}, ρ values for cubic materials and the corresponding polycrystalline elastic properties viz. B, G, E, υ, v_{L} and v_{T}, and the factors A^2, and A_f. For $A = 0$, i.e. an isotropic cubic crystal, the velocities reduce to

$$v_L = \sqrt{\frac{c_{11}}{\rho}}, \quad v_T = \sqrt{\frac{c_{44}}{\rho}} \tag{1.35}$$

These are the limiting values for the sound velocities for single crystals of Rb, K, and W close to the left side of the x-axis in Fig. 1.15, i.e. for $A \rightarrow 0$, see also the values in Table 1.2.

While (1.4) describes the proportionality between stress and strain, precise measurements show that this is not the case for high strains. As will be discussed later in detail, material changes in the microstructure by residual stresses, texture, plastic deformation, fatigue damage, creep damage, embrittlement, and presence of micro-cracks lead to deviations from Hooke's law. In addition to the second order elastic constants c_{ij}, higher-order terms have to be considered in the stress–strain relation by taking into account the third order elastic constants (TOEC) c_{ijk} and the fourth order elastic constants (FOEC) c_{ijkl}:

$$\sigma_i = c_{ij}\varepsilon_j + c_{ijk}\varepsilon_j\varepsilon_k + c_{ijkl}\varepsilon_j\varepsilon_k\varepsilon_l \tag{1.36}$$

The consequences and conditions for introducing higher-order elastic constants were worked out in detail long ago by Murnaghan [200]. In turn, as discussed in Sect. 1.1, the deviations from the parabolic part of the interatomic potential (Lennard–Jones-potential) can be taken care of by the introduction of the third and fourth order elastic constants. Table 1.1 lists the number of independent second, third, and fourth order elastic constants for different crystalline structures (Table 1.2).

Table 1.1 Number of independent second, third and fourth order elastic constants in different crystal symmetries

Number of independent elastic constants in different crystal symmetries			
Symmetry	SOEC (c_{ij})	TOEC (c_{ijk})	FOEC (c_{ijkl})
Triclinic	21	56	126
Monoclinic	13	32	70
Orthorhombic	9	20	42
Trigonal	7/6	20/14	42/28
Tetragonal	6/5	16/12	36/25
Hexagonal	5	12/10	24/19
Cubic	3	8/6	14/11
Isotropic	2	3	4

1.9 Ultrasonic Velocity in Polycrystalline Materials

1.9.1 Basic Relations

In polycrystalline materials, whether consisting of cubic crystallites or crystallites of other symmetries, an ultrasonic wave will propagate through the individual crystals in different directions, i.e. it is highly unlikely that the *k*-vector is parallel to one of the crystals main orientations. Neglecting scattering at the grain boundaries, an average velocity will result. Furthermore, it is easy to understand that in textured materials, resulting e.g. from mechanical and/or thermal treatments, three different bulk velocities can result due to potentially preferred directions of the polycrystalline structure as a result of the production process. Such processes can be evaluated by online velocity measurements, see Sect. 1.15.

As discussed in the preceding Sect. 1.8, one can measure the elastic constants of crystals by orienting the wave-vector into the various directions of the crystal structure and measure the time-of-flight of the ultrasonic waves, and hence the velocity. Knowing the density, one inverts the proper equation from (1.25) to (1.31). It does not matter that the ultrasonic pulse length ($\approx 10^{-3}$ m) is orders of magnitudes larger than the unit cell (length about 10^{-9} m). The measurement of the sound velocity in polycrystals, however, leads to averaged values for the elastic constants.

There are two well-known approximations to calculate the averaged elastic moduli for polycrystalline aggregates from their respective single crystal elastic constants. Voigt [298] averaged over all lattice directions assuming that the strain is uniform over a grain. Reuss [235], in contrast, assumed that the stress is uniform throughout

Table 1.2 Single crystal elastic constants and density values of cubic materials at room temperature. The polycrystal bulk and shear moduli values are calculated using the single crystal elastic constants employing Hershey-Kröner-Eshelby model [157]. Young's modulus, Poisson's ratio and ultrasonic velocities are calculated using bulk and shear moduli, and density. The table lists A^2 and A_f calculated according to (1.32) and (1.33). Elastic properties for a few non-crystalline and ceramic materials are also included. Further data for the cubic system can be found in [281]

Element	c_{11} [GPa]	c_{12} [GPa]	c_{44} [GPa]	ρ [g/cm^3]	B [GPa]	G [GPa]	E [GPa]	υ	v_L [mm/μs]	v_T [mm/μs]	A^2 [10^2] [GPa^2]	A_f	Reference
Ag	122.20	90.7	45.4	10.50	101.20	30.22	82.45	0.3642	3.6709	1.6965	35.165	2.88	[157]
Al	106.75	60.41	28.34	2.70	75.86	26.16	70.38	0.3454	6.4076	3.1142	1.069	1.223	[157]
Au	192.90	163.8	41.5	19.30	173.50	27.89	79.40	0.4237	3.3040	1.2020	29.052	2.852	[157]
C (Diamond)	1076.0	125	575.8	3.51	442.0	533.27	1140.96	0.0698	18.1193	12.3224	402.404	1.211	[157]
Cr	350.00	67.8	100.8	7.20	161.87	115.31	279.54	0.2122	6.6208	4.0018	64.964	0.714	[157]
Cu	169.60	122.4	75.4	8.93	138.13	48.25	129.66	0.3436	4.7616	2.3245	107.330	3.195	[157]
Fe	231.40	134.7	116.4	7.87	166.93	82.45	212.38	0.288	5.9305	3.2363	185.232	2.407	[157]
Ge	128.53	48.26	66.8	5.32	75.02	54.52	131.66	0.2075	5.2677	3.2003	28.441	1.664	[157]
Ir	599.47	255.82	268.82	22.52	370.37	224.95	561.23	0.2474	5.4557	3.1605	376.166	1.564	[157]
K	3.700	3.14	1.88	0.851	3.33	0.94	2.58	0.3709	2.3198	1.0509	0.1024	6.714	[157]
Li	13.500	11.44	8.78	0.533	12.13	4.05	10.94	0.3497	5.7349	2.7573	2.403	8.522	[157]
Mo	463.70	157.8	109.2	10.23	259.77	125.01	323.19	0.2926	6.4571	3.4960	76.565	0.714	[157]
Na	7.390	6.22	4.19	0.971	6.61	2.05	5.56	0.3597	3.1012	1.4517	0.520	7.162	[157]
Nb	240.19	125.58	28.22	8.57	163.78	37.65	104.92	0.3932	4.9969	2.0961	20.23	4.925	[157]
Ni	248.10	154.9	124.2	8.91	185.97	84.71	220.63	0.3023	5.7921	3.0834	240.87	2.665	[157]
Pb	49.660	42.31	14.98	11.34	44.76	8.88	24.98	0.407	2.2336	0.8847	5.112	4.076	[157]
Pd	227.10	176.04	71.73	12.04	193.06	48.28	133.69	0.3846	4.6244	2.0026	85.378	2.810	[157]
Pt	346.70	250.7	76.5	21.50	282.70	63.69	177.73	0.3952	4.1351	1.7212	32.490	1.594	[157]

(continued)

Table 1.2 (continued)

Element	c_{11} [GPa]	c_{12} [GPa]	c_{44} [GPa]	ρ [g/cm^3]	B [GPa]	G [GPa]	E [GPa]	υ	v_L [mm/µs]	v_T [mm/µs]	A^2 [10^2] [GPa^2]	A_f	Reference
Pu	36.280	26.73	33.59	15.75	29.91	16.12	40.99	0.2716	1.8066	1.0116	33.212	7.035	[157]
Rb	2.890	2.45	1.64	1.56	2.60	0.79	2.16	0.3616	1.5301	0.7123	0.0807	7.454	[157]
Si	165.78	63.94	79.62	2.33	97.89	66.62	162.91	0.2226	8.950	5.3462	32.948	1.564	[157]
Ta	260.23	154.46	82.55	16.63	189.72	69.22	185.14	0.3374	4.1185	2.0404	35.200	1.560	[157]
Th	75.30	48.9	47.8	11.70	57.70	29.05	74.62	0.2845	2.8716	1.5760	47.886	3.621	[157]
V	228.70	119.0	43.15	6.02	155.57	47.52	129.39	0.3614	6.0295	2.8091	5.476	0.787	[157]
W	522.39	204.37	160.83	19.26	310.38	160.10	409.83	0.2799	5.2156	2.8834	0.360	0.982	[157]
X6CrNi18.12	216	145	129	7.94	168.6667	78.41	203.66	0.2988	5.8659	3.1424	349.69	3.633	[156]
β-brass	120.00	105	78	8.4	110.0	34.10	92.72	0.3595	4.3021	2.0148	198.81	10.40	[133]
Steel				7.85	166.67	80.0	210.0	0.29	5.90	3.230			[156]
Graphite				1.40	11.70	5.60	14.49	0.2936	3.70	2.0			[156]
Al_2O_3 ceramic				3.90	242.41	159.70	392.70	0.23	10.80	6.40			[90]
HP-SIC				3.20	221.14	180.0	424.60	0.18	12.0	7.50			
HP-Si_3N_4				3.20	172.68	119.10	290.10	0.22	11.0	6.10			
Quartz glass				2.20	31.65	31.76	71.40	0.124	5.80	3.80			
Hard rubber				1.10	3.12	2.41	5.75	0.193	2.40	1.480			
Acrylic glass				1.09	3.11	1.78	4.48	0.26	2.18	1.250			
Polytetrafluor-ethylene				1.14	6.26	1.49	4.13	0.39	2.67	1.147			

(continued)

Table 1.2 (continued)

Element	c_{11} [GPa]	c_{12} [GPa]	c_{44} [GPa]	ρ [g/cm^3]	B [GPa]	G [GPa]	E [GPa]	υ	v_L [mm/μs]	v_T [mm/μs]	A^2 [10^2] [GPa^2]	A_f	Reference
Polyethylene				0.92	3.18	0.268	0.78	0.459	1.95	0.540			[128]

a grain. Considering that the strain energy in any isotropic material can be expressed in the terms of the principal stresses and the principal strains, Hill [95] showed that $B_R \le B \le B_V$, $E_R < E < E_V$, and $G_R \le G \le G_V$ hold. Here, $B_{R,V}$ are the bulk moduli, $E_{R,V}$ are the Young's moduli, and $G_{R,V}$ are the shear moduli in Reuss and Voigt approximations, respectively. For the shear velocity one obtains for these approximations [32, 78]:

$$v_{T,Voigt} = \sqrt{\frac{c_{11}-c_{12}+3c_{44}}{5\rho}} = \sqrt{\frac{c_{44}+A/5}{\rho}}; \quad v_{T,Reuss} = \sqrt{\frac{5(c_{11}-c_{12})c_{44}}{(4\,c_{44}+3(c_{11}-c_{12}))\rho}} \tag{1.37}$$

For the longitudinal sound velocity, we could only find the expression in the Voigt approximation [275]. Furthermore, using the Reuss values for the bulk and the shear modulus for a cubic lattice [78] and employing (1.42) which follows below, one obtains:

$$v_{L,Voigt} = \sqrt{\frac{c_{11}-2A/5}{\rho}}; \quad v_{L,Reuss} = \sqrt{\frac{(2c_{11}-c_{12})c_{44}+(3c_{11}-2c_{12})c_{12}+c_{11}^2}{(c_{44}+3(c_{11}-c_{12}))\rho}} \tag{1.38}$$

The differences between the two approximations amount to $\approx 10\%$. Hill [95] proposed to take the arithmetic mean of the Voigt and Reuss approximations, leading to a more satisfying agreement between measurements and theory. Further theoretical details can be found in [31, 78, 275] and more practical details and insight in [4, 157]. Finally, averaging the squared sound velocities for a cubic lattice in [100], [110], and [111]-directions, leads to

$$v_{L,poly} = \sqrt{\frac{c_{11}-7A/18}{\rho}}; \quad v_{T,poly} = \sqrt{\frac{c_{44}+7A/36}{\rho}} \tag{1.39}$$

In addition to the above approaches, several other models have been proposed for correlating single crystal elastic coefficients to the quasi-isotropic polycrystalline elastic constants [157]. There are only two independent elastic constants required to describe the elastic properties of an isotropic polycrystalline material. The natural choice are the bulk and shear moduli, as they represent extreme deformation cases. The bulk modulus represents volume change without a shape change, whereas, the shear modulus represents shape change without a volume change. As the bulk modulus B is a rotational invariant of the elastic-stiffness tensor, it is always given by: $B = (C_{11} + 2C_{12})/3$. Out of the various proposed models, [157] reported the closest agreement of the experimental polycrystalline value of the shear modulus with that obtained by the Hershey-Kröner-Eshelby model for cubic materials. As per this model, the shear modulus G is given by the positive root of the cubic equation $G^3 + \alpha G^2 + \beta G + \gamma = 0$ where $\alpha = (5C_{11} + 4C_{12})/8$, $\beta = -C_{44}(7C_{11} - 4C_{12})/8$

and $\gamma = -C_{44}(C_{11} - C_{12})(C_{11} + 2C_{12})/8$. Hence, Young's modulus, Poisson's ratio and ultrasonic velocities can be calculated using the bulk and the shear moduli, and the density. The magnitudes of the sound velocities according to (1.39) also deviate only a few percent from the values of (1.37) and (1.38). The consequence of these small but significant differences is that instead of using theoretically based averaged elastic constants or velocities, elastic constants are defined which are adapted to the measurements. The corresponding velocities are:

$$v_L = \sqrt{\frac{\lambda + 2\mu}{\rho}};\ v_T = \sqrt{\frac{\mu}{\rho}} \tag{1.40}$$

These averaged elastic constants are also known as the Lamé constants λ and μ. These equations are applicable for homogeneous quasi-isotropic polycrystalline solids of density ρ, as well as for homogenous isotropic solids. From the measured sound velocities for longitudinal and shear waves, the Lamé-constants λ, μ, Poisson's ratio υ, Young's modulus E, the shear modulus G, and the bulk modulus B are derived:

$$\lambda = \upsilon E/((1 + \upsilon)(1 - 2\upsilon)) = \rho\left(v_L^2 - 2v_T^2\right);$$
$$\lambda + 2\mu = \rho v_L^2;$$
$$\upsilon = \frac{\lambda}{2(\lambda + \mu)} = \frac{1}{2}\frac{v_L^2 - 2v_T^2}{v_L^2 - v_T^2} = \frac{1}{2}\frac{1 - 2\left(\frac{v_T}{v_L}\right)^2}{1 - \left(\frac{v_T}{v_L}\right)^2} \tag{1.41}$$

$$E = 2G(1 + \nu) = \frac{\mu(3\lambda + 2\mu)}{\lambda + \mu} = \rho v_T^2\left(\frac{4 - 3(v_L/v_T)^2}{1 - (v_L/v_T)^2}\right)$$
$$G = \mu = \frac{E}{2(1 + \upsilon)} = \rho v_T^2$$
$$B = \lambda + \frac{2\mu}{3} = \rho\left(v_L^2 - 4v_T^2/3\right) \tag{1.42}$$

Surface waves, also called Rayleigh waves, propagate along surfaces. Their displacement fields extend approximately one wavelength into the depth of the material, with an approximately exponential decay [295]. For the exact calculation of the Rayleigh wave velocity v_R, the secular equation (1.43) with $x = v_R/v_T$ must be solved [128, 231]. Equation (1.44) is an approximate solution for $0 < \upsilon < 0.5$. For $\upsilon = 0.25$ (1.45) holds. For other values of υ see Table 1.3.

$$x^6 - 8x^4 + 8x^2\left(3 - 2\left(v_T/v_L\right)^2\right) - 16\left(1 - \left(v_T/v_L\right)^2\right) = 0 \tag{1.43}$$

$$v_R = ((0.87 + 1.12\upsilon)/(1 + \nu))v_T \tag{1.44}$$

$$v_R \approx 0.92 v_T \tag{1.45}$$

Table. 1.3 Ratio of Rayleigh wave velocity to transverse velocity v_R/v_T for different Poisson's ratios υ based on the solution of the secular equation (1.43) [128, 231]

Poissons's ratio υ	v_R/v_T
0.20	0.912
0.25	0.920
0.30	0.928
0.35	0.935
0.40	0.941
0.45	0.948

If the single-crystal elastic constants are not known for a cubic crystalline material, they may be estimated from velocity and attenuation data obtained from the corresponding polycrystalline material, provided that the grain microstructure is sufficiently uniform, equiaxed and randomly-oriented. Haldipur et al. [83] used the longitudinal and transverse velocities for polycrystals in the Hill average (1.37 and 1.38) and in addition the attenuation for longitudinal waves. The attenuation depends both on the grain size d and the anisotropy factor $A = (c_{11} - c_{12} - 2c_{44})$ (1.32), see Sect. 1.17.2. Knowing the grain size, this allows one to estimate the three elastic constants (c_{11}, c_{12}, c_{44}) by a numerical search procedure which was applied on several billets of Inconel and Waspaloy. In order to test the procedure, a polycrystalline Ni sample with known single crystal elastic constants was examined, yielding a maximum deviation of 25%. For crystalline materials with structures other than cubic, the situation is more complex, because there are several anisotropy factors. Up to date no complete description of the situation is known.

1.9.2 *Monitoring of Microstructural Damage by Ultrasonic Velocity Measurements*

In the above part of this section, the ultrasonic velocities were derived for macroscopically isotropic polycrystalline materials, which are microscopically anisotropic due to their polycrystalline structure. Besides texture resulting in a macroscopically anisotropic material, there are many other effects which cause microscopically induced anisotropy and/or inhomogeneous behavior. Such microscopic changes may e.g. be caused by material damaging processes. With ultrasonic velocity measurements such processes can be monitored. This is discussed in what follows.

Based on ultrasonic velocity measurements in a wide range of alloy systems and data collected from literature, a polynomial correlation between velocity ratio and ultrasonic shear velocity was established for isotropic solid materials [138]. For specific materials/alloy systems, an increase in the velocity ratio R_{UV} is observed with increasing ultrasonic shear wave velocity, see Fig. 1.16. Here, the velocity ratio R_{UV} is defined as

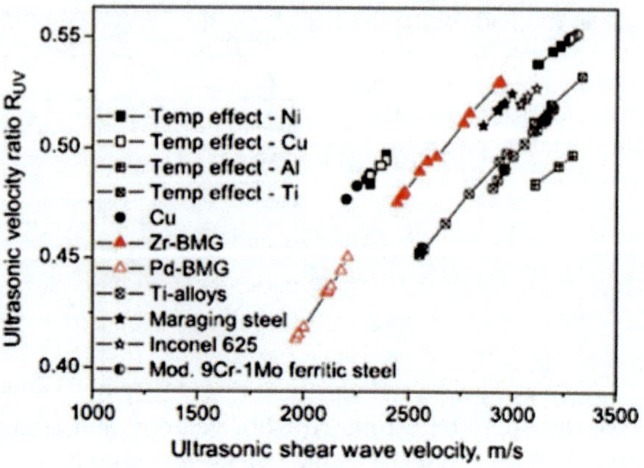

Fig. 1.16 Linear increase of the velocity ratio R_{UV} with increasing ultrasonic shear wave velocity in a wide range of materials (from [138], with permission from Hirzel Verlag)

$$R_{UV} = v_T / v_L \tag{1.46}$$

Because the velocity ratio does not require the thickness of the component to be measured physically, but only time-of-flight values, it has the potential for in-situ materials characterization applications rather than measuring ultrasonic velocity.

Different ultrasonic velocities were obtained by altering the microstructure by employing suitable heat treatments as indicated in Table 1.4. For example, specimens with various volume fractions of different phases in ferritic steel [144] and titanium alloys [143, 296] were produced by subjecting the specimens to heat treatments at different temperatures followed by quenching in oil and water, respectively. Data showing the influence of the temperature on the velocity of various pure metals [155], as well as corresponding data for a Pd-based bulk metallic glass [301], are also included. It can be observed that with microstructural changes in the alloy systems, R_{UV} increases with increasing shear ultrasonic velocity. For a certain range of R_{UV} and v_T, a linear correlation can be obtained as

$$R_{UV} = C + 1.02710^{-4} v_T \tag{1.47}$$

where C is a constant depending upon the alloy system. It can also be seen in Fig. 1.16 that the slopes are almost identical for all alloys. The largest variations in ultrasonic parameters are observed in bulk metallic glasses and Ti-alloys. Both R_{UV} and v_T exhibit minimum values in the as cast metallic glasses and increase with the extent of recrystallization [230, 301]. In case of Ti-alloys, the ultrasonic parameters decrease with increasing volume fraction of the β-phase [142, 143, 296]. A linear correlation between the ultrasonic velocities and the volume fraction of the α-/β-phase was observed in a β-Ti alloy for a range of the volume fraction of the α-phase from 0 to 50%, obtained with heat treatment at different temperatures [296]. In modified 9Cr–1Mo ferritic steel, ultrasonic parameters decreased with increasing heat treatment temperature in the inter-critical temperature region (α + austenite phase region), due to increasing volume fraction of the martensite phase upon quenching [144]. In

Table 1.4 Microstructural variations studied in different alloys and the corresponding range of variations of the ultrasonic velocity and the ultrasonic velocity ratio. The ultrasonic velocities could not be measured in Inconel 625 ammonia cracker tubes and boiler water tubes, due to lack of thickness values (from [138], with permission from Hirzel Verlag)

Materials	Microstructural variations	Range of shear velocity [m/s]	Range of R_{UV}
Ferritic steels	Volume fraction of ferrite and martensite	3272–3310	0.5500–0.5519
Maraging steel	Precipitation of intermetallic phases	2853–3000	0.5104–0.5248
$\alpha + \beta$ and β-Ti-alloys	Volume fraction of different phases	2570–2967	0.4546–0.4978
Zr-base metallic glass	Recrystallization of metallic glasses	2443–2936	0.4752–0.5303
Inconel 625	Precipitation of intermetallics		0.5152–0.5389
Boiler water tube	Hydrogen damage		0.5411–0.5964

M250 maraging steel, ultrasonic parameters increased with increasing precipitation of intermetallic phases (Ni_3Ti and Fe_2Mo) [233].

It is important to evaluate non-destructively the extent of degradation in mechanical properties of cracker tubes used in heavy-water plants, so that they can be subjected to solution heat treatment to regain their mechanical properties and thus any failure due to brittleness of the tube can be avoided. The nickel base alloy Inconel 625 is used in ammonia cracker tubes in the solution treated condition. In such cracker tubes, the velocity of longitudinal waves was found to increase with precipitation of intermetallic phases (Ni_2(Cr, Mo) and γ'') during long-term exposure to intermediate temperature ($\sim$ 873 K) and to decrease with their dissolution during resolution annealing [145]. In this condition, the yield strength is minimum (375 MPa) and the ductility is maximum ($\sim$ 60% uniform elongation). With the precipitation of intermetallic phases during service exposure, yield strength increases to about 1000 MPa along with a drop in ductility to about 6% uniform elongation.

Ultrasonic velocity measurements performed on the service-exposed and post-service heat-treated specimens exhibited that the ultrasonic velocities were minimum in the solution annealed condition and increased with the precipitation of intermetallic phases. However, due to the tube geometry, it was not possible to measure the thickness at the locations of ultrasonic measurements, and hence the absolute values of ultrasonic velocities could not be measured for non-destructive assessment of the tube. However, any change in the ultrasonic velocity due to microstructural changes is always associated with a change of the parameter R_{UV} [138]. Figure 1.17 shows the variations in R_{UV} with in-situ measured surface hardness values. It can be seen that the R_{UV} parameter exhibits minimum values in the virgin/resolution annealed conditions. It increases with increasing hardness upon service exposure. A large range of R_{UV} versus hardness values was observed in the tubes exposed to 120,000

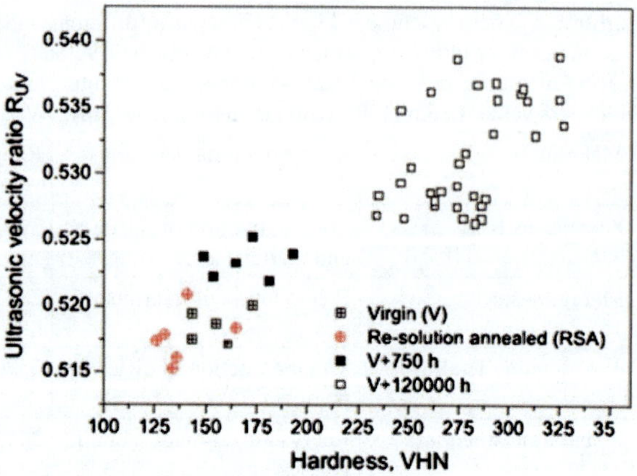

Fig. 1.17 Variation in ultrasonic velocity ratio with hardness for Inconel 625 ammonia cracker tubes in different service conditions (from [138] with permission from Hirzel Verlag)

h of service, when they were removed for resolution annealing treatment. Thus, even though the tubes were subjected to a similar duration of service exposure, the extent of degradation was not the same in different tubes. The scatter in the correlation of R_{UV} with hardness was attributed to the fact that different intermetallic precipitates affect the correlation between elastic and plastic properties differently, depending upon their chemical composition and induced coherency strain [145]. This example demonstrates the capability of R_{UV} for in-situ applications for the assessment of service induced degradation in materials properties, where only ultrasonic time-of-flight data can be measured.

1.10 Measurement of Elastic Properties by Dynamic Techniques

Elastic moduli of materials are determined in tensile or compressive test machines with special measures in order to take into account the machine compliance. However, it is often more convenient to obtain the elastic moduli with dynamic techniques. In Sects. 1.8 and 1.9, it was discussed that by measuring time-of-flights of ultrasonic signals and knowing their path-length, their orientation, their polarization, and the density of the material or component, one can derive the elastic constants using (1.25–1.31, 1.35, 1.41 and 1.42). The measurements can be made at different positions in order to collect information about the inhomogeneous distribution of the elastic properties due to texture, stresses or other inhomogeneities such as micro-cracks [210], and thus provide a measure of quality.

There are dynamic testing techniques allowing one to measure the eigenresonances of simply shaped components made of the material of interest in order to obtain their Young's and shear moduli, and their Poisson's ratios [206, 216, 277].

Furthermore, there are ASTM standards as guidelines for carrying out such measurements [6]. Experimental and commercial apparatuses were developed which can be operated up to about 2000 °C [237]. The components are usually excited with an impact force and they are simply supported at their rims in order not to influence the resonance frequencies. The frequencies of the ensuing oscillations as well as their amplitude decay are often measured with microphones. From these parameters the elastic moduli and the internal friction constants are determined. The frequencies are in the sonic range and hence this technique is sometimes called sonic testing. An overview of the various applications can be found in [311] and the theoretical foundations for the eigenmodes of the plates in [61, 74, 197, 204] and other textbooks of mechanical engineering. Some eigenfrequencies of differently shaped bodies are given here:

(1) Thin rectangular plates with thickness h and side-lengths a and b, Young's modulus E: if the rims of the plates are simply supported, their resonance frequencies $\omega_{n,p}$ are given by

$$\omega_{n,p} = \pi^2\left(\frac{n^2}{a^2} + \frac{p^2}{b^2}\right)\sqrt{\frac{D}{\rho h}} \tag{1.48}$$

where n and p indicate mode number and mode shape, respectively with $D = Eh^3/(12(1-\nu^2))$ also called as flexural plate stiffness. The first fundamental frequency of vibration corresponds to the value $n = p = 1$ yielding

$$\omega_{1,1} = \pi^2\left(\frac{1}{a^2} + \frac{1}{b^2}\right)\sqrt{\frac{D}{\rho h}} \tag{1.49}$$

which can be easily inverted to obtain E from the natural frequency of the fundamental mode. Every frequency is connected with a corresponding mode shape [61]. The resonance frequencies of circular plates can be found in [44].

(2) From the fundamental flexural mode of a rectangular bar, Young's modulus can be obtained [6]:

$$E = 0.9465\frac{mf^2}{b}\frac{L^3}{t^3}T_{corr} \tag{1.50}$$

where f is the resonance frequency, m is the mass of the bar, L is its length, t is its thickness, b is its width, and T_{corr} is a correction factor of the order of 1 depending on Poisson's ratio and the ratio L/t.

(3) The shear modulus G is determined from torsional resonances of bars, either rectangular or cylindrical. The fundamental torsional resonance of a cylindrical bar yields [6]:

$$G = 16mf_t^2(L/\pi d^2) \tag{1.51}$$

where m is the mass of the cylindrical bar, L is its length, and d is its diameter. The torsional resonance frequency is f_t.

(4) For a rectangular bar of length L, width t, and thickness b, the shear modulus G is related to the eigen-frequency of the first torsional mode as [6]:

$$G = f_t^2 \frac{4mL}{bt}(B_c/(1 + A_{corr})) \tag{1.52}$$

with

$$B_c = \left(\frac{b/t + t/b}{4(t/b) - 2.52(t/b)^2 + 0.21(t/b)^6} \right) \tag{1.53}$$

and A_{corr} is a correction factor of the order of 10^{-2}.

From an inverse problem view-point, monitoring the eigen-frequency of complicated shaped components becomes a quality assurance tool in order to determine deviations from their geometries and/or homogeneity [91].

The dependence of the resonance frequencies of small bars on the boundary conditions was the key idea to exploit the eigenmodes of the cantilevers in atomic force microscopes as probes to measure elastic constants, damping, and friction properties on a nanoscale [222, 229, 256, 314, 324].

Some results of elastic constants obtained by dynamic tests for Fe–Ni alloys manufactured by powder metallurgy are shown in Fig. 1.18 in comparison with ultrasonic measurements based on (1.41 and 1.42) [216]. The purpose of this study was to monitor the sintering process by measuring the elastic parameter. In focus were in particular the effects of porosity and pore anisotropy.

The difference of elastic data obtained by the various techniques was at most 1%. From a fundamental point-of-view, there is always a small difference in elastic data measured with high-frequency ultrasonics and static methods such as tensile testing with the dynamic techniques in between. It originates from the fact that absorption leads to a softening of the material which becomes noticeable always below the frequencies where the absorption is maximum. This is a consequence of the Kramers–Kronig relations, see Sect. 1.24.

In this context, it should be mentioned that there is also a difference between the elastic moduli defined and measured for an adiabatic deformation where no heat flows out of a deformed volume, i.e. at constant entropy, and the elastic constants measured at a constant temperature T, i.e. isothermal conditions. However, the differences between adiabatic and isothermal elastic moduli are not significant in the context of this book, and the reader is referred to [78] for a thorough discussion.

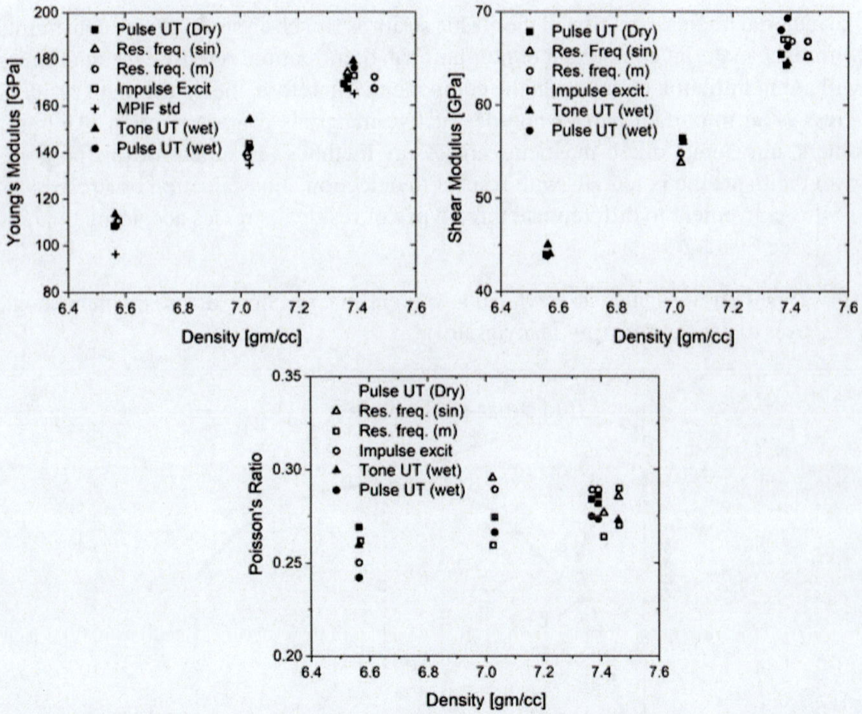

Fig. 1.18 Young's modulus, shear modulus, and Poisson's ratio as a function of density for an iron-nickel powder-metallurgical alloy. Ultrasonic measurements were made by pulse excitation of a transducer with a negative pulse ("Impulse excit"), dry-coupled to the sample, and by a tone-burst generator ("Tone UT") wet-coupled to the sample under test. The dynamic measurements were made with an impulse excitation ("Impulse excit"), random noise excitation ("m") and sinusoidal wave excitation ("sin") (from [216] with permission from Springer Nature)

1.11 Measurement of Stresses using Ultrasonic Velocity

1.11.1 Strains and Stress in Materials

Materials in use have to withstand mechanical loads leading to tensile or compressive stresses and/or shear stresses. The material's response to the applied load is strain, which leads to stress in the material. The strain and stress amplitudes are described in the linear elastic regime by Hooke's law (1.4) and in the nonlinear regime by its extension, taking into account the higher-order elastic constants, see (1.36). After removing the external load, the elastic strains will disappear completely, also the non-linear part. However, when the stress values exceed the elastic limit, i.e. the yield strength Y_U, the material deforms plastically and permanent deformation sets in (Fig. 1.19). The mechanical stress in the material continues to increase with strain beyond the elasticity limit due to strain hardening. Upon removal of the applied load, the material retains a fraction of the elastic strain as frozen-in strain, due to constraints within the material/component or non-uniform deformation. A tensile residual stress will aid in initiation of cracks in the component. Therefore, the detection of residual stress is an important part of non-destructive materials characterization. Amongst others, ultrasonic, micro-magnetic, and X-ray methods are suited for this purpose, with their specific potentials with respect to detection and evaluation of stress.

It is convenient to differentiate three types of residual stresses according to [175, 263], see Fig. 1.20:

(1) Type I stresses, also abbreviated as σ_I, are macroscopic stresses which extend over distances covering many grains;

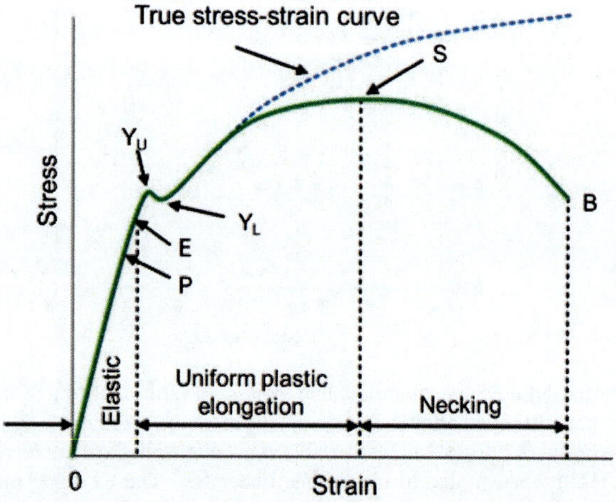

Fig. 1.19 Stress-strain-diagram for mild steel. Here, P is the proportional limit beyond which the linearity of stress–strain graph ceases, E is the elastic limit beyond which macroscopic plastic deformation sets in, Y_U is the upper elastic limit or yield stress, and Y_L is the lower elastic limit. In modified from www.substech.com; licensed under a Creative Commons Attribution 3.0 License

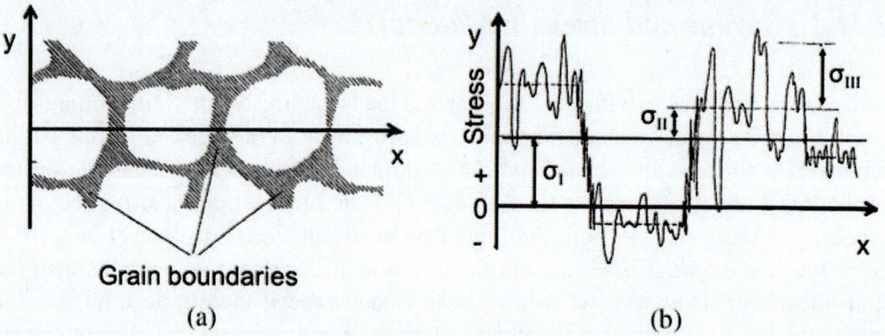

Fig. 1.20 **a** Schematic view of grain boundaries; **b** definition of stresses of the first (σ_I), second (σ_{II}), and third kind (σ_{III}) (in modified form from [176], with permission from Springer Nature)

(2) Type II stresses, σ_{II}, are microscopic stresses within single grains.
(3) Type III stresses, σ_{III}, are localized stresses caused by grain boundaries, small inclusions, and dislocations.

The locally given total residual stress σ_{RS} is the sum of the above defined stresses:

$$\sigma_{RS} = \sigma_I + \sigma_{II} + \sigma_{III} \tag{1.54}$$

While the measurement of stresses using X-rays is restricted to near surface regions due to their limited penetration depth, typically about 10–20 μm (see Chap. 2, Sect. 2.4.3), ultrasonic methods are suited for surface as well as for volume stress measurements. With ultrasound due to its finite beam diameter and its finite pulse length, both of which are large compared to the scale length of the microstructure of a material, only macroscopic stresses of the first kind (σ_I) can be measured. Higher resolution techniques, such as acoustic microscopy or micro-Raman scattering [56]), allow one with some effort to measure second order stresses (σ_{II}) [45], or in the case of atomic force acoustic microscopy even third order stresses (σ_{III}) [115, 222, 228]. Although Raman scattering is an optical technique, the scattering process involves an optical phonon, i.e. high-frequency sound waves whose frequencies are in the optical range. When photons scatter at the so-called optical phonon branch of the lattice oscillations, their momentum and energy must be conserved [129]. The oscillation frequency is stress dependent and there is a close analogy to the stress dependent sound velocity [48].

In the following sections, we discuss the measurement of stresses by ultrasonic techniques. The ultrasonic velocity is sensitive to stresses, because they strain the lattice which changes the curvature of the interatomic potential and the equilibrium position of the atoms, see Fig. 1.1. This entails that the velocity is determined by a combination of the second order and third order elastic constants. Besides the articles cited in the next section, review articles which discuss various material and solid-state physics aspects of stress measurements by ultrasound, can be found in [2, 212]. We discuss third-order elastic constants and their determination in the next section.

1.11.2 *Third-Order Elastic Constants of Polycrystalline Materials and Sound Velocity*

The non-linear stress–strain relations discussed above (1.36) for single crystal materials apply to polycrystalline materials as well. The higher-order elastic constants for the ensemble of grains in the microstructure are then averaged values, like for the second order constants. Starting from the extended Hooke's law and taking into account the second (c_{ij}) and third order elastic constants (TOEC) (c_{ijk}) for crystalline materials, we get:

$$\sigma_i = c_{ij}\varepsilon_j + c_{ijk}\varepsilon_j\varepsilon_k \tag{1.55}$$

Hughes and Kelly [113] found a solution for the equation of motion for ultrasonic waves for a polycrystalline ensemble. For a quasi-isotropic polycrystalline material, the TOEC reduce to the three constants *l*, *m*, and *n*, the so-called Murnaghan constants. They obtained three ultrasonic velocities for a quasi-isotropic polycrystalline ensemble, one longitudinal wave velocity v_{11} and two shear wave velocities v_{12} and v_{13}, allowing one to calculate the stress related strains ε_1, ε_2 and ε_3 with sound velocity measurements. Here, 1, 2, and 3 are the three orthogonal directions in space; λ and μ are the Lamé constants, see (1.40–1.42), and $\varepsilon_1 + \varepsilon_2 + \varepsilon_3 = \theta$ is the volume change caused by the strains:

$$\rho v_{11}^2 = \lambda + 2\mu + (2l + \lambda)\theta + (4m + 4\lambda + 10\mu)\varepsilon_1 \tag{1.56}$$

$$\rho v_{12}^2 = \mu + (\lambda + m)\theta + 4\mu\varepsilon_1 + 2\mu\varepsilon_2 - 0.5n\varepsilon_3 \tag{1.57}$$

$$\rho v_{13}^2 = \mu + (\lambda + m)\theta + 4\mu\varepsilon_1 + 2\mu\varepsilon_3 - 0.5n\varepsilon_2 \tag{1.58}$$

The direct measurement of stress by measuring the strains using (1.56–1.57), is not possible due to several reasons. At the point of measurement v_L and v_T have to be known with appropriate accuracy. Then, there are velocity variations caused by inhomogeneities due to the manufacturing processes. Finally, the length of the sound path must be known with the same accuracy as the time-of-flight for the ultrasonic signals. In practice this is not achievable. However, stresses can be measured based on relative ultrasonic time-of-flight data. Before we discuss the details, let us consider again the one-dimensional case. In this case, the stress–strain relation for materials possesses one higher-order elastic constant c_{112}, see (1.36 and 1.55):

$$\sigma_1 = c_{11}\varepsilon_1 + c_{112}\varepsilon_1^2 \tag{1.59}$$

Here, ε_1 is the strain and σ_1 is the stress. From the wave equation:

$$\rho\frac{\partial^2 u}{\partial t^2} = \frac{\partial \sigma_1}{\partial x} \tag{1.60}$$

we get with (1.59):

$$\frac{\partial \sigma_1}{\partial x} = c_{11}\frac{\partial \varepsilon_1}{\partial x} + 2c_{112}\varepsilon_1\frac{\partial \varepsilon_1}{\partial x} = (c_{11} + 2c_{112}\varepsilon_1)\frac{\partial \varepsilon_1}{\partial x} \tag{1.61}$$

$$\frac{\partial^2 u}{\partial t^2} = \left(\frac{c_{11} + 2c_{112}\varepsilon_1}{\rho}\right)\frac{\partial \varepsilon_1}{\partial x} = \left(\frac{c_{11} + 2c_{112}\varepsilon_1}{\rho}\right)\frac{\partial^2 u}{\partial x^2} \tag{1.62}$$

This is a nonlinear wave equation where the sound velocity depends on strain and hence on stress:

$$\rho v_1^2 = c_{11} + 2c_{112}\varepsilon_1 \tag{1.63}$$

The strain ε_1 consists of a dynamic part, due to the ultrasonic wave and a static part caused by the external or residual stress. The strain amplitude of an ultrasonic wave is usually smaller than 10^{-6} and hence much smaller than the external or residual strains. Therefore, the stress dependence of the sound velocity in (1.63) can be linearized provided that $c_{112}\varepsilon_1$ is small enough compared to c_{11}. In this case one can expand the velocity as a function of static strain to first order, which yields

$$v = ((c_{11} + c_{112}\varepsilon_1)/\rho)^{0.5} = v_0\left(1 + \frac{2c_{112}\varepsilon_1}{c_{11}}\right)^{0.5} \approx v_0\left(1 + \frac{c_{112}}{c_{11}}\varepsilon_1\right) = v_0\left(1 + \frac{c_{112}}{c_{11}^2}\sigma_1\right) \tag{1.64}$$

Here σ_1 is the stress caused by the strain ε_1. This equation shows that the stress dependence of the sound velocity contains second and higher-order elastic constants. Higher-order elastic constants possess about the same magnitude as the second order constants, see Table 1.5. Therefore, the approximation made for (1.64) is fulfilled in most applications of NDMC.

The TOEC can be obtained if test specimens are made from the material of interest, which can be subjected to a defined strain ε upon measurements of the sound velocities. Such measurements are carried out in a tensile test machine in samples subjected to an uniaxial strain ε, as shown in Fig. 1.21a [55]. Five independent velocities must be measured, and from the $\Delta v_{ij}/v_{ij}$ versus strain curves the full set of λ, μ, l, m, and n can be determined. For a uniaxial load in 1-direction, as shown in Fig. 1.21a, $\varepsilon = \varepsilon_1$, $\varepsilon_2 = \varepsilon_3 = -\,\upsilon\varepsilon$ holds. Equations (1.56–1.58) then become [20, 55]:

$$\rho v_{11}^2 = \lambda + 2\mu + (4(\lambda + 2\mu) + 2(\mu + 2m) + 2\mu\upsilon(1 + 2l/\lambda))\varepsilon \tag{1.65}$$

$$\rho v_{12}^2 = \rho v_{13}^2 = \mu + (4\mu + n\upsilon/2 + m(1 - 2\upsilon))\varepsilon \tag{1.66}$$

The velocities of the plane waves propagating perpendicular to the uniaxial stress are:

$$\rho v_{22}^2 = \rho v_{33}^2 = \lambda + 2\mu + (2l(1 - 2\upsilon) - 4\upsilon(m + \lambda + 2\mu))\varepsilon \tag{1.67}$$

$$\rho v_{21}^2 = \rho v_{31}^2 = \mu + ((\lambda + 2\mu + m)(1 - 2\upsilon) + n\upsilon/2)\varepsilon \tag{1.68}$$

$$\rho v_{23}^2 = \rho v_{32}^2 = \mu + ((\lambda + m)(1 - 2\upsilon) - 6\mu\upsilon - n/2)\varepsilon \tag{1.69}$$

The first index indicates the propagation direction, i.e. the direction of the k-vector, and the second the direction of polarization. From Eqs. (1.65–1.69) the expressions for the relative change of the velocity as a function of strain $(dv_{ij}/v_{0,ij})/d\varepsilon$ can be

Table 1.5 Elastic constants λ, μ, l, m, and n determined by ultrasonic time-of-flight measurements and tensile tests

Second and third order elastic constants λ, μ, *l*, *m*, and *n*						
Material	λ [GPa]	μ [GPa]	*l* [GPa]	*m* [GPa]	*n* [GPa]	References
Armco-iron	110 ± 0.4	82 ± 1	− 348 ± 65	− 1030 ± 70	1100 ± 1100	[113]
Steel ferritic-pearlitic	110	81	− 270	− 580	− 710	[259]
St 42	110	81	− 48	− 503	− 652	[259]
Steel 0.2% C	115	82	− 301 ± 37	− 666 ± 6.5	− 716 ± 4.5	[126]
St E 355	109	82	− 192	− 565	− 724	[328]
Rail steel	116 ± 2	80 ± 2	− 248 ± 3	− 623 ± 4	− 714 ± 3	[55]
Rail steel	110 ± 2	82 ± 2	− 302 ± 3	− 616 ± 4	− 724 ± 3	[55]
Rail steel	112 ± 1	81 ± 1	− 358 ± 5	− 650 ± 3	− 721 ± 3	[260]
Steel Hecla 37	111 ± 1	82	− 461 ± 65	− 636 ± 46	− 708 ± 32	[271]
Steel Hecla 17	111	82	− 328 ± 30	− 595 ± 32	− 668 ± 24	[271]
Steel Hecla 138A	109	82	− 426 ± 55	− 619 ± 50	− 708 ± 40	[271]
Steel Hecla austenite	87	72	− 535 ± 90	− 752 ± 100	− 400 ± 40	[271]
Steel A 533 B	119	79	− 218	− 486	− 564	[198]
Steel A 471	120	79	− 179	-496	-628	[198]
17 Cr Ni Mo 6	109	81	− 58 ± 38	− 517 ± 14	− 718 ± 18	[260]
24 CrMo 5V	112	83	− 350	− 624	− 702	[260]
24 CrMo 5V	112	82	− 440 ± 10	− 600 ± 2	− 670 ± 2	[260]
30 Cr Mo Ni V 511	109	83	− 357	− 574	− 670	[260]
22 Ni Mo Cr 37	109	82	− 185	− 503	− 652	[260]
22 Ni Mo Cr 37	109 ± 2	82 ± 1	− 196 ± 10	− 520 ± 2	− 657 ± 2	[260]
22 Ni Mo Cr 37	111 ± 2	82 ± 1	− 190 ± 10	− 555 ± 2	− 659 ± 2	[260]
24 Ni Cr Mo V 145	110	80	− 90	− 439	− 546	[260]
Ni-Steel	109	82	−328	−578	−676	[260]
Ni-Steel S/NTV	109 ± 1	82	−56 ± 20	−671 ± 6	−785 ± 7	[46]
Ni-Steel Rex 535	91	78	− 46	−590	−730	[46]
Ni-Steel Rex 535	109	82	− 327 ± 75	− 578 ± 80	− 676 ± 60	[271]

(continued)

Table 1.5 (continued)

Second and third order elastic constants λ, μ, l, m, and n

Material	λ [GPa]	μ [GPa]	l [GPa]	m [GPa]	n [GPa]	References
15 Mn Ni 63,	110 ± 2	81 ± 2	– 270 ± 10	– 580 ± 2	– 710 ± 2	[260]
X6 Cr Ni 1811	101	75	– 370	– 532	– 236	[260]
Roll H-Cr R 78/80	129 ± 2	84 ± 2	– 573 ± 8	– 775 ± 3	– 996 ± 3	[260]
WC–Co sintered	178	256	– 464	– 1390	– 2108	[260]
WC–Co sintered	178	257	– 552	– 1453	– 2153	[260]
Al 99	61 ± 1	25	– 47 ± 25	– 342 ± 10	– 248 ± 10	[46]
Al 9905	64	27	– 319 ± 10	– 373 ± 5	– 354 ± 3	[126]
Al 99.3	57	28	– 311 ± 12	– 401 ± 138	– 408 ± 136	[271]
Al B 53 S	62	26	– 201 ± 23	– 305 ± 27	– 300 ± 24	[271]
Al B 53 S	58	26	– 223 ± 40	– 237 ± 20	– 276 ± 20	[271]
Al D 54 S	49	26	– 387 ± 125	– 358 ± 15	– 320 ± 12	[271]
Al JH 77 S	58	27	– 337 ± 25	– 395 ± 24	– 436 ± 28	[271]
$AlMg_3$	56 ± 0.6	28 ± 0.3	– 212 ± 21	– 309 ± 6	– 369 ± 12	[260]
Al Mg_3	58	27	– 245	– 313	– 350	[260]
AlCu2Mg1	45	31	– 218	– 378	-435	[260]
AlZn5.5MgCu(A)	57 ± 1	27 ± 0.3	– 160 ± 80	– 305 ± 30	– 345 ± 35	[260]
AlCuMn	58 ± 1	27 ± 0.3	– 155 ± 54	– 283 ± 14	– 305 ± 6	[260]
AlMgSi0.7	56 ± 1	28 ± 0.3	– 240 ± 84	– 303 ± 30	– 336 ± 13	[260]
AlZn7Cu2Mg2.5	60	27	– 324	– 397	– 403	[260]
Cu 99.85	107	47	– 8378 ± 131	– 2429 ± 16	– 527 ± 4	[126]
Cu 99.9	104	46	– 542 ± 30	– 372 ± 5	– 401 ± 5	[46]
Al_2O_3 (ceramic; ρ = 3.9 g/cm^3)	395 ± 12	160 ± 4.8	– 68 ± 10.2	– 970 ± 97	– 1300 ± 130	[259]
Al_2O_3 (ceramic; ρ = 3.5 g/cm^3)	284 ± 8.5	119 ± 3.6	– 70 ± 10.5	– 600 ± 90	– 865 ± 87	[259]

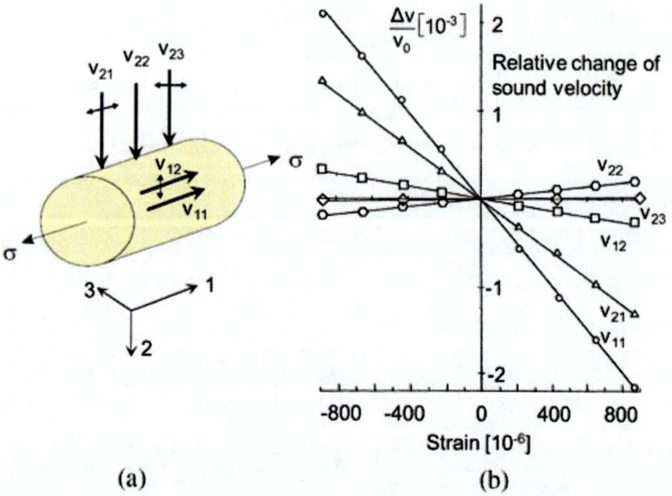

Fig. 1.21 **a** Stressed cylinder in order to measure the relative change $\Delta v_{ij}/v_{ij}$ of sound velocities for longitudinal and shear waves as a function of tensile strain in the elastic regime, here in 1-direction; **b** The change of velocity is largest for a longitudinal wave polarized and propagating in tensile direction (v_{11}). For a strain of 10^{-3}, $\Delta v_{11}/v_{11}$ is of the order of -2×10^{-3}, here for a rail steel. That means that the corresponding AEC is of the order -2. Slightly smaller is $\Delta v_{21}/v_{21}$ for a shear wave polarized in stress direction 1 and with the *k*-vector in 2-direction. The effect of stress is smallest for shear waves polarized perpendicular and propagating perpendicular to the 1-direction, i.e. for the velocities v_{23} and v_{32}. From such measurements, *l*, *m*, and *n* can be obtained by measuring all $\Delta v_{ij}/v_{0,ij}$ versus strain ε in a tensile test machine for different directions and polarizations, see Eqs. 1.75–1.77. As can be seen from the figure, stresses due to an external load entail changes in the ultrasonic velocities up to about 0.5% for metals. They are up to a few % for polymeric and similar materials. Note that $\Delta v_{12}/v_{0,12} \neq \Delta v_{21}/v_{0,21}$, as can be seen from Fig. 1.21b and Eqs. 1.71–1.73. The velocities in the stress direction were measured using angled wedges [(**b**) from [55], with the permission of the Acoustical Society of America]

calculated:

$$AEC_{11} \equiv \left(dv_{11}/v_{0,11}\right)/d\varepsilon = 2 + \frac{2m + \mu + \upsilon\mu(1 + 2l/\lambda)}{(\lambda + 2\mu)} \tag{1.70}$$

$$AEC_{12} \equiv (dv_{12}/v_{0,12})/d\varepsilon = 2 + \frac{\upsilon n}{4\mu} + \frac{m}{2(\lambda + \mu)} \tag{1.71}$$

$$AEC_{22} \equiv (dv_{22}/v_{0,22})/d\varepsilon = -2\upsilon\left(1 + \frac{m - \mu l/\lambda}{\lambda + 2\mu}\right) \tag{1.72}$$

$$AEC_{21} \equiv (dv_{21}/v_{0,21})/d\varepsilon = \frac{\upsilon n}{4\mu} + \frac{\lambda + 2\mu + m}{2(\lambda + \mu)} \tag{1.73}$$

$$AEC_{23} \equiv (dv_{23}/v_{0,23})/d\varepsilon = \frac{m - 2\lambda}{2(\lambda + \mu)} - \frac{n}{4\mu} \tag{1.74}$$

Figure 1.21b shows the relative change of sound velocity, $dv_{ij}/v_{0,ij}$, for the different wave polarizations as a function of strain. The slopes of $dv_{ij}/v_{0,ij}$ *versus* ε, i.e. $(dv_{ij}/v_{0,ij})/d\varepsilon$, are also called acoustoelastic constants (*AEC*). The zero in the subscript of $v_{0,ij}$ stands for the corresponding velocity at zero stress. Most sensitive are longitudinal waves with the wave vector parallel to the stress direction, followed by shear waves with polarization in stress direction. Least sensitive are waves with the polarizations perpendicular to the stress directions. Tensile stresses in general decrease the sound velocity, whereas compressive stresses lead to an increase. In the case of the v_{23} and v_{22} velocities, the increase is due to Poisson's ratio. Solids with negative Grüneisen constants can show the opposite behavior in some glasses and crystals [78, 226]. The phenomenon is relatively rare and therefore it is not discussed here.

The three TOEC constants *l, m, and n* can be determined from the acoustoelastic constants defined in Eqs. (1.70–1.74) [55]:

$$l = \frac{\lambda}{1-2\nu}\left(\frac{(1-\upsilon)}{\upsilon}\frac{dv_{22}/v_{0,22}}{d\varepsilon} + \frac{2}{1+\upsilon}\left(\frac{dv_{21}/v_{0,21}}{d\varepsilon} + \upsilon\frac{dv_{23}/v_{0,23}}{d\varepsilon}\right) + 2\upsilon\right) \tag{1.75}$$

$$m = 2(\lambda+\mu)\left(\frac{\upsilon}{1+\upsilon}\frac{dv_{23}/v_{0,23}}{d\varepsilon} + \frac{1}{1+\upsilon}\frac{dv_{21}/v_{0,21}}{d\varepsilon} + 2\upsilon - 1\right) \tag{1.76}$$

$$n = \frac{4\mu}{1+\upsilon}\left(\frac{dv_{21}/v_{0,21}}{d\varepsilon} - \frac{dv_{23}/v_{0,23}}{d\varepsilon} - 1 - \upsilon\right) \tag{1.77}$$

1.11.3 Bulk Stresses and Acoustical Birefringence

We first consider polarized shear waves with their *k*-vectors insonifying perpendicular to the surface of a component, i.e. parallel to the surface vector and with a reflection path back to the exciting probe which is now acting as a receiver in pulse-echo or single-ended mode. For convenience, the flight path is in direction 1. The polarization or displacement direction of the shear waves can be selected freely in the surface plane and shall be perpendicular to each other. The velocities v_{13} and v_{12} have the same value if the material under test is homogeneous and quasi-isotropic, i.e. without any preferred orientations due to texture, internal or external stresses. If there are stresses in 3- and 2-directions with the corresponding strains ε_3 and ε_2, there are differences between the two velocities and the material is acoustically birefringent in analogy to the same effect in optics. By orientating one polarization direction in the stress direction and the other perpendicular to this direction, the time-of-flight difference between the two polarizations becomes maximum. To obtain an equation for the corresponding stress differences $\sigma_3 - \sigma_2$ (1.58) is subtracted from (1.57) which yields:

$$\rho\left(v_{12}^2 - v_{13}^2\right) = \frac{(4\mu + n)(\varepsilon_2 - \varepsilon_3)}{2} \tag{1.78}$$

Using $(v_{12} + v_{13}) \approx 2v_{12} \approx 2v_{13} \approx 2v_T$, and according to Hooke's law $(\varepsilon_2 - \varepsilon_3) = (\sigma_2 - \sigma_3)/2\mu$ and $\varepsilon_i = \left(\sigma_i - \upsilon\left(\sigma_j + \sigma_k\right)\right)/E$, (1.78) can be rewritten as [258, 260]:

$$\frac{v_{12} - v_{13}}{v_{13}} = \frac{t_{13} - t_{12}}{t_{12}} = (\sigma_2 - \sigma_3)\frac{4\mu + n}{8\mu^2} \tag{1.79}$$

Equation (1.79) contains only one third order elastic constant in order to determine the stress difference $(\sigma_2 - \sigma_3)$. Here, $v_{12} \approx v_T$ and $t_{12} \approx t_T$ are the stress-free velocity and time-of-flight, respectively. The relative measurements remove the need to know the velocities v_{13} and v_{12} at zero stress. The inaccuracies for measuring absolute time-of-flights and sound paths data are thus avoided by using two perpendicularly shear wave polarizations. The advantage of this procedure is that the knowledge of the sound path length is not needed. Equation (1.79) is called the birefringence equation. In an ultrasonic pulse-echo train, one can directly observe the splitting of individual echoes in two partial echoes caused by the different velocities due to stress-differences in the 3- and 2-directions, see Fig. 1.22b. In order to obtain quantitative information of the stress amplitude in a component in a practical situation, the relative time-of-flight differences must be measured with an accuracy of at least 10^{-4}. There are analog expressions to Eqs. (1.78 and 1.79) for shear waves of other polarization and propagation directions [258]. They allow to determine the three unknown stresses σ_i $(i = 1, 2, 3)$. However, measurements for all three sides of a component are not always feasible, but at least the stress differences can be determined for the directions of interest. To measure out-of-plane stresses, ratios of TOF's of longitudinal and transverse velocities must be used.

1.11.4 Hydrostatic Pressure

In case a material is strained by a hydrostatic pressure, $\varepsilon_i = -\varepsilon$ (i = 1, 2, 3) holds and hence only two independent velocities exist. The strain ε is related to the pressure p by the bulk modulus B:

$$3\varepsilon = \Delta V/V = -p/B \tag{1.80}$$

where $\Delta V/V$ is the relative volume change. Two independent sound velocities exist as a function of strain or pressure:

$$\rho v_L^2 = (\lambda + 2\mu) - (7\lambda + 2\mu + 6l + 4m)\varepsilon \tag{1.81}$$

$$\rho v_T^2 = \mu - (3(\lambda + 2\mu) + 3m - n/2)\varepsilon \tag{1.82}$$

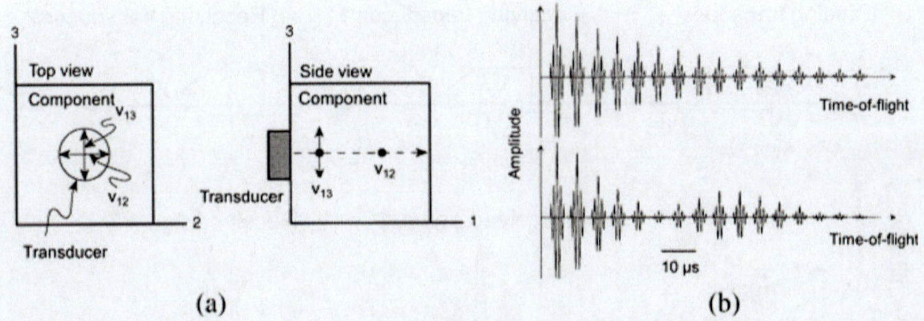

Fig. 1.22 **a**: Transducer orientation in order to obtain the principal stress directions in a component. The arrows indicate the polarization directions; **b** Ultrasonic back-wall echo train. Top: the polarization of the incident shear wave is parallel to a principal stress axis [v_{13} or v_{12} in (**a**)]; bottom: polarization under 45° to this axis. In this case the wave packet can be decomposed into two partial wave packets. One is polarized in the stress direction and the other perpendicular to the stress direction. Each packet propagates with slightly different velocity according to the stress level it experiences. After some time, they are 180° out-of-phase and interfere destructively and thus cause the minimum [(**b**) in modified form from [258], with permission from Elsevier]

1.11.5 Near Surface Stresses and Surface Waves

In order to measure in-plane surface stresses, the use of waves propagating along a surface is a natural choice. With a transmitter–receiver set-up as shown in Fig. 1.23, where two receiving transducers are arranged at a pre-defined fixed distance, time-of-flight differences for different measurement directions are proportional to the surface-stress differences [260]. Two types of waves are suitable for such measurements: Surface waves, also called Rayleigh waves, and surface skimming longitudinal waves also called as critically refracted longitudinal waves (L_{CR}-wave).

Rayleigh waves contain two displacement components perpendicular to each other, one parallel to the *k*-vector, i.e. in propagation direction and the other perpendicular to the surface, see Sect. 1.3. This entails that Rayleigh waves are sensitive to the two stress components of the displacement directions and also to the perpendicular direction due to Poisson's ratio, see Fig. 2.3.6 in [128]. Rayleigh waves also allow to make measurements on curved surfaces of components as long as the radius of curvature is large compared to the wavelength.

Instead of using Rayleigh waves for surface stress measurements, it is also possible to use L_{cr}-waves with the same arrangement as shown in Fig. 1.23, except that the incident angle in the transmitter and the receivers are adjusted to excite and receive longitudinal waves. These waves are not pure modes because longitudinal waves skimming along a free surface do not fulfill the boundary conditions. Therefore, they radiate a head wave into the component inspected, i.e. they are damped [321]. The damping distance is of the order of decimeters to meters and hence this is not a limitation for many applications. The penetration of the L_{cr}-wave was examined by [152] and by [38] in a numerical study.

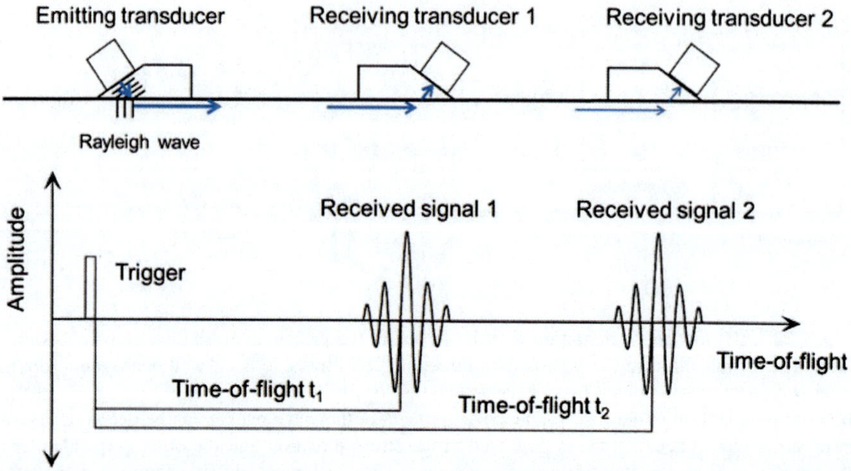

Fig. 1.23 Device for measuring surface stresses using Rayleigh waves (or critically refracted longitudinal surface waves, so-called L_{cr}-waves). Both can be generated by an angled-transducer which matches the wavelength in the plastic wedge to either the Rayleigh wavelength or the skimming longitudinal wave by varying the incidence angle (Snell's law, see Fig. 1.8). They are received by two transducers at fixed distances 1 and 2 using the same principle. In case of the L_{cr}-wave there is geometrical damping. If there is a change of sound velocity due to surface stresses, the stress variations as a function of the direction of the *k*-vector can be measured [260]

Various aspects of the theory of the acoustoelastic effect for surface waves are discussed in a number of publications. Three early authors are [64, 89, 117]. If the extension h of the stresses into the depth of the component becomes comparable to the wavelength λ (which is also its penetration length), i.e. $2\pi h/\lambda = kh \cong 1$, the velocity of the SAW becomes dispersive allowing one to differentiate between the stresses in different depth regions (Fig. 1.24) [96, 195]. By the same token, surface contaminants hinder the application of stress measurements. Similar measurements were carried out with three point-bending tests by [167], so that the surface underwent either tensile stress or compressive stress close to the loading nose of the assembly proving the relatively small sensitivity of Rayleigh waves to surface stresses.

Lee et al. [160] examined the stress dependence of Rayleigh waves in detail by employing scanning acoustic microscopy operating at 225 MHz and equipped with a line-focus lens (L-SAM). A line-focus lens having an opening angle θ_L allows to generate Rayleigh waves and surface skimming longitudinal waves, provided θ_L encompasses their critical angle [147]. Suppose that the surface waves propagate in the 1-direction (Fig. 1.25). The relative velocity change for propagation along the surface in direction 1 is given by

$$\frac{v_1 - v_0}{v_0} = H_1\varepsilon_{11} + H_2\varepsilon_{22} = K_1\frac{\sigma_{11}}{E} + K_2\frac{\sigma_{22}}{E} \tag{1.83}$$

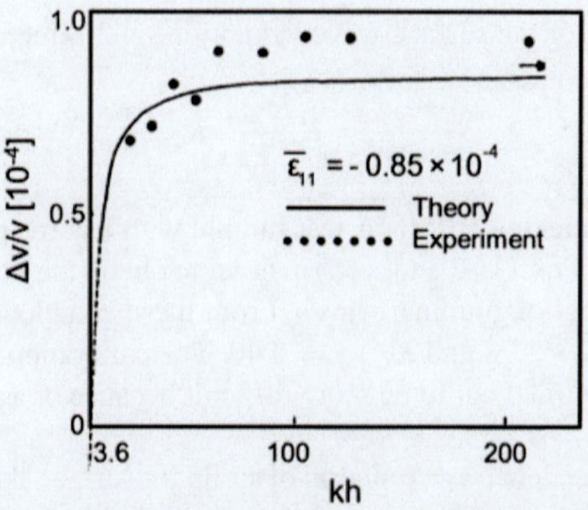

Fig. 1.24 Change of Rayleigh velocity in a plate of mild steel as a function of *kh* where *k* is the wave-vector and *h* is the plate thickness. The plate was bent in order to produce a surface strain and hence surface stress. The Rayleigh wave propagated with the *k*-vector parallel to the direction of uniaxial tensile strain $\varepsilon_{11} = -0.85 \times 10^{-4}$. For this situation $\Delta v/v_0 = -0.31\varepsilon_{11}$ held. The factor -0.31 is the corresponding acoustoelastic constant (Figure replotted with data from [96]), with permission from American Society of Mechanical Engineers

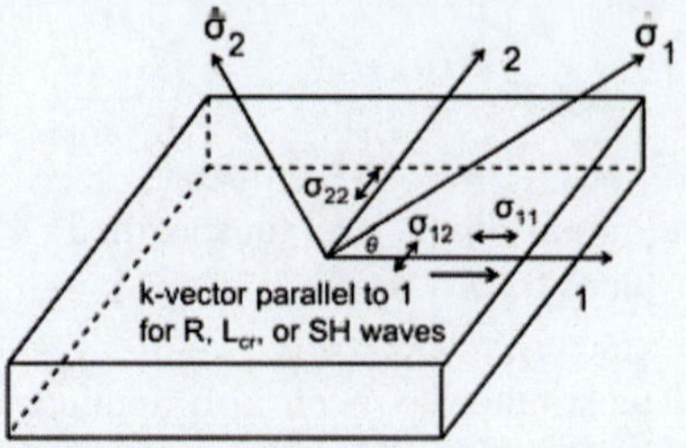

Fig. 1.25 Coordinate system for stress measurements using Rayleigh, surface skimming longitudinal waves, or shear-horizontal (SH) waves. The *k*-vector of the wave mode employed is along the 1-direction

where v_0 is the velocity at zero strain or stress. Here, $\varepsilon_{11}, \varepsilon_{22}, \sigma_{11},$ *and* σ_{22} are the strains and stresses in 1 and 2-directions, respectively. The parameters H_1 and H_2 are the acoustoelastic constants, E is Young's modulus, and K_1 and K_2 are constants. H_1 and H_2 can be expressed in terms of the second and third order elastic constants [64, 117] (see also Eqs. 1.70–1.74). Furthermore, it is assumed that the stress components σ_{13}, σ_{23}, and σ_{33} are zero and that σ_{11}, σ_{22}, and σ_{12} do not vary within the penetration depth of the surface or Rayleigh wave. Finally, stresses σ_{12} do not affect the Rayleigh velocity.

Equation (1.83) can be rewritten in terms of the principal stresses $\hat{\sigma}_1, \hat{\sigma}_2$ and principal directions [160]:

$$\frac{v_1 - v_0}{v_0} = \frac{1}{2}(K_1 + K_2)\left(\frac{\hat{\sigma}_1 + \hat{\sigma}_2}{E}\right) + \frac{1}{2}(K_1 - K_2)\left(\frac{\hat{\sigma}_1 - \hat{\sigma}_2}{E}\right)\cos(2\theta) \qquad (1.84)$$

Applying (1.84) for the surface wave velocity v_2 in 2-directions yields:

$$\frac{v_2 - v_0}{v_0} = K_1 \frac{\sigma_{22}}{E} + K_2 \frac{\sigma_{11}}{E} \tag{1.85}$$

Lee et al. [160] stressed a PMMA test sample with a corresponding device in an L-SAM and applied Eqs. (1.83 and 1.85) in order to obtain the expression for σ_{11}, σ_{22} for surface skimming longitudinal waves. From these equations Lee et al. obtained the factors $K_{Lcr,1} = -1.78$ and $K_{Lcr,2} = 0.40$. The calibration for surface waves in an aluminum alloy turned out to be more difficult because of a rolling texture in the sample.

The velocity dependence as a function of strain for surface skimming longitudinal waves is the same as for bulk waves and thus (1.70) holds:

$$(dv_{11}/v_{11})/d\varepsilon = 2 + \frac{2m + \mu + \nu\mu(1 + 2l/\lambda)}{(\lambda + 2\mu)} \equiv L_{11} \tag{1.86}$$

where L_{11} is the acoustoelastic constant. Like in (1.83), this can be converted into an equation for stress as a function of relative time-of-flights $\Delta t/t_0$ [21, 266]:

$$\Delta\sigma = \frac{E(dv_{11}/v_{11})}{L_{11}} = \frac{E}{L_{11}} \frac{\Delta t}{t_0} \tag{1.87}$$

For three different aluminum alloys, steel (unspecified), 4140 steel, 316L stainless steel, and rail steel, the values for L_{11} are -2.7, -3.1, -3.55 (aluminum), -2.26, -2.2, -2.1, and -2.38 (steel), respectively.

The L_{cr}-wave based technique has been also applied to measure surface/sub-surface residual stresses in carbon steel weld joints. It was found to be very sensitive to small variations in the residual stress profiles across the weld joints induced by marginally different heat inputs during welding [121]. In this case the calibration of L_{11} in (1.86) yielded $L_{11} = -1.52$, slightly smaller than the values above.

Like for other wave modes, the acoustoelastic constants for L_{cr}-waves are obtained by calibration in a tensile-test machine and by measuring the slope of the relative variation of the time-of-flight as a function of stress $\Delta t/(\Delta\sigma t_0)$ (1.87). In the literature, the parameter L_{11}/E is often also called the acoustoelastic constant, but now having the unit [MPa^{-1}] [118]. Likewise, one may omit the reference of the absolute time-of-flight t_0 and then the unit for the acoustoelastic constant is [ns/MPa] [209]. There seems to be no standard definition for the acoustoelastic constant followed by everyone in the field.

The application of L_{cr}-waves propagating along the tread of a rail has been used to evaluate the longitudinal stress in new rails [259]. In order to discriminate the influence of microstructural changes along the length of the rail, the simultaneous application of L_{cr}-waves in the upper part of the web of the rail is recommended. The longitudinal stress is negligibly small in that part and the time-of-flight data represent the value for the stress-free state. Ultrasonic data, evaluated continuously

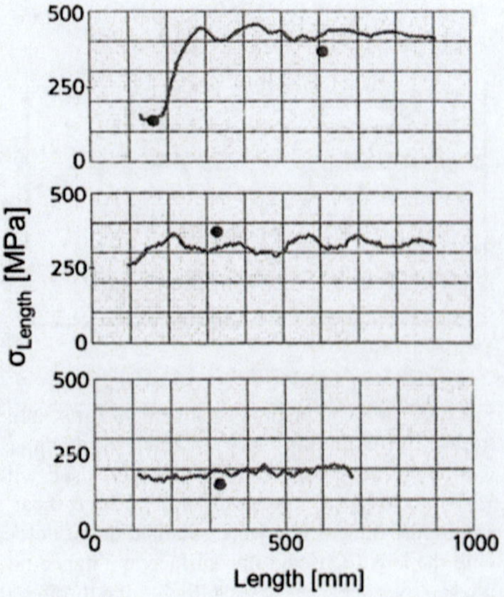

Fig. 1.26 Longitudinal stress in the tread of three freshly manufactured rails as a function of length. The data points were obtained by the time-of-flight technique. The solid lines represent data using the ring-core technique (in modified form from [259], with permission form Elsevier)

along 1 m of rail length are found to be in good agreement with the results of the partly destructive ring-core technique, see Fig. 1.26.

Beside the generation of surface acoustic waves and L_{cr}-waves in an L-SAM, the generation of obliquely generated longitudinal and shear waves can also be used for stress measurements in a scanning acoustic microscope (SAM) using a spherical lens, see Fig. 1.27a and Sect. 1.6. If stresses are present in a component, the various mode-converted waves from the impinging longitudinal wave undergo phase shifts due to the stresses which in turn determine the contrast in the images (Fig. 1.27b–g) [53]. Using the appropriate acoustoelastic constants, the contrast can be calibrated [52]. Likewise, after a control calibration with X-ray diffraction, surface skimming longitudinal waves were used in a SAM in order to quantitatively image stress-fields at a crack tip and other surface discontinuities [251].

1.11.6 Plate Stress and Shear Horizontal Waves

The sensitivity of shear horizontal waves (SH_0-mode) for surface stress measurements in a plate is practically the same as for transverse waves in an infinite bulk volume similarly to the case of surface skimming longitudinal waves. The SH-modes can be excited and detected by electromagnetic transducers (EMAT's), using again the fixed transmitter–receiver arrangement (Fig. 1.23). Texture parallel to the surface vector does not influence the result because of $v_{12,texture} = v_{21,texture}$. The normalized time-of-flight differences between the two different polarization and propagation directions yield directly the stress difference between the two directions [69, 260, 289]:

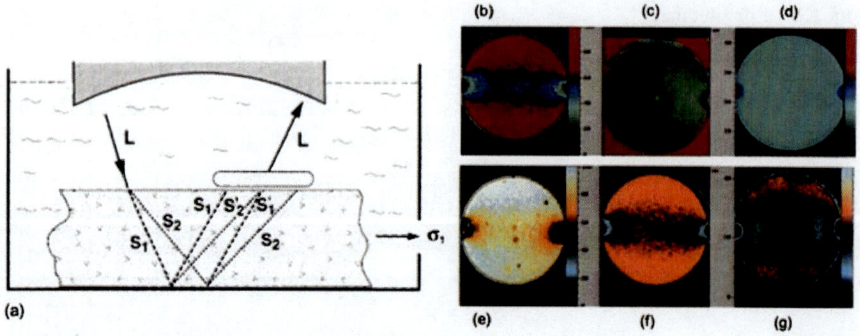

Fig. 1.27 **a** Experimental arrangement for scanning acoustic microscope images of a compressed (to 2002 N) aluminum disk 3 mm thick and 40 mm Ø. A low-frequency scanning acoustic microscope with frequency range 1–150 MHz was used with a 15-MHz transducer with a focal length of 12.7 mm; **b** shear wave peak-amplitude; **c** shear waves time-of-flight; **d** longitudinal wave peak amplitude. Images **b**, **c** were obtained under defocused conditions, whereas image **d** was obtained with the lens focused at the surface and hence no contrast variation can be seen; **e**–**g** stress-maps imaged by a shear pulse travelling twice through the sample thickness at loads of 1745, 2002, and 2436 N (in modified form from [53], with permission from Springer Nature)

$$(\sigma_1 - \sigma_2)_{SH} = \frac{8\mu^2}{8\mu + n}(t_{12} - t_{21})/t_{21} \tag{1.88}$$

1.11.7 Uniaxial Stresses

The load stress in screws and bolts determines the quality of a fastened connection. There are a number of devices on the market which exploit the stress dependence of the velocity of a longitudinal wave in a bolt. An improvement of this technique can be obtained by using longitudinal and shear waves simultaneously. This allows one to measure the stress levels even after the screw has been built-in for a long time and the original length is not known. One exploits that stresses influence the shear velocity less than the longitudinal velocity if the shear wave is polarized perpendicular to the stress direction, here the thickness direction of the screw or the bolt. Furthermore, the path-lengths for the shear and longitudinal waves are identical and hence the strain for both waves is the same for all loads. As can be seen from Fig. 1.21, an increasing velocity difference of the longitudinal and the shear wave velocity develops with increasing strain. Based on Eqs. 1.65 and 1.66 for an uniaxial load, one obtains for the absolute strain in a bolt [69]:

$$\varepsilon_i = \frac{t_L - t_T Q_S}{X_T t_T Q_s - X_L t_L} \tag{1.89}$$

where t_L and t_T are the time-of-flights for the longitudinal and the shear waves and the parameter Q_s is a function of Poisson's ratio only:

$$Q_s = \left(\frac{\mu}{\lambda + 2\mu}\right)^{1/2} = \left(\frac{1 - 2\upsilon}{2(1 - \upsilon)}\right)^{1/2} \tag{1.90}$$

The parameter X_L and X_T contain the second and the third order elastic constants:

$$X_L = 2+\frac{\mu + 2m}{\lambda + 2\mu}+\lambda\mu\left(\frac{1 + 2l/\lambda}{2(\lambda + \mu)(\lambda + 2\mu)}\right) \tag{1.91}$$

$$X_T = 2+\frac{\lambda n}{8\mu(\lambda + \mu)}+\frac{m}{2(\lambda + \mu)} \tag{1.92}$$

1.11.8 Biaxial Surface Stresses

Similar to uniaxial stress states, biaxial stress states can also be measured by using surface skimming longitudinal waves [221]. If, for example, direction 1 is the main stress-direction parallel to the surface plane, the measurement of the longitudinal wave velocity v_{11} allows the determination of the stress σ_1 in a near surface region of some mm depth. The velocity of the longitudinal wave is given by

$$\rho v_{11}^2 = \lambda + 2\mu + A\sigma_1+B(\sigma_2+\sigma_3) \tag{1.93}$$

Here, A and B are combinations of the SOEC and TOEC (λ, μ, l, m, and n respectively), see (1.56). Because A is about one order of magnitude larger than B, one can neglect the last term in (1.93) and one obtains:

$$\frac{(v_{11}-v_{11,0})}{v_{11,0}} = \frac{(t_{11,0}-t_{11})}{t_{11}} = \frac{A}{2(\lambda + 2\mu)}\sigma_1 \tag{1.94}$$

Here, the approximation $v_{11} + v_{11,0} \approx 2v_{11}$ is made because the stress dependence of the velocity is small. Indices with the subscript "0" refer to velocities and time-of-flights in the stress-free case.

If the stresses in thickness direction 3 are small [in the coordinate system of Fig. 1.25 the − 3 direction)], the stresses in directions 1 and 2 can be determined by the measurement of the shear velocities v_{31}, v_{32} and the longitudinal velocity v_{33}. By exchanging the indices of Eqs. 1.79 and 1.56 in the proper way, one obtains:

$$(v_{33}+v_{32}+v_{31})/v_{33} = 1+\sqrt{\mu/(\lambda + 2\mu)}(2 + C(\sigma_1+\sigma_2)) \tag{1.95}$$

$$(\sigma_1+\sigma_2)_{biaxial} = (1/C)\left(\frac{((v_{33}+v_{32}+v_{31})/v_{33}-1)}{(\sqrt{\mu/(\lambda+2\mu)})}\right) - 2 \tag{1.96}$$

Here, C is again a combination of second and third order elastic constants and has the dimension of an inverse stress, see discussion above.

1.11.9 Sound Velocity, Temperature, and Stresses

Starting at very low temperature, the sound velocity decreases in solids with increasing temperature. This is caused by the anharmonicity of the interatomic potential, which is formally taken into account by the higher-order elastic constants, see Eqs. 1.3, 1.36, and 1.55. Thermal expansion is also a consequence of the anharmonicity. The thermal excitations by phonons and other elementary excitations also contribute to the decrease of sound velocity in a solid, as a function of increasing temperature. An approximate expression for this behavior is of the form

$$E(T) = (1 - bTexp(-T_0/T))E_r \approx (1 - b(T - T_0))E_r \tag{1.97}$$

where b is a material dependent parameter, T_0 is the reference temperature and $E(T)$ is the temperature dependent Young's modulus with E_r being its value at the reference temperature. The reference temperature is about half the Debye temperature of the material, see [78]. Analog relations hold for all elastic constants. Furthermore, it has been found that the slope dv/dT of the temperature dependent sound velocity varies with the applied stress, see Fig. 1.28 [39]:

$$\left(\frac{\Delta v}{\Delta T}\right)_{\sigma,T} = \frac{dv/dT_\sigma - dv/dT_0}{dv/dT_0} = \pm\ K\sigma \tag{1.98}$$

The amplitude of dv/dT depends on the stress amplitude and its orientation relative to the polarization of the wave. The proportionality constant K may be positive or negative, see Fig. 1.28, showing data for the steel grade A 533 B. The effect is of the order of $\approx 10^{-3}$ for a temperature change of 10 °C in ferritic and austenitic steels for longitudinal as well as for shear waves, and of the order of 2×10^{-3} for aluminum and its alloys. The temperature dependence of the acoustoelastic effect must be taken into account when carrying out stress measurements by using corresponding correction factors. Further details are discussed in [259]

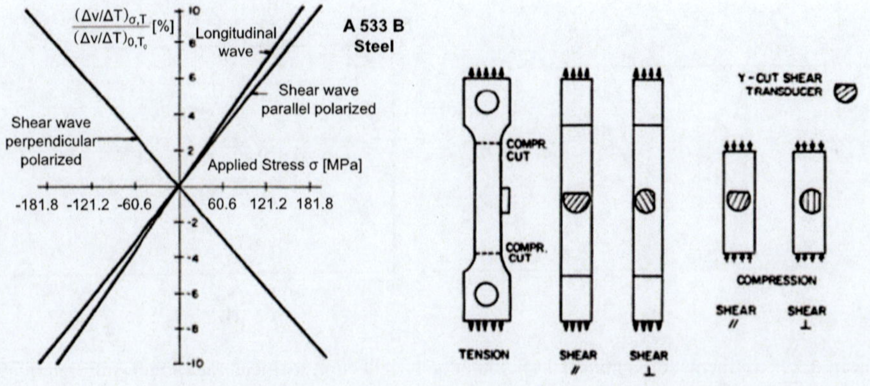

(a) (b)

Fig. 1.28 **a** Relative change of the temperature dependence, $(\Delta v/\Delta T)_\sigma,_T$ of longitudinal and shear velocity as a function of applied stress, σ, in an A 533B steel sample; **b** Schematic of the loading direction with the arrangement of the transducer orientation and hence polarization of the shear, respectively, longitudinal waves (in modified form from [39], with permission from Springer Nature)

1.12 Critical Angle Reflectometry

When ultrasonic waves are reflected at an interface under an oblique angle, mode conversion occurs. An incident longitudinal wave is converted into a longitudinal and a shear wave. The amplitudes and energy balances depend on the boundary conditions. This has been discussed in detail in many textbooks, see for example [25, 128] and therefore this will not be repeated here. A special case has been discussed for incident longitudinal waves in a liquid with velocity v_0 onto a solid surface with the Rayleigh velocity v_R under the critical angle θ_R for Rayleigh wave generation:

$$\sin\theta_R = \frac{v_0}{v_R} \tag{1.99}$$

If the material does not attenuate, the Rayleigh wave is still damped, because it radiates back into the liquid and therefore it is called a leaky wave, and there is a total reflection of the incident wave. If the material possesses internal friction, for example due to an agglomeration of dislocations, the reflection coefficient shows a dip and its width as a function of incident angle gets wider, see Fig. 1.29 [40]. This is caused by the fact that the energy of the ultrasonic wave absorbed in the surface layer can no longer be radiated back into the liquid and hence the reflection coefficient becomes smaller. This effect has been modeled and possible applications have been discussed in detail by [11]. The penetration depth is approximately one Rayleigh wavelength (see Sect. 1.3). It is useful for (i) when the surface layer adequately represents the bulk properties of the material; (ii) when the component geometry does not allow the use of pulse-echo techniques or the various through-transmission techniques; (iii)

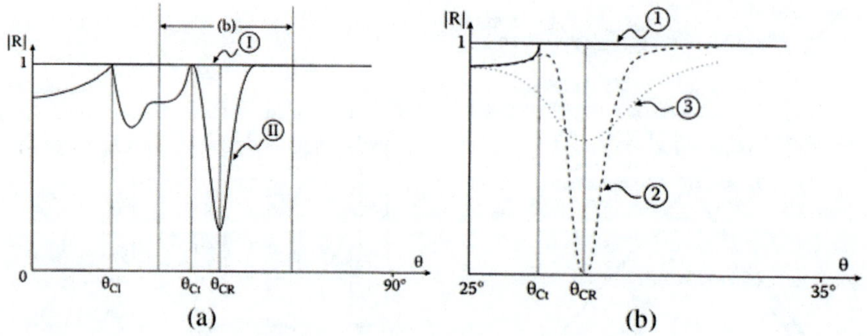

Fig. 1.29 **a** Schematic diagram of the modulus of the reflection coefficient at a liquid–solid interface as a function of the incident angle: (I) perfect, non-attenuating solid; (II) finite value of attenuation in the solid gives rise to a reduction of the reflection coefficient at the critical angle; **b** Blowup of (**a**) around the Rayleigh angle: (1) Zero attenuation; (2) small attenuation in the solid; (3) high attenuation. Increasing attenuation progressively washes out the Rayleigh dip (from [40], with permission from Francis & Taylor. Concerning the measurement of the attenuation of shear waves see [11])

when material attenuation or thickness is too large; and (iv) in order to evaluate the differences between bulk and surface layer properties.

These effects must be taken into account when measuring stresses by critical reflectometry using the SAM or the L-SAM, see Sects. 1.6 and 1.11.5 [160] or by using a goniometer. With a goniometer, it is not necessary to measure the time-of-flight of the ultrasonic pulses but exploit (1.99).

1.13 Acoustic Nonlinearity Parameter

Residual stresses can be caused in a material's microstructure by inhomogeneous plastic deformation. Examples are thermal loading of metal-matrix composites with globular SiC-particles of size 1–5 μm [190] due to the different expansion coefficients of the constituents, shot peening of aluminum components [170], rolling of copper, which changes the microstructure due to an increase of the dislocation density also leading to a change of the hardness [327] and many other examples. Finally, residual stresses contribute to the anisotropy of the material and to the nonlinearity parameter, as discussed in what follows.

The nonlinear interatomic forces, (1.1), lead to the generation of higher harmonics in ultrasonic propagation. To elucidate the main effects, let us consider the equation of motion for an ultrasonic wave, for simplicity in a one-dimensional form:

$$\rho\left(d^2u/dt^2\right) = d\sigma/dx \tag{1.100}$$

where u is the particle displacement, t is the time, and $d\sigma/dx$ is the stress gradient across a small material element. We now take into account a nonlinear stress–strain relation due to the deviation from the parabolic interatomic potential

$$\sigma = E\,\varepsilon + \frac{1}{2}D\,\varepsilon^2 \tag{1.101}$$

where E is the second order elastic constant, i.e. Young's modulus, and D represents the appropriate third order elastic constant (TOEC) in a one-dimensional form, (1.59). Solving Eqs. (1.100 and 1.101) for a sinusoidal input wave of amplitude A_1, leads to the generation of harmonics due to the (1/2 $D\varepsilon^2$) term:

$$u = A_1\cos(kx_0 - \omega t) + A_2\cos 2(kx_0 - \omega t) \tag{1.102}$$

where A_1 is the amplitude of the fundamental wave, A_2 is the amplitude of the first harmonic, and k is the wave vector. The first harmonic builds up during propagation of the fundamental wave. A measure of this built-up is the so-called β parameter:

$$\beta = 8A_2/(kA_1)^2 x_0 \tag{1.103}$$

where x_0 is the propagation distance after which the accumulated amplitude A_2 of the first harmonic and the amplitude of the fundamental mode A_1 are measured. In case the attenuation at the double frequency is much larger than at the fundamental frequency, a correction has to be made when using (1.103) [171].

The parameter β is dominantly determined by the deviation of the interatomic potential from a parabola, see Fig. 1.1. This part is often called "classical nonlinearity". For solids with a cubic structure as well as for two-phase materials, β is given in [35, 36]. For example for longitudinal waves in [100] direction:

$$\beta = -\left(3 + \frac{c_{111}}{c_{11}}\right) \tag{1.104}$$

The parameter β can also be expressed in terms of the Lamé and Murnaghan constants, as discussed in [190, 191]:

$$\beta_L = -\left(3 + \frac{2l + 4m}{\lambda + 2\mu}\right) \tag{1.105}$$

Due to symmetry conditions for third order elastic moduli, the β-parameter for pure transverse waves in isotropic solids is identical to zero. For transversely symmetric materials β is also zero for transverse waves propagating both parallel and perpendicular to the symmetry axis [76, 203].

The effect of harmonic generation using longitudinal, shear, and Rayleigh waves was exploited to measure the higher-order elastic constants in a large class of materials, in particular by Prof. Breazeale and collaborators [23, 24, 322]. It is interesting to note that linear behavior for ultrasonic propagation for longitudinal waves is not provided when the higher-order elastic constants are zero, but when $c_{111}/c_{11} = -3$ for a cubic crystal in [100] direction and $(2l + 4m)/(\lambda + 2\mu) = -3$ for an isotropic material. For other directions see [22].

A vast number of experiments and applications have been reported to use the β parameter for NDMC. For example, β changes with the content of SiC-particles in the 7064 and 8091 aluminum alloy metal-matrix composites (MMC). The particles had a more or less globular shape, ranging from 1 to 5 μm in size, with a random distribution of the reinforcement in the plane normal to the extrusion direction and a particle alignment along the extrusion direction. The β parameter is found to decrease with increasing volume % of particles in metal-matrix composites, whereas the same is observed to increase sharply with volume % of particles in a precipitation hardened aluminum alloy (Fig. 1.30) [191].

Due to increased interfacial strain, an increase in acoustic nonlinearity with increasing volume fraction of second phase has been reported in various studies [296]. The normalized NLU parameter (= β (1.103)) exhibited a sharp increase with the α-phase volume fraction in a beta Ti-alloy during the initial increase in the α-phase volume fraction from 0 to 25% (Fig. 1.31). With further increase in the α-phase volume fraction, the normalized NLU parameter appeared to saturate. It was intuitively stated that, with further increase in the α-phase volume fraction beyond 50%, the normalized NLU parameter would decrease due to the decrease in the α/β interface.

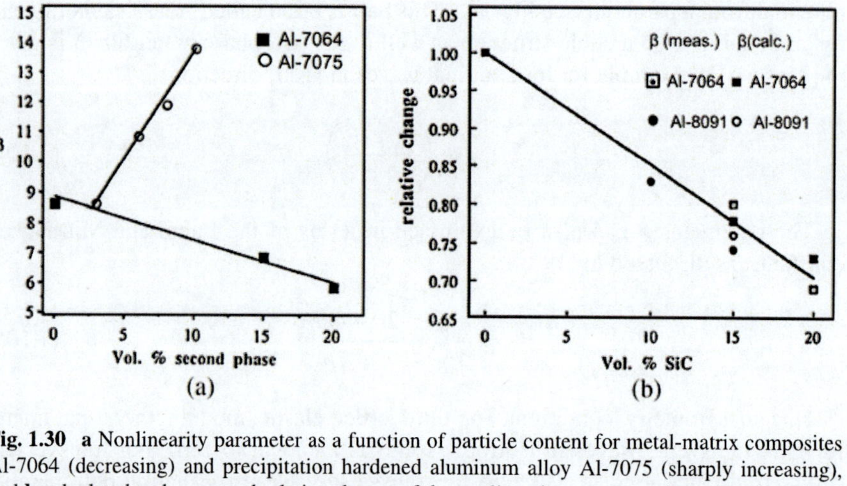

Fig. 1.30 **a** Nonlinearity parameter as a function of particle content for metal-matrix composites Al-7064 (decreasing) and precipitation hardened aluminum alloy Al-7075 (sharply increasing), and **b** calculated and measured relative change of the nonlinearity parameter (based on the rule of mixing) as a function of particle content (from [191] Copyright © 1991; The American Society for Nondestructive Testing Inc; reprinted with permission)

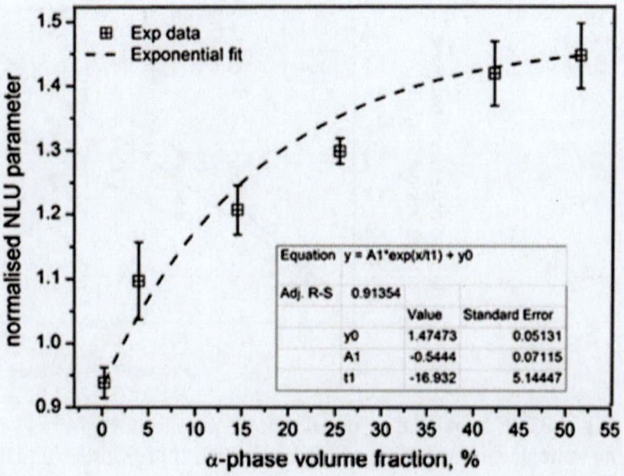

Fig. 1.31 Variation in the normalized NLU parameter with α-phase volume fraction in a beta Ti-alloy (from [296]), with permission from Springer Nature)

Shot-peening is a cold working process to produce residual compressive stresses in a component by shooting hard spheres onto its surface, see for example [63, 280]. Figure 1.32 shows the parameter $A_2/A_1^2 \propto \beta$ (see (1.103)) versus propagation distance measured with surface waves on three plates made of an aluminum alloy. One plate was not shot-peened and served for comparison with the other two plates [169]. Figure 1.32 clearly shows that the slope increases for samples with higher shot-peen intensity; the slopes for the 8 and 16 A samples are about 81% and 115% larger, respectively, than the as-received reference sample. Here, A stands for Almen intensity, see [63]. The large change in the measured acoustic nonlinearity parameter is due to the combined effect to residual stress from the shot-peening and its associated plastic deformation and any additional surface roughness. It is postulated in [169] that the decrease or downturn of A_2/A_1^2 with increasing propagation distance is due to higher attenuation at higher frequencies of the Rayleigh waves.

The growth of the first harmonic cannot continue forever, because its wave energy stems from the fundamental wave and therefore the fundamental wave gets depleted. There is another limit. The generation of the harmonics leads to a saw-tooth like wave shape and when the spatial derivative of the displacement is infinite, the so-called discontinuity length L_d is reached:

$$L_d = \frac{K_2/\rho}{((3K_2+K_3)/K_2)\left(\omega^2/2\right)A_1} \tag{1.106}$$

Here, K_2 and K_3 are second and third order elastic constants, respectively. For example, in cubic crystals in [100] direction $K_2 = C_{11}$ and $K_3 = C_{111}$. As said above, for a "linear solid" for which $L_d \to \infty$, the condition $K_3 = -\ 3K_2$ must be fulfilled and not $K_3 = 0$ [22].

There are a number of review articles discussing experiments and theories as to how micro-damage of materials can be detected and monitored by non-linear ultrasound, see for example [119, 168, 182]. In our view, the most important aspect is the

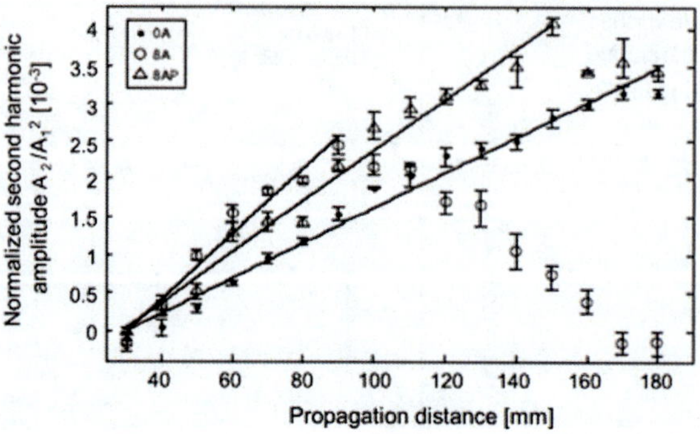

Fig. 1.32 Normalized second harmonic amplitude A_2/A_1^2 as a function of propagation distance for an untreated aluminum alloy plate, and two shot-peened Al plates at Almen intensities 8 and 16 A. The SAW frequency was 2.25 MHz. The error bars represent the maximum measured variations from ([169], with permission from Elsevier). For surface waves the β-parameter is evaluated analog to (1.103), with a pre-factor which depends on v_T, v_L, and v_R and which is of the order of 1, see [93]

calibration of these techniques, so that the data can be inverted in order to extract the damage level. Furthermore, research groups often restrict their analysis to determining the β-parameter only. This is not sufficient, because micro-cracks, dislocation networks, fatigue and creep damage introduce also defects on larger, mesoscopic scale into a material influencing not only the higher-order elastic constants. They introduce hysteresis and discrete memory in the stress–strain relations. To first approximation, the one-dimensional constitutive relation between the stress σ and the strain ε, which describes the dynamic behavior of solids, can be expressed as follows:

$$\sigma = \int K(\varepsilon, \dot{\varepsilon}) d\varepsilon \tag{1.107}$$

with

$$K(\varepsilon, \dot{\varepsilon}) = K_0\big(1 - \beta\varepsilon - \delta\varepsilon^2 - \alpha\big[\Delta\varepsilon + \varepsilon(t) sign(\dot{\varepsilon})\big] + \ldots..\big) \tag{1.108}$$

which is an extension of (1.101). K_0 is the linear modulus, $\Delta\varepsilon$ is the local strain amplitude, and $\dot{\varepsilon} = d\varepsilon/dt$ is the strain rate (sign $(\dot{\varepsilon}) = 1$ if $\dot{\varepsilon} > 1$ and sign $(\dot{\varepsilon}) = -1$ if $\dot{\varepsilon} < 1$. A review of the various effects is given by [1]; Van Den [292]. A comparison between experimental data and theoretical simulations provides essential information as to what type of nonlinearity is dominant in the material.

Solodov et al. [273] studied the nonlinearity parameter β of polymers and glass fiber-reinforced composite plates at tensile stresses up to the fracture limit by air-coupled ultrasound. Local noncontact measurements of the flexural wave velocity as a function of static strain were used to calculate the nonlinearity parameter. Molecular untangling and crazing phenomena were identified in the material as causes for the non-linearity. The hysteresis in velocity variation during loading–unloading cycle was used as an indicator of residual defect accumulation.

Matlack et al. [182] measured a normalized β-parameter in order to monitor radiation damage in two reactor pressure vessel steels using longitudinal waves. The microstructural changes associated with radiation damage includes changes in the dislocation density and the formation of precipitates. Six test samples were previously irradiated in a nuclear power reactor with two fluence levels, up to 10^{20} n/cm^2 with energies E higher than MeV. The results showed an increase in the β-parameter from the unirradiated state, which served for normalization, to the medium dose (50×10^{18} n/cm^2), and then a decrease from medium dose to the highest doses.

1.14 Acoustoelastic Constants (AEC)

Young modulus E, the shear modulus G, the density ρ, and Poisson's ratio υ determine the longitudinal velocity, v_L, the shear velocity, v_T, and the Rayleigh velocity v_R, see Eqs. 1.41–1.42. When stress is present and hence strain, the materials velocities change, see Fig. 1.21b. The slopes of the velocities as a function of strain are the so-called acoustoelastic constants AEC, see Eqs. 1.70–1.74 and Sect. 1.11.2. They are combinations of the second and third order elastic constants. For example, for the slope of the velocity v_{11} as a function of strain ε, (1.70) holds:

$$AEC_{11} = \frac{1}{v_L}\frac{dv_{11}}{d\varepsilon} = \frac{1}{v_L}\frac{(v_{11}-v_L)}{d\varepsilon} = 2+\frac{\mu + 2m + \nu\mu(1 + 2l/\lambda)}{(\lambda + 2\mu)} \tag{1.109}$$

Beyond the measurements of stresses, the AEC itself can be used for NDMC. The interatomic potential, (1.1), not only responds to strains on a macroscopic scale, but also to localized strains acting on the atomic lattice caused by dislocation networks, by solid solution phases, i.e. by modifications of the microstructure.

For aluminum alloys, the correlations of yield strength, Brinell hardness, and AECs with the content of different solid-solution phases were determined experimentally by [262], see Fig. 1.33. In this case, the acoustoelastic constant was determined by dividing the change of stress by the change in normalized longitudinal wave velocity and therefore has the unit [Pa] and not dimensionless like in Eq. 1.109. The data were obtained for 10 MHz longitudinal waves propagating perpendicular to the stress direction. The yield strengths were determined from tensile tests. Likewise, the hardness values of the specimens were evaluated using conventional techniques.

Razvi et al. [234] examined two different aluminum alloys, one heat treatable, and another work hardenable. They behaved differently in respect to β and the AEC. The work-hardenable material showed no change in the parameter β and the AEC as a function of aging time, while for the heat treatable alloy the AEC and the parameter β increased as a function of the volume content of the second phase. Furthermore,

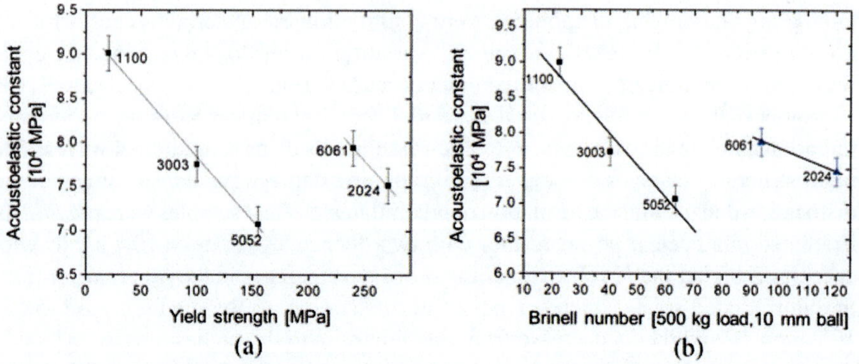

Fig. 1.33 **a** Acoustoelastic constant of several Al alloys versus yield-strength with solid solutions of Si, Fe (1100, annealed), Mn (3003, strain-hardened), Mg (5052, strain-hardened and stabilized), Si, Mg (6061, solution heat-treated and then artificially aged), Cu, Mg (2024, solution heat-treated, cold worked and stress relieved); **b** Acoustoelastic constant of the same alloy versus Brinell number. In modified form from [262], with permission from Springer Nature

the heat treatable alloys showed saturation in AEC above a certain percentage of the second phase (Fig. 1.34).

The dependence of the AEC as a function of carbide content was measured by [36] in different steel grades, as shown in Fig. 1.35. In addition, theoretical calculations were performed in order to explain the influence of precipitates on the acoustoelastic constants and on the nonlinearity parameter β.

Summarizing this section, there is a strong need for additional research work in order to put the experimentally determined correlations of hardness, tensile strength

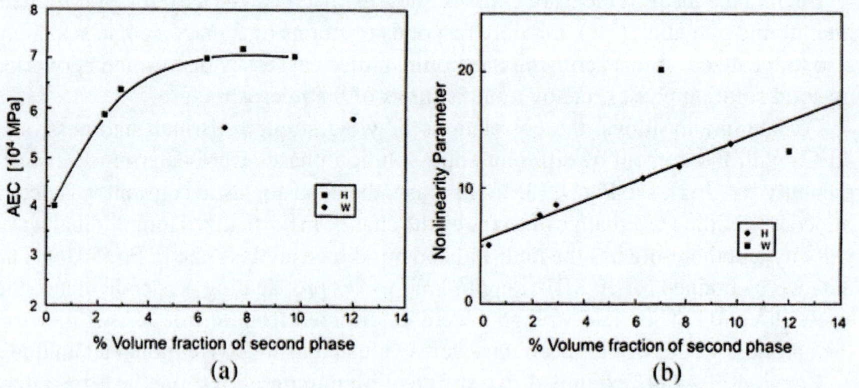

Fig. 1.34 **a** Acoustoelastic constant for the heat-treatable (H) Al alloy 7075 and the work-hardenable (W) Al alloy 5086-H111 Al alloy as a function of the second phase content; **b** nonlinearity parameter for the same alloys. The data were obtained with 10 MHz longitudinal waves and the definition of the AEC is the same as for Fig. 1.33 (from [234], with permission from Springer Nature)

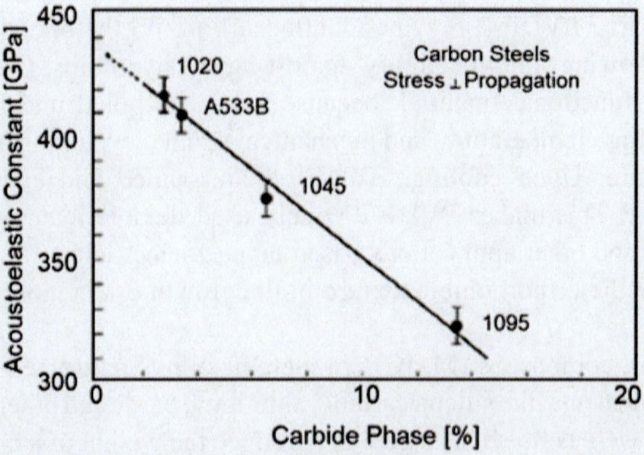

Fig. 1.35 Acoustoelastic constant as a function of carbide content for four different carbon steels with different volume percent of carbide [AISI 1020: 0.2% C; 1045 steel: 0.45%; C 1095 steel: 0.95% C; and ASTM 533B (not given)]. The measurements were performed using longitudinal waves at 10 MHz propagating perpendicular to the applied load (from [36], with permission from Francis & Taylor)

(R_e, R_m, R_f), fracture strain A, and fracture toughness κ_1 with the acoustoelastic constants on a solid footing.

1.15 Texture

Macroscopic elastic/plastic anisotropies of a material are called textures. The anisotropy can be desired or undesired, however, it must be controlled in each case. Texture may originate from the production process, e.g. in metal sheets and slabs by rolling or thermo-mechanical treatment and in fiber-reinforced structures by the orientation of the fibers. It results in direction-dependent orientation of the microstructure's constituents. It is an important parameter in order to optimize the properties of technical materials.

- If there is a homogeneous microstructure without any texture, the sound velocities are given by Eqs. (1.40–1.42). If there is a texture due to a preferred orientation of the grains, the sound velocity also becomes directionally dependent. In this case the anisotropy factor A, Eqs. (1.32–1.34), enters in the expression for the sound velocities. It contains now a contribution from the texture.
- Metals like steel, aluminum, and other cubic crystallizing materials easily develop an orthorhombic texture. They belong to the crystal groups with three axes, perpendicular to each other, including primitive as well as body and face-centered lattice structures [31, 223].
- Reinforced composites are transverse isotropic if they are unidirectional or more general show an orthorhombic texture [92].

- Elastomers like PVDF (Polyvinylidenfluorid) are on the one side construction materials showing a high stability against aggressive fluids. On the other side, PVDF is a functional material, because it can be poled under a strong electrical field, high temperature, and mechanical strain, developing an orthorhombic microstructure. Upon cooling, PVDF remains poled and exhibits piezoelectric behavior. This makes PVDF a widely used flexible material for ultrasonic transducers and other applications based on piezoelectricity.
- Welds show the orthorhombic texture in the growth orientation of the dendrites [31].
- Metal-matrix composites (MMC) are metallic alloys reinforced by adding low-weight components like silicon carbide, aluminum oxide and others, in the form of particles as well as fibers. The goal is to reduce the weight of a component while its strength increases extending the capacity to carry high stresses and strains. Between the "harder" and the "tougher" components residual stresses as well as texture can develop during the production processes like casting, extrusion, and isostatic pressing [123]. The typical texture is orthorhombic.
- Transformer sheets are preferably made of Fe–Si steels which are soft magnetic materials. A texture is produced by heating and rolling processes and secondary recrystallisation. The aim is to reduce heating by minimizing the combined losses due to hysteresis, eddy currents, and domain motions [120]. To this end, the so-called Goss texture is preferred [51, 72, 73]. It is characterized by a (110) ⟨001⟩crystal orientation in rolling direction, with a perpendicular face-diagonal direction of the body-centered cubic structure. This sharp texture develops due to a discontinuous or abnormal grain growth during a high-temperature annealing at the end of the production process.

In the Voigt approximation the ultrasound velocities for longitudinal (L) and transverse waves (T) without texture are given by Eqs. (1.37 and 1.38):

$$\rho v^2_{L,without\ texture} = c_{11} - \frac{2}{5}A \tag{1.110}$$

and

$$\rho v^2_{T,without\ texture} = c_{44} + \frac{2}{5}A \tag{1.111}$$

with the anisotropy factor A defined in Sect. 1.8, (1.32).

Texture is mathematically described by the orientation distribution function (ODF) both for elastically induced anisotropies or plastic anisotropies. For elastic anisotropy of the crystallites in a polycrystalline ensemble, the ODF describes the probability to find a single crystallite of a certain orientation with respect to the sample coordinate system. After Bunge [30, 31] the ODF can be written as a series expansion of symmetrical generalized spherical harmonics.

Let us consider texture in a rolled material. The rolling direction is the 1-direction, the perpendicular direction is the 2-direction, and the normal direction is the 3-direction. Texture entails directionally dependent sound velocities, which are related to the texture expansion coefficients w_{400}, w_{420} and w_{440} [236, 253]:

$$\rho v_{11}^2 = c_{11} - \frac{2}{5}A + \frac{12\pi^2\sqrt{2}}{35}A\left(w_{400} - \frac{2}{3}\sqrt{10}w_{420} + \frac{1}{3}\sqrt{70}w_{440}\right) \quad (1.112)$$

$$\rho v_{22}^2 = c_{11} - \frac{2}{5}A + \frac{12\pi^2\sqrt{2}}{35}A\left(w_{400} + \frac{2}{3}\sqrt{10}w_{420} + \frac{1}{3}\sqrt{70}w_{440}\right) \quad (1.113)$$

$$\rho v_{33}^2 = c_{11} - \frac{2}{5}A + \frac{32\sqrt{2}\pi^2}{35}Aw_{400} \quad (1.114)$$

$$\rho v_{12}^2 = c_{44} + \frac{1}{5}A + A\frac{4\sqrt{2}\pi^2}{35}\left(w_{400} - \sqrt{70}w_{440}\right) \quad (1.115)$$

$$\rho v_{23}^2 = c_{44} + \frac{1}{5}A - A\frac{16\sqrt{2}\pi^2}{35}\left(w_{400} - \sqrt{\frac{5}{2}}w_{420}\right) \quad (1.116)$$

$$\rho v_{31}^2 = c_{44} + \frac{1}{5}A - A\frac{16\sqrt{2}\pi^2}{35}\left(w_{400} - \sqrt{\frac{5}{2}}w_{420}\right) \quad (1.117)$$

A different notation for the texture coefficients, C_{lmn}, was developed by [31, 32]. The origin of the differences and the relations to the fourth-order coefficients in the Roe notation is discussed in [329]. These are:

$$w_{400} = C_{411}d\sqrt{2}, \quad w_{420} = C_{412}d, \quad w_{440} = C_{413}d \quad (1.118)$$

where $d = \sqrt{7/12}/24\pi^2$. Using Eq. 1.118, Eqs. 1.112 to 1.117 can be written as [274]:

$$\rho v_{11}^2 = c_{11} - \frac{2}{5}A + \frac{1}{70}\sqrt{\frac{7}{3}}A\left(C_{411} - \frac{2}{3}\sqrt{5}C_{412} + \frac{1}{3}\sqrt{35}C_{413}\right) \quad (1.119)$$

$$\rho v_{22}^2 = c_{11} - \frac{2}{5}A + \frac{1}{70}\sqrt{\frac{7}{3}}A\left(C_{411} + \frac{2}{3}\sqrt{5}C_{412} + \frac{1}{3}\sqrt{35}C_{413}\right) \quad (1.120)$$

$$\rho v_{33}^2 = c_{11} - \frac{2A}{5} + \frac{1}{70}\frac{8}{3}\sqrt{\frac{7}{3}}AC_{411} \quad (1.121)$$

$$\rho v_{12}^2 = \rho v_{21}^2 = c_{44} + \frac{A}{5} + \frac{1}{70}\sqrt{\frac{7}{3}}A\left(\frac{1}{3}C_{411} - \frac{1}{3}\sqrt{35}C_{413}\right) \quad (1.122)$$

$$\rho v_{23}^2 = \rho v_{32}^2 = c_{44} + \frac{A}{5} - \frac{1}{70}\sqrt{\frac{7}{3}}A\left(\frac{4}{3}C_{411} + \frac{2}{3}\sqrt{5}C_{412}\right) \tag{1.123}$$

$$\rho v_{31}^2 = \rho v_{13}^2 = c_{44} + \frac{A}{5} - \frac{1}{70}\sqrt{\frac{7}{3}}A\left(\frac{4}{3}C_{411} - \frac{2}{3}\sqrt{5}C_{412}\right) \tag{1.124}$$

There are numerous possibilities to determine the various expansion coefficients C_{411}, C_{412}, and C_{413} from ultrasonic velocity data. In order to reduce error propagation, relationships were used which do not require absolute velocities but only their ratios canceling out path-length errors. Together with other measures, the following equations were developed by [274]:

$$C_{411} = \frac{210}{8A}\sqrt{\frac{3}{7}}\left(2\mu - (c_{11} + 2c_{44})\frac{(v_{31}^2 + v_{32}^2)}{(v_{31}^2 + v_{32}^2 + v_{33}^2)}\right) \tag{1.125}$$

$$C_{412} = \frac{210}{4A\sqrt{5}}\sqrt{\frac{3}{7}}\left((c_{11} + 2c_{44})\frac{(v_{31}^2 - v_{32}^2)}{(v_{31}^2 + v_{32}^2 + v_{33}^2)}\right) \tag{1.126}$$

$$C_{413} = \frac{210}{8\sqrt{35}A}\sqrt{\frac{3}{7}}\left(6\mu - \lambda + (c_{11} + 2c_{44})\frac{(v_{33}^2 - 8v_{12}^2)}{(v_{31}^2 + v_{32}^2 + v_{33}^2)}\right) \tag{1.127}$$

The agreement of the values of the expansion coefficients C_{41i} $(i = 1, 2, 3)$ determined by ultrasonic velocities of longitudinal and shear waves, and with those determined by X-ray measurements, reassures the correct interpretation of the underlying physical principles for determining the C_{41i}, see Fig. 1.36 [275].

Up to here, the assumption was made that the ultrasonic waves propagate along the macroscopic directions *x*, *y*, *z*, which are also the texture directions. However, if the *k*-vector is oriented under an angle θ relative to the rolling direction, the above relations have to be modified [159]. In this case one starts with the equation:

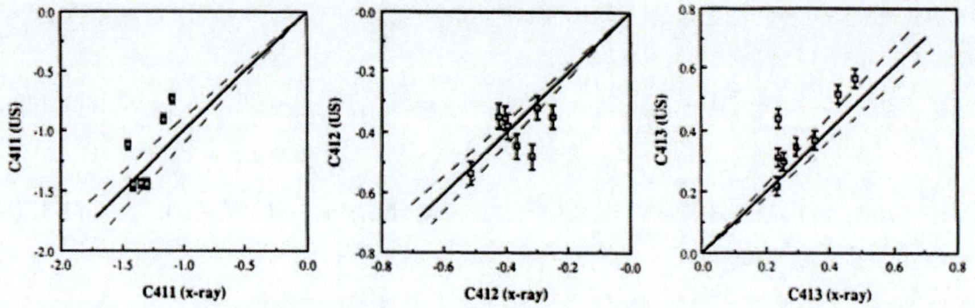

Fig. 1.36 Comparison of the expansion coefficients determined by ultrasonic time-of-flight measurements and by X-ray diffraction in reflection, from [275], with permission under (*Creative Commons Attribution License*)

$$\rho v_{L,\theta}^2 = c_L + \frac{\beta_1}{2} c_L \cos 2\theta - \frac{\beta_2}{2} c_T (1 - \cos(4\theta)) \tag{1.128}$$

$$\rho v_{T,\theta}^2 = c_T + \frac{\beta_2}{2} c_T (1 - \cos(4\theta)) \tag{1.129}$$

For the case of the lowest-order symmetric Lamb and shear-horizontal modes in a plate having an orthorhombic texture, the coefficients β_1 and $\beta_2$were derived by Lee et al. [159]. These equations were then used in order to determine w_{420} and w_{440} for aluminum and copper plates.

For a material with an orthorhombic texture, the wave equation for Rayleigh waves is given by [49]:

$$v_{R,\theta} = v_{R_0} \left(1 + \frac{A}{c_{44}} (\alpha_1 C_{411} + \alpha_2 C_{412} \cos(2\theta) + \alpha_3 C_{413} \cos(4\theta)) \right) \tag{1.130}$$

where v_{RO} is the Rayleigh-wave velocity in the material free of texture. The parameters α_1, α_2, α_3, and the conditions for obtaining the texture coefficients C_{411} and C_{413} by RW velocity measurements are discussed in [49].

Guided waves like Lamb waves and shear-horizontal (SH) waves in plates are well suited for texture evaluation. Solutions for the velocities were developed for symmetric plate waves [286] and shear horizontal (SH) waves [3, 166]. For SH-waves propagating with angle θ to the rolling direction, the following expression holds:

$$\rho v_{SH}^2(\theta) = \mu + \frac{4\sqrt{2}}{35}\pi^2 \; A w_{400} - \frac{8\pi^2}{\sqrt{35}} A w_{440} + \\ + \frac{64\pi^2}{\sqrt{35}} A w_{440} \cos^2\theta \sin^2\theta \left(1 - \left(\frac{16\pi^2 A}{\sqrt{35}(\lambda + \mu)} \right) w_{440} (1 - 2\cos^2\theta)^2 \right) \tag{1.131}$$

Further examples demonstrate the value of ultrasonic guided waves in order to characterize anisotropy and texture:

- Lamb-wave velocity-plots were used to characterize laminate orientation in composites [244].
- Rayleigh-waves were used to characterize the surface texture of aluminum plates allowing one to deduce w_{420} [189].
- The inclusion of SiC fibers in aluminum forms a metal-matrix composite. The orientation of the fibers leads to texture which was theoretically evaluated [103]. Due to the elongated form of the fibers, the sound velocity and attenuation become directionally dependent.
- An ultrasonic pole figure, similar to the X-ray equivalent (see Chap. 2, Sect. 2.4.11), can be obtained by direction-dependent velocity plots. While

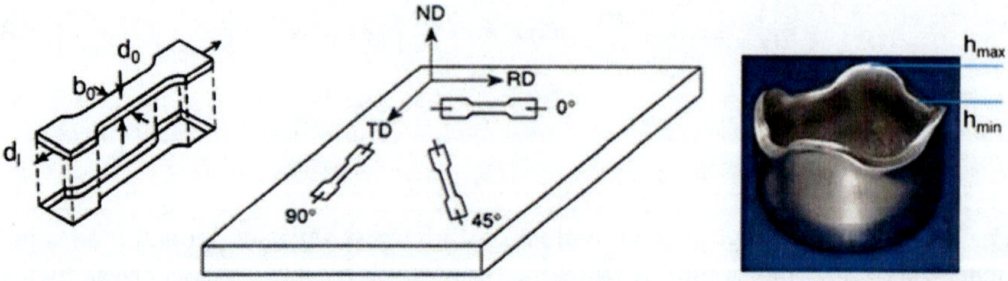

Fig. 1.37 Specimen shapes and orientations required in order to determine the *R*-value by loading them in a tensile test machine in length direction, so that the samples are plastically deformed within the regime of uniform elongation

Rayleigh waves allow to characterize the surface-near area, Lamb waves allow to characterize the texture over the thickness of a metal sheet [28].

The deep-drawing capability of metal sheets which are characterized by the Lankford parameter *R* [123, 153], which are usually measured by destructive tests. Tensile specimens are cut out of a metal sheet (Fig. 1.37) and subjected to strains depending on the yield limit. Then, their dimensions are measured and compared to the dimensions before loading. The *R* parameter is defined as:

$$R = ln(b_0/b_\varepsilon)/ln(d_0/d_\varepsilon) \tag{1.132}$$

Here, b_0 is the width of the test specimen before the test, b_ε is the width after plastic deformation of 15 to 20% straining (ε), and d_0 and d_ε are the corresponding thicknesses. The *R*-parameter is also called normal anisotropy or plastic anisotropy. If $R > 1$, the material flows more in width direction and less in thickness direction, and for $R < 1$ vice versa. Using the volume constancy for plastic deformation $\varepsilon_d + \varepsilon_b + \varepsilon_l = 0$, Eq. 1.132 can be rewritten as

$$R = ln(b_0/b_\varepsilon)/ln(d_\varepsilon l_\varepsilon/d_0 l_0) \tag{1.133}$$

where l_o and l_ε are the changes in length direction.

If a metal sheet is planar anisotropic, a mean value for *R*, called R_m is used, which is composed of three values measured in three directions in the plane of the metal sheet with each value determined according to Eq. 1.132, see (Fig. 1.37):

$$R_m = \left(\frac{R(0^\circ)+2R(45^\circ)+R(90^\circ)}{4}\right) \tag{1.134}$$

Finally, the planar anisotropy value of the plate, ΔR, is defined as

$$\Delta R = \left(\frac{R(0^\circ)-2R(45^\circ)+R(90^\circ)}{2}\right) \tag{1.135}$$

$$Earing\ Z_e(\%) = 100\left(\frac{h_{\max} - h_{\min}}{h_{max}}\right) \tag{1.136}$$

Large anisotropy of metal sheets leads to earing when deep-drawn, see Fig. 1.37. In directions of large R-values humps are generated, for directions of small R-values valleys. Cup-tests yield the ear heights $\Delta h = (h_{max} - h_{min})$, the earing parameter Z_e (Eq. 1.136) and the concurrent determination of R_m and ΔR in tensile tests. They allow one to obtain the parameters which result in minimum earing [297]. From practice, it turns out that $R_m > 1$ is favorable for deep-drawing, but this entails in general that $\Delta R > 0$, which leads to the occurrence of ears, see Fig. 1.37. Their number (4, 6, or 8) depends on the polycrystalline crystallographic structure of the metal sheet. In order to minimize the occurrence of ears in cup forming, the planar anisotropy value should be zero, i.e. $\Delta R = 0$. Within certain limits this can be controlled by alloying, temperature of drawing, heat cycles, and alternate drawing in length and thickness directions, see [123].

Using various ultrasonic wave modes under different angles relative to the rolling direction, the texture coefficients C_{411} and C_{413} were measured in a series of cold-rolled ferritic steel sheets by Spies and Schneider [275, 276] and were compared to X-ray diffraction or to destructively obtained data showing excellent agreement. The coefficients were then correlated to the Lankford parameters R_{m} and ΔR. It was shown that the relations

$$R_m = a + bC_{411} \tag{1.137}$$

$$\Delta R = c + dC_{413} \tag{1.138}$$

hold. The parameters a, b, s, and d were determined by calibration tests [275, 276].

Following Sayers [253], Serebryany [265] obtained the texture coefficients in sheets of low carbon steel and aluminum alloys using the velocities of longitudinal and two shear waves (Eqs. 1.112 to 1.118). The ultrasonic data for C_{440} and C_{402} were then used to calculate from the angle-dependent Talyor factor $M(q,\theta)$ [265]

$$M(q, \theta) = \frac{\sum_{l=0}^{6} \sum_{n=0}^{l} \sum_{p=0}^{N} m_{0,nlp} C_{0ln} q^p \cos n\theta}{2l + 1} \tag{1.139}$$

the anisotropy factor R. Here $m_{0,nlp}$ are the coefficients of the expansion into a series of spherical functions, C_{0ln} are the coefficients obtained by the ultrasonic measurements, θ is the angle between the rolling and the transverse directions, and q is the orientation-dependent Taylor factor for single crystals. The R-value is a function of the expansion coefficients $m_{0,nlp}$ up to the sixth order ($l = 6$). This leads to R by

$$R = \frac{q_{\min}}{1 - q_{\min}} \tag{1.140}$$

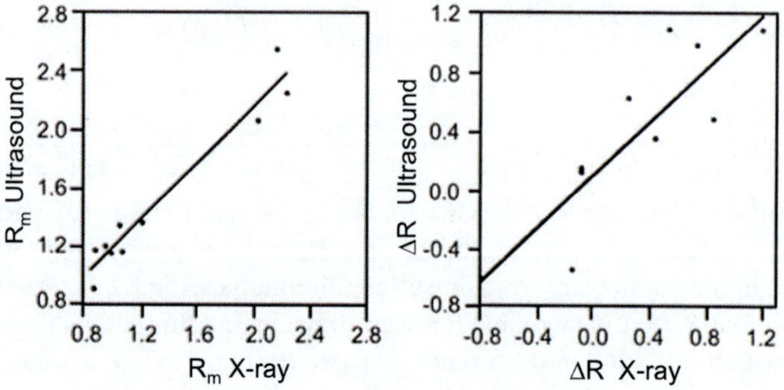

Fig. 1.38 Correlation between the values of R_m and ΔR determined by ultrasonic and X-ray diffraction based on the Taylor model (in modified form from [37], with permission from Springer Nature)

where q_{min} is the so-called contraction ratio, the value of q in Eq. (1.140), when $M(q,\theta)$ achieves its minimum value. Based on this formalism, Serebrany [265] determined the R- values and the coefficients C_{400} and C_{402} for steel and aluminum test samples and obtained good agreement with X-ray data.

For the analysis of the ultrasonic data there are the Hill and Taylor models relating the texture anisotropy to the R_m and ΔR values, see Fig. 1.38 [37]. Such measurements allow one to develop techniques for obtaining the parameters R_m and ΔR in situ in a production line [19, 196, 199]. Eventually such measurements can be enhanced by modeling efforts [245].

An interesting approach was pursued by Sakata et al. [249]. In this work, the higher-order orientation distribution function (ODF) coefficients up to $l = 12$ in cold-rolled and annealed sheet steels were evaluated and calculated from the anisotropy of the ultrasonic velocities of the lowest order symmetrical Lamb and shear horizontal waves propagating in the rolling plane. The elastic energy method was employed, together with a decomposition of the texture into the principal preferred orientations, following a procedure originally developed for Young's modulus data obtained destructively. Plastic strain ratios were estimated using the series expansion method, in conjunction with a relaxed constraint grain interaction model. The R-value predictions based on the ODF coefficients obtained were compared with tensile test data and with empirical predictions. Then, the anisotropy was expressed in terms of an angle dependent Young's modulus relative to the rolling direction (RD). Figure 1.39 displays experimental data (open circles), theoretical work both for the angle dependent Young's modulus, and the corresponding r-values for three steel types (AKDQ = batch annealed Al-killed drawing quality steel; rimmed: commercial grade rimmed steel; HSLA: high-strength low-alloy steel).

Goebbels and Salzburger [71] used a transmitter-receiver arrangement with fixed distance in order to measure the group velocity of the first asymmetric Lamb mode a_1 in 13 different cold and warm-rolled steel plates as a function of angle between

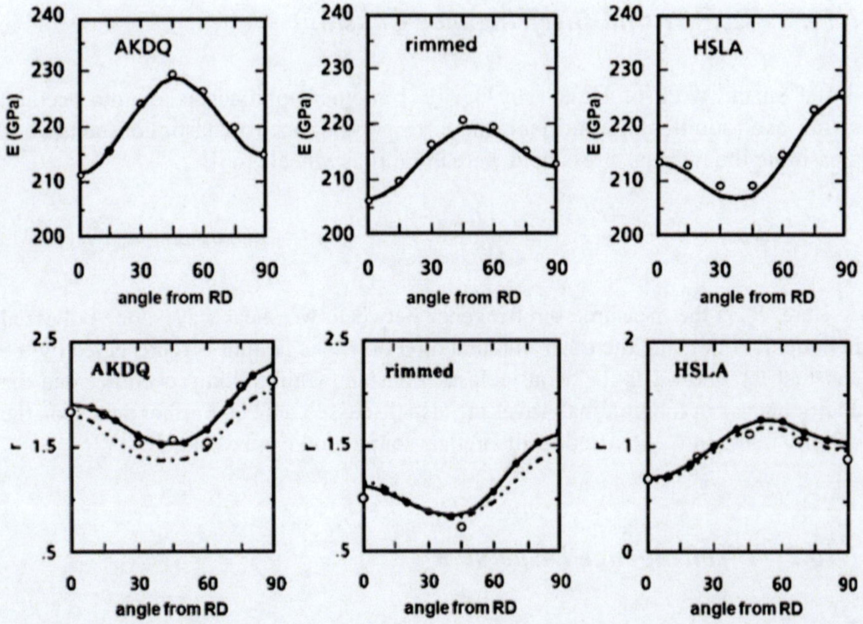

Fig. 1.39 Angle dependent Young's modulus for three types of steel and corresponding *r*-values (*R*-values) obtained by ultrasonic velocity measurements (RD: rolling direction). Data points represented by open circles and solid curves are according to theory (in modified form from [249], with permission from Springer Nature)

the rolling direction and the propagating direction. The relative change of the time-of-flight data $\Delta t/t$ obtained in the direction of maximum planar anisotropy were correlated with the earing parameter Z_e (1.136). They corresponded almost in a 1:1 relation to Z_e, i.e. $\Delta t/t \approx Z_e$.

As discussed in Sect. 1.10, the elastic constants of materials, if they are cast into simply shaped bodies like cubes, can be determined from their eigenresonances. The experimental technique is called RUS. The theoretical base of such measurements can be extended to obtain information on the presence of texture in the materials [59, 116].

1.16 Stress and Texture

Stress and texture may be present simultaneously in a material's microstructure, and especially texture may influence the sound velocities stronger than stress, as was noticed quite some time ago [62]. Therefore, there is a need for a procedure to separate both or at least to take the effect of texture into account. One approach is to use the different spatial symmetries of the textures developed for example by rolling in a plate geometry relative to the symmetries of the stress field [270].

1.16.1 Texture and Birefringence Constant

Based on the work of Hirao and Ogi [97] another approach takes into account texture as an additional parameter in the corresponding acoustoelastic constants when measuring the residual stress field, here in railroad wheels [66].

$$B_b = \frac{v_{s,radial} - v_{s,hoop}}{v_0} \cong \frac{t_{hoop} - t_{radial}}{\left(t_{hoop} + t_{radial}\right)/2} = B_0 + C_A \sigma_{hoop} \tag{1.141}$$

Here, B_b is the measured birefringence between two shear waves, one polarized in hoop direction and the other in radial direction. The texture is represented by the constant B_0, and C_A is the acoustoelastic constant. Calibration procedures and the additional use of longitudinal waves are also discussed in [66]. Further details on the validity range of (1.141) and its theoretical foundation can be found in [97].

1.16.2 Birefringence Dispersion

Texture caused by a preferred orientation of an ensemble of grains in a polycrystalline microstructure entails a change of dispersion of the sound velocity. This opens the possibility to separate stress and texture. The dispersion itself is caused by grain scattering, see Sect. 1.17.4. Hence, if there is an orientation dependent scattering, the dispersion becomes orientation dependent as well. This was recognized quite some time ago [177] and analyzed in detail both experimentally [70, 108] and theoretically by [101]. Because stresses of first order extend over distances much larger than the grain size, the birefringence due to stress is frequency independent, while birefringence due to texture is frequency dependent, as shown schematically in Fig. 1.40. The velocity contains a dispersion-free part, which corresponds to the velocity for $ka \to 0$ (i.e. $\omega \to 0$), where k is the wave-vector and a is the grain size. For shear waves, with one polarization propagating in the rolling direction (R, $v_{T,R}$) and with the other in the perpendicular direction (T, $v_{T,T}$), the relative difference is given by:

$$\left(\frac{v_{T,R} - v_{T,T}}{v_{T_0}}\right)_{stress+texture} = \frac{t_{T,T} - t_{T,R}}{t_{T_0}} = \frac{\Delta t}{t} = c_1 + c_2 \left(\frac{2\pi a}{v_{T_0}}\right)^2 f^2 \tag{1.142}$$

where v_{T0} is the shear wave velocity in the material free of texture and c_1 and c_2 depend on the single crystal elastic constants of the individual grains and on the fourth order expansion coefficients C_{411} and C_{412} of the grain structure. Furthermore, c_2 depends also on the sixth order expansion coefficients C_{611} and C_{612}. The correlation between the c_1 and c_2 are shown in Fig. 1.41. The dispersive part of (1.142), which is proportional to f^2, holds for the Rayleigh regime [99, 100]. Let us now consider in addition the relative time-of-flight differences due to stresses (Eqs. (1.78–1.79)):

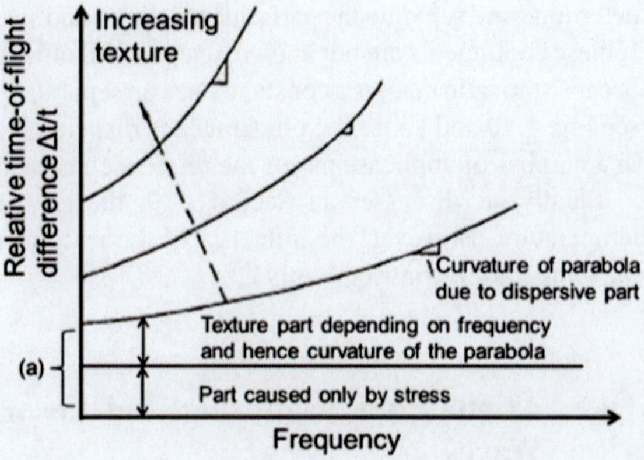

Fig. 1.40 Principle of separation of stress and texture by using ultrasonic dispersion due to texture. Part (a) is determined by stress and texture. The texture dependent part becomes larger at larger frequencies because of its increasing larger curvature thus opening the possibility to separate the two (according to [69])

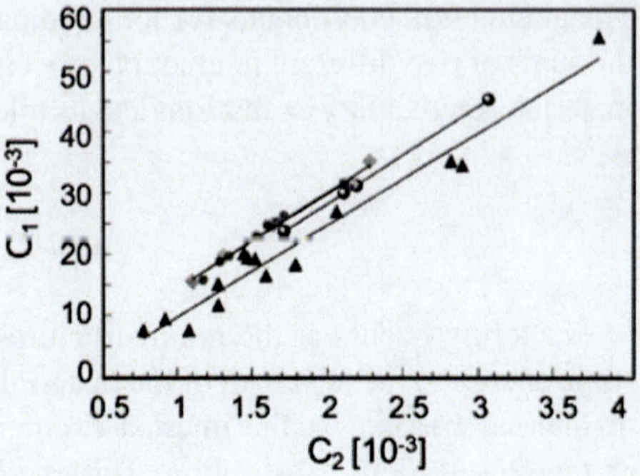

Fig. 1.41 Calculated and experimentally determined correlation between c_1 and c_2, which enter in the relative time-of-flight difference of two perpendicularly polarized shear waves for several rolled steel plates (from [108], in modified form with permission from Springer Nature)

$$\left(\frac{\Delta t}{t}\right)_{stress} = \frac{(\sigma_2 - \sigma_3)(4\mu + n)}{8\mu^2} \tag{1.143}$$

by replacing in (1.142) c1 by c_1' so that

$$c_1' = c_1 + \left(\frac{\Delta t}{t}\right)_{stress} \tag{1.144}$$

one obtains

$$\frac{\Delta t}{t} = \left(\frac{\Delta t}{t}\right)_{stress} + \left(\frac{\Delta t}{t}\right)_{texture} = c_1' + c_2\left(\frac{2\pi a}{v_0}\right)^2 f^2 \tag{1.145}$$

The advantage of this expression is that only relative time-of-flight differences have to be measured. If the relevant constants entering c_1 and c_2 were all determined beforehand by mechanical tests, including the expansion coefficients of the sixth order by ultrasonic or X-ray measurements, the application of (1.142) allows to determine and separate the parts due to texture and stress in a straightforward manner. If these coefficients are not known, one can still obtain from the dispersive part c_2 and because the ratio c_1/c_2 is a constant, one can separate the stress dependent component, see Fig. 1.40 and [70]. The birefringence dispersion method has been demonstrated in a number of applications see the reference given above.

Finally, as discussed in Sect. 1.11.9, the acoustoelastic constants depend on temperature. However, the influence of the texture remains dominant compared to the temperature contribution [41].

1.17 Attenuation, Scattering, and Absorption of Ultrasonic Waves

The attenuation coefficients for longitudinal as well as shear waves, α_L and α_T, are the sums of two different interaction processes between the ultrasonic waves and the material's elementary excitations and its microstructure, absorption α_A and scattering α_S:

$$\alpha = \alpha_A + \alpha_S \tag{1.146}$$

Scattering occurs at the grain structures whereas absorption transfers the ultrasonic energy to the heat bath of the material. Throughout this book, we call damping or internal friction by the internal excitations of a material, such as dislocations or thermo-elasticity, absorption. The expressions attenuation, absorption, internal friction, and damping are often used concurrently without further differentiation. However, damping by scattering is often called attenuation, in particular in the publications from Fraunhofer IZFP. The use of the terms of attenuation and absorption for other wave modes, like surface waves, Lamb waves, SH-waves, is analog. For all cases, damping is formally described by an imaginary part $i\alpha$ of the k-vector of the ultrasonic waves, i.e. the k-vector becomes complex

$$k^* = k + i\alpha \tag{1.147}$$

The phenomenology and definitions of the parameters of internal friction or ultrasonic absorption are described in many textbooks and in review articles, for example by [60].

1.17.1 Absorption

Ultrasonic absorption or internal friction, as it is also called, probes the elementary excitations which exist in a material by inelastic interaction processes, resulting in ultrasonic absorption. The coupling to the elementary excitations occurs usually by the dynamic elastic strain. There are many processes, such as the interaction with dislocations and with spin waves, magneto-elastic effects, thermoelastic effects, interaction with the free electrons in a metal, with tunneling sites in disordered materials and their associated relaxation phenomena, interaction with interstitials and a number of other interactions. These are discussed in different books: [14, 15, 47, 173, 204, 290, 291] and in the book series on Physical Acoustics (Academic Press, now Elsevier). For materials characterization in metals, two effects are in particular of importance: thermo-elastic dissipation in polycrystalline materials, and dislocation damping.

Dislocations are line defects or irregularities in the crystallographic structure of materials (Fig. 1.42a, b). They may extend over many atoms with lengths of tens to hundreds of atoms. The two primary types of dislocations are edge dislocations and screw dislocations. Mixed dislocations are intermediate between these. Mathematically, dislocations are topological defects. Dislocations can behave as stable defects and are mobile by forces acting in their slip plane. Two dislocations of opposite orientation can cancel each other when brought together, but a single dislocation cannot “disappear” on its own within a crystalline material. A dislocation may be pinned by other dislocations or by defects like vacancies, inclusions or interstitials. Under load dislocations may move and they start to vibrate by oscillatory dynamic loads. Higher loads will break them away from their pinning points, and they can move along energy barriers like grain boundaries, or if the loads are high enough, move across such barriers. Some types of dislocation pinning points even act as sources of dislocations by multiplying the number of dislocations in the material under externally applied load (e.g. Frank-Read sources, Orowan mechanism). The Burgers vector b of a dislocation is the displacement it generates. There are many textbooks in material science detailing the physics of dislocations, also in compact form, see for example [73]. The presence of dislocations strongly influences many of the material’s properties.

The dynamic strain of propagating ultrasonic waves excites the dislocations to oscillations. They in turn modulate the distribution of electrons and thermal excitations or phonons in a metal, which lead to resonance and relaxational effects resulting in the damping of the ultrasonic waves [75, 184]. The resonance frequencies of dislocation strings are of the order of 1 GHz for a dislocation length L of 1 μm and for shorter L correspondingly higher and vice versa. However, for engineering metals, the dislocation lengths are much shorter and the dislocation motion is strongly damped. The relaxation part of the damping due to dislocation motion remains, which leads to [290]:

$$\alpha_{A,dis} = \frac{1}{v}\frac{16}{\pi^4}\frac{Gb^2Bf^2}{C^2}\Lambda L^4 \tag{1.148}$$

The absorption due to dislocations entails also a relative change of velocity (or the elastic modulus):

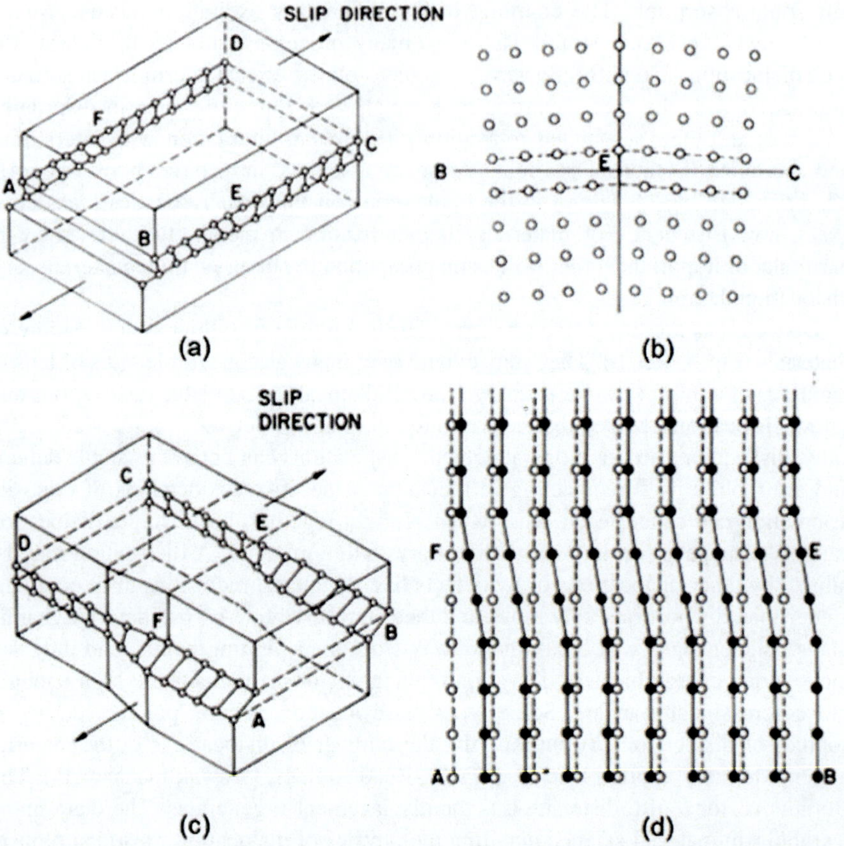

Fig. 1.42 Structure of an edge dislocation in a simple cubic crystal; **a** Slip plane ABCD is divided by dislocation line EF into the slipped region ABEF and un-slipped region FECD; **b** View perpendicular to the line EF showing atomic arrangement; **c** Structure of a screw dislocation. Again, the slip plane ABCD is divided by dislocation line EF into a slipped region ABEF and an un-slipped region FECD; **d** view of plane ABCD from above, showing atomic arrangement in the upper plane (●) and lower plane (○) (from [204], with permission from Elsevier)

$$\frac{\Delta v}{v_0} = \frac{4Gb^2}{\pi^4 C} \Lambda L^2 \tag{1.149}$$

Here, Λ is the dislocation density (total length of the dislocation lines per unit volume. Hence the unit of Λ is [m/m^3] = [m^{-2}]), L is the dislocation length, f is the ultrasonic frequency, $C = 2Gb^2/\pi(1 - \upsilon)$ is the line tension of the bowed-out dislocations, G is the shear modulus, v is the sound velocity, and υ is Poisson's ratio. If there is a distribution of dislocation lengths, the numerical pre-factor in (1.149) changes, but the overall dependences remain. The damping constant B (of the order of $\approx 10^{-4}$ [Ns/m^2]) is caused by the interaction of the oscillating dislocations, with the phonon and electron bath modulating their energy distribution, which leads to relaxation and hence to an imaginary part in (1.148). Equations (1.48 and 1.49) are derived for the case, when the oscillating ultrasonic stress-field is in the glide direction of the dislocations. Otherwise amplitudes of the dispersion and absorption are reduced, which is taken into account by an orientation factor Ω. Likewise, the distribution of the dislocation lengths can be taken into account by defining an effective length, see [47] for a detailed discussion and for further references for experimental data. Using contactless laser ultrasonics, the absorption according to the Granato-Lücke theory could be clearly established in a copper single crystal, with a dislocation density of $\Lambda = 10^{11}$ [m^2] [82].

In materials with many point defects which act as pinning points for dislocation, two contributions determine the ultrasonic absorption, namely thermal activated processes and athermal hysteretic behavior leading to linear frequency dependence of absorption [77], which is indeed observed in many metals [68]. A number of experimental verifications of dislocation relaxation absorption have been carried out [124, 132, 208], also separating attenuation due to grain scattering and inelastic absorption or internal friction [131].

There are clear signs that ultrasonic absorption and the velocity depend on accumulated plastic deformation and hence on the fatigue cycles a material has undergone [57, 98, 125, 250]. In addition, at high enough ultrasonic strains, the so-called β-parameter (nonlinearity parameter) can be used as a measure of fatigue in the material (see Sect. 1.13). Based on Eqs. (1.148 and 1.149), an absorption increase due to accumulated dislocation densities and the corresponding decrease of the sound velocity resulting from plastic deformation becomes perfectly understandable.

In an annealed iron specimen with $G = 80$ GPa, $v_L = 6000$ m/s, $f = 5$ MHz, $B = 10^{-4}$ Ns/m^2, $b = 0.25$ nm, $\Lambda = 2 \times 10^{13}$ m^{-2} and average $L = 0.1$ µm, increase in attenuation due to dislocation is $\alpha_{A,dis} = 0.033$ Np/m $= 0.29$ dB/m and the decrease in velocity is $\Delta v/v_0 \approx 0.01$ as per Eqs. (1.148 and 1.149), respectively. It should be noted that with increasing Λ, L is expected to decrease.

Besides internal friction caused by dislocation motion, there are several other mechanisms which contribute to the overall absorption coefficient of ultrasonic waves in engineering materials, for example domain wall motion in magnetic materials, inter-crystalline thermal currents in polycrystalline materials, electron–phonon interactions, grain-wall relaxation, and thermoelastic interaction. There is relatively scarce information discussing these issues in materials of complicated microstructures, whereas there is a wealth of information in crystalline and amorphous materials, see the reference books cited in this chapter.

To determine the absorptive part α_A independent of the scattering part α_S, the so-called reverberation technique can be used. A material of finite size, for example a cube of some cm side-length, is insonified at a certain point by a transducer or by a laser source. First a back-wall echo sequence develops. Due to scattering at the grain structure and reflections with associated mode conversion at the sample walls, a reverberation field develops which decays only due to absorption [307]. This technique uses the fact that an incident ultrasonic beam loses spatial coherence when it is repeatedly scattered at grain and phase boundaries and reflected at the sample surfaces, see Fig. 1.43. After a few microseconds, the elastic energy is distributed over the whole sample volume, creating a diffuse sound field, the decay of which is determined only by absorption—no losses due to scattering or diffraction play a role. The distribution of energy among the various modes (longitudinal, shear, and surface waves) depends on the size and shape of the specimen and on the wave speeds of the material [208]. The various contributions to absorption have been examined in detail in a master thesis carried out at the Fraunhofer IZFP and are summarized in [81, 218].

For steels, aluminum, polymers, nanocrystalline materials, and the aluminum alloy Al3Mg in the frequency range from 1 to 50 MHz, a direct proportionality between α_A and f was found [82, 86, 151]. This entails that $1/Q = \alpha\lambda/2\pi$ is constant and a characteristic value for the material and its state. For steels $\alpha_A\lambda \approx 0.007 \pm 0.005$; and for aluminum $\alpha_A\lambda \approx 0.003$ [68]. It is also known that not only the absorption increases with increasing fatigue level or creep damage [124, 208], but also the non-linearity parameter β (see Fig. 1.44) introduced above (Sect. 1.13) [139, 146].

It has been suggested that the origin of $1/Q$ being a constant is a hysteretic behavior due to micro-inhomogeneities in the solid. For the ultrasonic absorption caused by dislocations, their interaction with pinning-points leads to hysteretic behavior [77]. In polymers, the inhomogeneities might be the polymer units, which re-arrange under cyclic stress in the minima of the potential energy landscape and which also cause β-relaxations [86]. Furthermore, it has been proposed by C. Zener and co-workers that in polycrystalline solids thermal inter-crystalline currents are modulated by cyclic stresses leading to a broad thermoelastic absorption peak as a function of frequency,

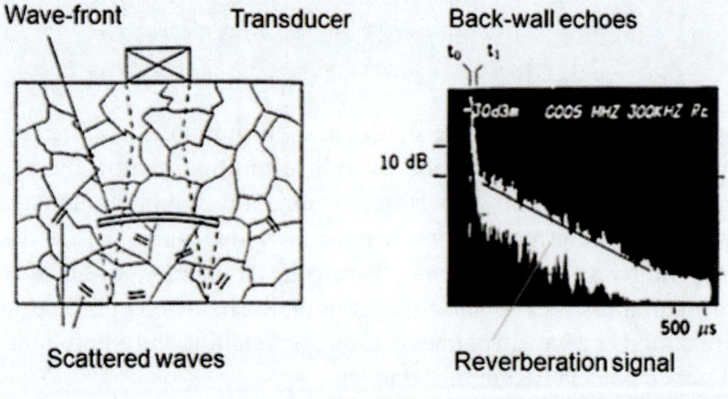

Fig. 1.43 Reverberation signal in a volume of finite size. The decay of the signal is eventually only determined by inelastic absorption

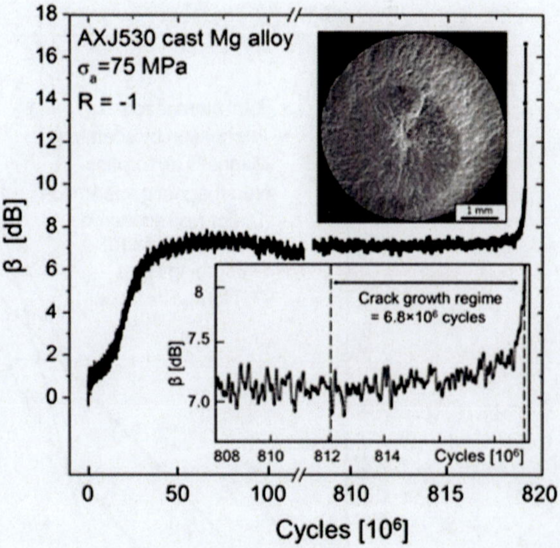

Fig. 1.44 Nonlinear ultrasonic parameter β in a cast Mg alloy as a function of fatigue cycles during a test at 75 MPa stress amplitude. The fatigue-crack growth took place during the last 6.8×10^6 cycles before macroscopic failure occurred (insert). The SEM image shows the fractured surface (in modified form from [139], with permission from Elsevier)

sec [204]. This absorption mechanism gets smeared out in frequency by grain size distributions and has to be taken into account in the lower MHz range for grain sizes of the order of 10 μm [150]. A similar mechanism has been suggested if micro-cracks are present, like in rocks [252]. Extending this picture to micro-inhomogeneous solids, thermal currents are modulated locally, leading to an increased absorption also by friction in the crack walls [325] or micro-contact points [1]. This mechanism also leads to non-linear stress–strain curves which generate higher harmonics. It is interesting to note that micro-slip friction in single asperity contacts lends support to this idea, because it leads to an increased damping and a nonlinear behavior in local internal friction as measured recently [174]. Furthermore, magneto-elastic damping of Bloch walls in the ultrasonic strain field leads to noticeable absorption in the MHz range if the material is ferromagnetic [81].

The energy absorbed by internal friction leads to a temperature increase in a material. This holds also for the energy transferred by ultrasonic waves to the thermal bath by the inelastic absorption mechanism discussed above. The temperature increase can be made visible by aid of an infrared camera (IR). Despite its amplitude, the signal of the IR-camera is measurable if synchronous detection is used. The signal is proportional to the ultrasonic absorption coefficient and other parameters which are determined by calibration. This technique has been used to separate the various contributions to the total absorption, like inter-crystalline thermal currents, magnetoelastic interactions and the expected dislocation absorption in a ferritic-pearlitic steel [194] (Fig. 1.45).

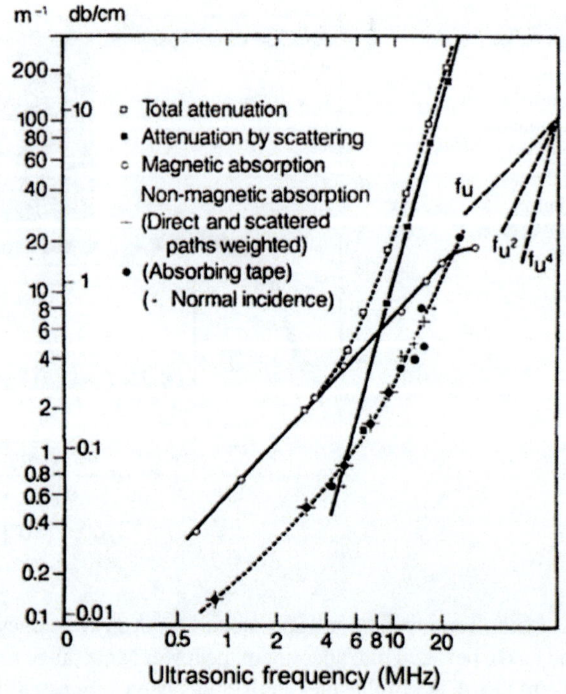

Fig. 1.45 Experimentally measured total ultrasonic attenuation as a function of frequency in a ferritic-pearlitic steel by measuring the increase of temperature increase of a sample allowing to separate the different contributions to internal friction or absorption and to scattering. The lines f_u f_u^2 and f_u^4 are guidelines for the eye in order to judge the slope of the attenuation where f_u represents the frequency; from [194], with permission from Springer Nature

1.17.2 Attenuation Due to Scattering – Basic Concepts

Ultrasonic scattering occurs at grain boundaries, inclusions, defects of various shapes, and voids. Scattering also occurs in liquids with suspended solid particles, aerosols, emulsions, and liquids which contain gas bubbles. Therefore, scattering has many applications in medical imaging as well as in material science and engineering. The ultrasonic field excites the particles to oscillations which radiate as ultrasonic waves in all directions and which are scattered again. This effect is called multiple scattering. If the amplitudes of the secondary waves are small enough, the effects of multiple scattering can be neglected.

Scattering occurs at heterogeneities, because the boundary conditions of continuity for the amplitude as well as for the strain must be fulfilled when ultrasonic waves cross their boundaries. Here, reflection, refraction and mode conversion occur. In solids, the mode conversion encompasses shear waves as well. For extended interfaces, large compared to the extent of the ultrasonic beam and wavelength, this can be expressed in terms of changes in acoustic impedance. For an ensemble of grains in a microstructure, it is quite helpful to consider changes of impedance at the grain boundaries in order to estimate the strength of scattering. However, that is not sufficient and additional parameters must be taken into account. These are the shape of heterogeneities, their size, size distribution, and orientations, i.e. texture. In case of a single size, spherical shape and statistical orientation of the heterogeneities, one considers the ratio of the grain diameter to the wavelength, d/λ. This ratio allows one to distinguish three scattering regimes in polycrystalline materials consisting of randomly oriented anisotropic grains: Rayleigh scattering ($d/\lambda \ll 1$), stochastic scattering ($d/\lambda \approx 1$), and geometric scattering ($d/\lambda \gg 1$).

In order to lay out the main points, let us consider scattering at a single sphere of radius R. The scattered waves are spherical waves emanating from the sphere, see for example [281]. For $kR \ll 1$, i.e. in the Rayleigh regime, one can expand the potentials describing the scattered waves as a function of the parameter kR. The calculations for this process are simplest for an infinitely stiff immobile sphere. This leads to a scattering cross-section of the form

$$\sigma_{eff} = (7/9)\pi R^2 (kR)^4 = (7/9)\pi R^6 k^4 \tag{1.150}$$

In analogy to scattering phenomena in nuclear physics, this equation tells us that the "cross section" for scattering at a single sphere is adjusted by the factor $(kR)^4$, compared to its geometrical cross-section πR^2 (see also Fig. 2.5 in Chap. 2).

Let us assume that the grain structure of a material consists of an ensemble of such stiff spheres. As long as the scatterers are independent, which is equivalent to neglecting multiple scattering, the loss in intensity I of the ultrasonic wave at a point x is given by [290]:

$$dI = n_0 \sigma_{eff} I dx = 2\alpha_s I dx \tag{1.151}$$

where n_0 is the number of scatterers per volume V. Here, n_o is proportional to $1/d^3$ and $d = 2R$. One arrives at

$$\alpha_S = \frac{1}{2} n_0 \sigma_{eff} \propto \frac{1}{d}\left(\frac{d}{\lambda}\right)^4 = const.Vf^4 \tag{1.152}$$

From this equation we conclude that for the same ratio d/λ, attenuation due to scattering in a fine-grained material is higher than for a coarse-grained material. However, for a given wavelength or frequency, the scattering part increases with the third power of the grain size, i.e. fine-grained microstructures with constituent sizes in the μm range become transparent for ultrasonic waves in the MHz range.

In reality, the grains in a microstructure have finite elasticity or stiffness. For multi-phased materials, the situation becomes more complex, leading to scattering equations which contain the elastic differences of the grains in the direction of the wave-vector of the ultrasonic waves relative to the host material as the origin of the scattering contrast. Multiple scattering has to be taken into account for strongly scattering materials. Scattering also leads to dispersion, and has been exploited for differentiating between birefringence caused by texture and by stresses (see Sect. 1.16). The work of Stanke and Kino [279] contains a detailed discussion of the historic developments, with the corresponding physical approximations and references up to 1984.

1.17.3 Scattering in Single-Phase Polycrystalline Materials

In single-phase polycrystalline materials the crystallites consist of the same crystal structure. Scattering in this type of material has been examined theoretically and experimentally and is summarized in [68, 215]. For the Rayleigh regime, $k\overline{R} \ll 1$, where k is the wave-vector and $2\,\overline{R} = \overline{d}$ is the mean grain diameter. The attenuation coefficient is of the form:

$$\alpha_{S_L,S_T} = S_{L,T}\overline{d}^3 f^4 = S_{L,T} V f^4 \tag{1.153}$$

Here, S_L and S_T stand for the scattering strength for longitudinal and shear waves, respectively. The parameters S_L and S_T depend on the elastic constants, anisotropy factors, and the longitudinal and shear sound wave velocities. They are tabulated in [68, 215] for a number of metallic polycrystalline materials. For ceramics one finds theoretical and experimental data in [58]. In (1.153) the pre-factor $\overline{d}^3$ or V is the so-called appropriate grain volume for which various expressions exist in the literature. If there is only one grain size of radius R, $V = (4/3)\pi R^3$ [99]. If there is a distribution of grain sizes with many large grains, they will dominate the attenuation coefficient, because the cross-section for scattering in the Rayleigh regime is proportional to V^2 of the scatterers, see (1.150). Therefore, the appropriate V is rather given by [213]:

$$V \rightarrow \frac{4\pi}{3} \frac{\langle R^6 \rangle_{av}}{\langle R^3 \rangle_{av}} \tag{1.154}$$

For longitudinal (L-) and transverse or shear waves (T-), the scattering parameters S_L and S_T are given by [68, 215]:

$$S_L = \frac{8\pi^3}{375} \left(\frac{A}{\rho v_L^2} \right)^2 \frac{1}{v_L^4} \left(2 + 3 \left(\frac{v_L}{v_T} \right)^5 \right) \tag{1.155}$$

$$S_T = \frac{6\pi^3}{375} \left(\frac{A}{\rho v_T^2} \right)^2 \frac{1}{v_T^4} \left(3 + 2 \left(\frac{v_T}{v_L} \right)^5 \right), \tag{1.156}$$

Here, $A = c_{11} - c_{12} - 2c_{44}$ is the anisotropy factor, see (1.32). The values of S_L and S_T calculated from the elastic properties given in Table 1.2 are listed for different materials in Table 1.6. A typical value for the sound velocity ratio v_L/v_T is ≈ 2.1. This value yields for the ratio S_L/S_T:

Table 1.6 Scattering coefficients S_L, S_T, Σ_L, Σ_T, and ratios S_L/S_T and Σ_L/Σ_T for a series of cubic polycrystalline materials calculated with the parameters given in Table 1.2

Elements	$S_L[10^{-4}]$ [μs/mm]^4	S_T $[10^{-4}]$ [μs/mm]^4	S_L/S_T	Σ_L $[10^{-2}]$ [μs/mm]^2	Σ_T $[10^{-2}]$ [μs/mm]^2	Σ_L/Σ_T $[10^{-2}]$
Ag	923.4255	7016.9566	0.1316	0.3921	25.1548	1.559
Al	3.8535	25.174	0.1531	0.0064	0.3029	2.109
Au	1717.3616	26,747.56	0.0642	0.1804	48.6094	0.371
C(Diamond)	0.042	0.1002	0.4194	0.0028	0.0175	15.829
Cr	8.7977	29.8795	0.2944	0.0448	0.5736	7.802
Cu	371.2967	2393.6622	0.1551	0.3473	16.0401	2.166
Fe	82.6895	381.6271	0.2167	0.2067	4.891	4.225
Ge	42.8284	143.2453	0.299	0.1413	1.7563	8.045
Ir	30.0047	115.7519	0.2592	0.0846	1.3996	6.047
K	17,756.3804	143,250.142	0.124	2.7287	197.3273	1.383
Li	567.8502	3831.9099	0.1482	0.715	36.1788	1.976
Mo	10.6495	50.3235	0.2116	0.0304	0.7535	4.031
Na	5776.1269	42,235.5811	0.1368	1.8645	110.7631	1.683
Nb	182.6552	1855.9567	0.0984	0.089	10.2117	0.872
Ni	114.3412	568.4274	0.2012	0.2417	6.6371	3.642
Pb	13,140.1067	158,629.355	0.0828	0.9621	155.8001	0.618
Pd	370.7892	3424.2316	0.1083	0.1812	17.1717	1.055
Pt	131.6871	1369.5586	0.0962	0.0423	5.0825	0.832
Pu	44,097.5168	188,372.244	0.2341	11.5842	234.8708	4.932
Rb	101,583.335	755,063.15	0.1345	7.7699	476.9711	1.629
Si	4.0381	14.2094	0.2842	0.0355	0.4882	7.268
Ta	104.313	643.3547	0.1621	0.0785	3.3173	2.366
Th	3118.5086	14,147.7736	0.2204	1.8786	42.9608	4.373
V	7.9283	58.7976	0.1348	0.0095	0.5776	1.636
W	0.0259	0.1151	0.2253	0.000053	0.00117	4.568
X6CrNi18.12	183.1973	893.703	0.205	0.4095	10.8289	0.0378
β-brass	2146.8253	15,674.5781	0.137	1.3368	79.1832	1.688

$$S_L/S_T \approx 0.144 \quad (1.157)$$

which means that the scattering of shear waves is much stronger than for longitudinal waves of the same frequency. Finally, for the scattering of surface waves in the Rayleigh regime, S_R is only slightly larger than S_T, e.g. for steel [310]:

$$S_R/S_T \approx 1.02 \quad (1.158)$$

Equations (1.155–1.157) allow to estimate the attenuation due to scattering, using the values of c_{11}, c_{12}, c_{44}, v_L v_T, and ρ from Table 1.2. For $R = 0.025$ mm and $f = 10$ MHz:

	S_L [10^{-4}] [μs/mm]^4	S_T [10^{-4}] [μs/mm]^4	S_L/S_T	α_{SL} [Np/mm]	αSL [dB/mm]	αST [Np/mm]	αST [dB/mm]
Al	3.85	25.2	0.153	2.52×10^{-4}	2.19×10^{-3}	1.65×10^{-3}	1.43×10^{-2}
Fe	82.7	382	0.216	5.41×10^{-3}	4.70×10^{-2}	2.5×10^{-2}	0.217
X6CrNi18-12 (austenitic stainless steel)	183	894	0.205	1.20×10^{-2}	0.104	5.85×10^{-2}	0.508

For the case $\lambda \cong 2\pi \overline{d}$, also called the intermediate, stochastic or resonance scattering regime, the attenuation is given by [213]

$$\alpha_{S_L,S_T} = \Sigma_{L,T}\overline{d}f^2 = \Sigma'_{L,T}f\left(\frac{\overline{d}}{\lambda}\right) \quad (1.159)$$

The scattering parameters Σ_L and Σ_T for longitudinal (*L*-) and transverse (*T*-) waves are given by

$$\Sigma_L = \frac{16\pi^2}{525 v_L^2}\left(\frac{A}{\rho v_L^2}\right)^2 \quad (1.160)$$

$$\Sigma_T = \frac{4\pi^2}{210 v_T^2}\left(\frac{A}{\rho v_T^2}\right)^2 \quad (1.161)$$

The scattering strength $\Sigma_{L,T}$ is given in Table 1.6 for various cubic polycrystalline materials. In case of a grain size distribution, the substitution

$$\overline{d} \rightarrow 2\langle R \rangle_{av} \tag{1.162}$$

should be made [213]. It is not straightforward to derive this distribution from the attenuation versus frequency curve if the transition from the Rayleigh region is concerned, see [202, 319].

Finally, if $\lambda \ll 2\pi\,\overline{d}$, i.e. in the geometric regime, the ultrasonic attenuation becomes frequency independent [15]:

$$\alpha_{S_L,S_T} = \frac{\langle R_{refl} \rangle_{av}}{\overline{d}} \tag{1.163}$$

Remember from above that the amplitude of ultrasonic wave gets reflected from grain to grain, due the change of impedance in the direction of propagation. For calculations, we may use the reflection coefficient for plane waves, $R_{refl} = (Z_2 - Z_1)/(Z_1 + Z_2)$, where Z_1, Z_2 are the acoustic impedances of two neighboring grains, i.e. the grains at which the wave is incident and refracted, respectively. An averaging procedure has to be carried out in order to obtain $\langle R_{refl} \rangle_{av}$, see [181]. The scattering process at the grain boundaries involves mode conversions which gets stronger with increasing anisotropy. For an incident *L*- or *T*-wave, the scattered energy is distributed into *L*-, *T*- and surface waves. From (1.157) it can be seen that attenuation due to scattering is about 7 times stronger for shear waves than for longitudinal waves in the Rayleigh regime. That is even more pronounced (~ 50 times) in the stochastic regime, see the ratios of Σ_T/Σ_L in Table 1.6. The scattered ultrasonic wave field diffuses through the microstructure and loses its spatial coherence. The diffusive field is characterized by a normalized diffusivity parameter *D*. The built-up of the diffusive field occurs in all three scattering regimes. Intuitively, *D* must be large when scattering is small, i.e. in the Rayleigh regime, because then the diffusive field propagates further due to the small scattering power of the grains. Conversely, *D* becomes small when a strong scattering regime sets in due to multiple scattering in the microstructure and high anisotropy. The diffusivity *D* has been calculated by [305] and for small frequency $D \propto 1/f^4$ holds, see Fig. 1.46. Measurements of the diffusion coefficient in a frequency range from 1 to 10 MHz of an ultrasonic diffusive field caused by multiple scattering in pearlitic and austenitic steels [80], with grain sizes between 35 and 124 μm fit well to the theoretical curve, see Fig. 1.46.

1.17.4 Advanced Theories of Ultrasonic Scattering

Hirsekorn [99, 100] presented an ultrasonic scattering theory which allowed to calculate the scattering coefficients and phase and group velocities of plane longitudinal and transverse waves in polycrystals as a function of *ka* (*k*: wavevector, *a* grain radius) without limitation to the Rayleigh regime. The theory includes mode conversion and multiple scattering and was used to describe ultrasonic propagation in polycrystals

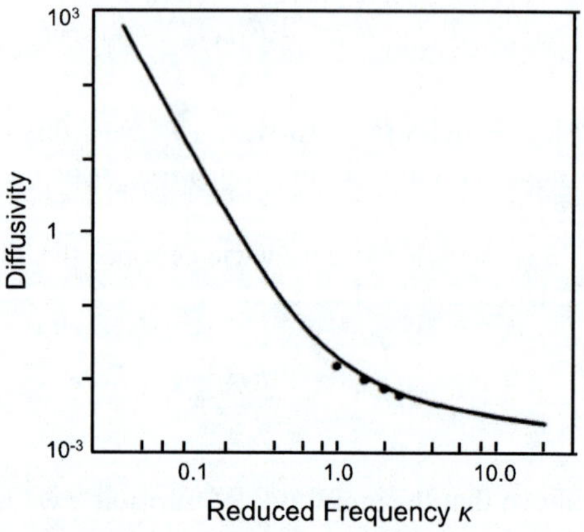

Fig. 1.46 Dimensionless diffusivity as a function of reduced frequency for the case of a cubic crystalline microstructure. The low frequency part follows an inverse fourth power law and represents the Rayleigh regime. The high frequency stochastic regime is characterized by an inverse logarithmic frequency dependence. The four data points represent the experimental work of [80]. The reduced frequency is $\kappa = \omega \overline{d}/v_T$, where $\overline{d}$ is the mean distance between two adjacent grains (from [305], with permission from Elsevier)

with randomly orientated grains as well as in those with preferred grain orientation. The calculations were first carried out for compressional [99] and for shear waves [100] for polycrystals of cubic symmetry, with randomly orientated grains in second-order perturbation theory, using the assumption that the polycrystals are weakly anisotropic, i.e. that the deviations from the average of the elastic constants and of the density are small. Furthermore, the grains were assumed to be approximately spheres of the same size. The asymptotic values for the attenuation coefficient at low *ka* (Rayleigh scattering) correspond to (1.153) which was obtained earlier by [16]. Besides recovering these equations, the theory predicted that the velocity becomes dispersive. The amplitude of the dispersion is of the order of 1–2%, depending on the anisotropy factor squared A^2. For $kR \ll 1$ it decreases like

$$v_{L,T}(f) = Bv_{L,T}\left(1 - Df^2\right) \tag{1.164}$$

B is slightly smaller than 1. This means that also the velocity for frequency $f \to 0$ is already reduced compared to the case, without taking scattering into account. The parameter B and D are constants which depend on the sound velocities, density and above all on A^2 (1.32). For $A = 0$, $D = 0$, $B = 1$ and $v_L = v_g$, where v_g is the group velocity. At $kR \approx 1$ the velocity passes through a minimum and then increases again, see [99, 100]. An example is shown in Fig. 1.74 and discussed in Sect. 1.24 on the Kramers–Kronig relations. Eventually, Hirsekorn extended the theory to take into account rolling texture [101]. As discussed in Sect. 1.16.2, this theory was used to develop a methodology to separate the texture influence and stresses in sound velocity measurements.

Furthermore, a unified theory of elastic wave scattering was developed by [279, 305] in order to describe scattering in the Rayleigh, stochastic, and geometric regimes in polycrystalline materials consisting of randomly oriented anisotropic grains. The crystallites possess spatially dependent elastic properties, due to orientation differences from grain to grain. The spatially variant elastic properties, weakly perturbing the elastic wave propagation, cause dispersion in the wave speed and attenuation by scattering at the grain boundaries. An effective reference medium was built up by the Voigt averaging of the elastic constants of the polycrystalline ensemble. Numerically solving the wave equation by a perturbation approach, [279] obtained the attenuation and dispersion for all three scattering regimes, see Figs. 1.47 and 1.48, where the logarithm of the normalized attenuation coefficient $\alpha\,\overline{d}$ is plotted for polycrystalline aluminum and iron versus $(\varepsilon\, x_0)$ in a double logarithmic plot. Here $x_0 = k_0\overline{d} = 2\pi\,\overline{d}/\lambda_0 = \omega\,\overline{d}/v_0$ contains the ratio between the grain diameter and the wavelength, which is also equivalent to a normalized frequency. The parameter ε represents the inhomogeneity of the propagation constant

$$\varepsilon^2 \equiv \left\langle [k(r) - k_0]^2 \right\rangle / k_0^2 \tag{1.165}$$

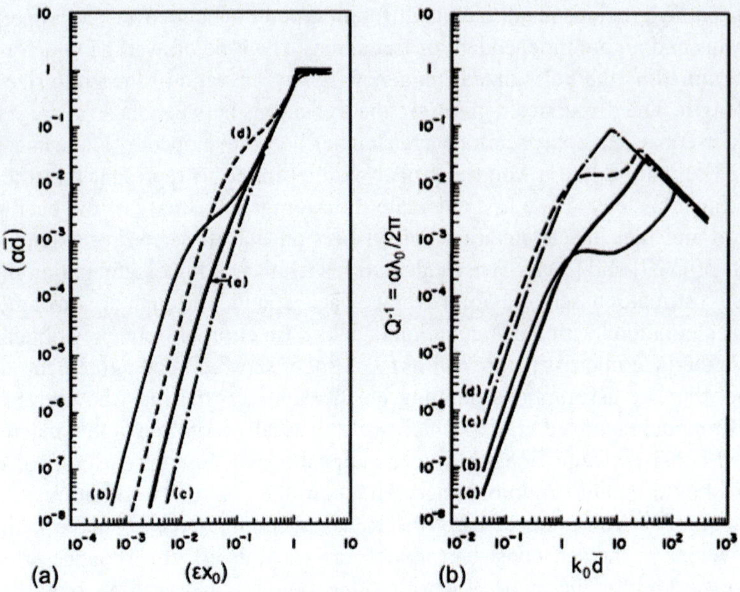

Fig. 1.47 **a** Plot of normalized attenuation $\left(\alpha\,\overline{d}\right)$ versus (εx_0), $\overline{d}$ is the mean grain diameter, and $x_0 = k_0\overline{d}$. The parameter ε is the inhomogeneity of the polycrystalline structure proportional to the anisotropy factor; **b** Plot of the same data like in **(a)**, however plotted as $Q^{-1} = (\alpha\lambda_0/2\pi)$ versus normalized grain size $k_0\overline{d}$. In both figures the letters (a), (b), (c), and (d) designate curves for shear and for longitudinal waves in aluminum, and for shear and for longitudinal waves in iron, respectively (from [279], with the permission of the Acoustical Society of America)

where $k(r)$ and k_0 are the wavevectors at position r and at position zero. The symbol $\langle\rangle$ stands for spatial averaging. The parameter ε is proportional to the anisotropy factor. It is assumed that ε is small and that all Euler rotations of the grains are equally likely, i.e. there is no texture. Stanke and Kino [279] executed their theory for cubic-class crystals. In Fig. 1.47a the graphs (a) and (b) respectively designate the attenuation for shear waves and longitudinal waves in aluminum; and the graphs (c) and (d) the attenuation for shear waves and longitudinal waves in iron. Figure 1.47b shows the same data in different normalized coordinates. These plots emphasize the transitions between Rayleigh, stochastic, and geometrical regimes and the dependence on the grain size.

Figures 1.48a–c show the phase velocity for the same cases that were used to calculate the attenuation. In the figures, the normalized variation in phase velocity is plotted relative to the Voigt velocity $(v_{ps} - v_0)/v_0$ versus the logarithm of normalized frequency. The Rayleigh limit, the Born approximation, and the lowest order term in ε of the geometric asymptote is shown (an analog (e) to electromagnetic waves is also shown, which is not discussed in this book). In all cases the unified theory shows regions that are virtually non-dispersive at low and high frequencies, where the phase velocity can be approximated by simpler theories. The Born approximation predicts a non-dispersive stochastic asymptote. Only the case of longitudinal waves in aluminum, which has the lowest ε, has a corresponding region. Each of these non-dispersive regions result from a different type of average over the properties of the grains and all are independent of frequency. The type of average which is valid is determined by the normalized frequency, i.e. by the ratio of the grain size to the wavelength. The dispersive regions are the transitions between these averages.

A self-consistent approach for scattering has been developed by Kube and Turner [137]. They argue that a Voigt approach is too simple to represent the true polycrystalline elasticity. They first calculate the averaged elasticity of the background medium and then the attenuation and dispersion due to scattering extending the models of [279] and [305]. These calculations result in a lower attenuation than the previous estimates based on a Voigt average background. The comparison of the theoretical calculations with the attenuation data as a function of frequency obtained for well-defined Cu microstructures with $\overline{d} = 18\ \mu\text{m}$ showed good agreement, despite the fact that the assumptions defining an ideal polycrystal must be obeyed, such as uniform and equiaxed grains, which are statistically isotropic on the macroscale. The self-consistent approach is physically appealing, because the elasticity of a polycrystal should include grain boundary effects in the continuity conditions of stress and strain. The Voigt, respectively the Reuss averaging over- or underpredict the phase velocity. The self-consistent model removes this effect and hence presents a step forward within the scope of wave propagation in polycrystalline metals.

Furthermore, several authors focused on a more accurate representation of scattering in complex polycrystalline microstructures by accounting for grain elongation in cubic polycrystalline materials [317], in equiaxed hexagonal [316] and in elongated hexagonal grain-structures [318]. Furthermore, scattering in microstructures were discussed with arbitrary crystallite and texture symmetries [163, 164], and in polycrystals of any symmetry class [136]. Also, the effect of the presence

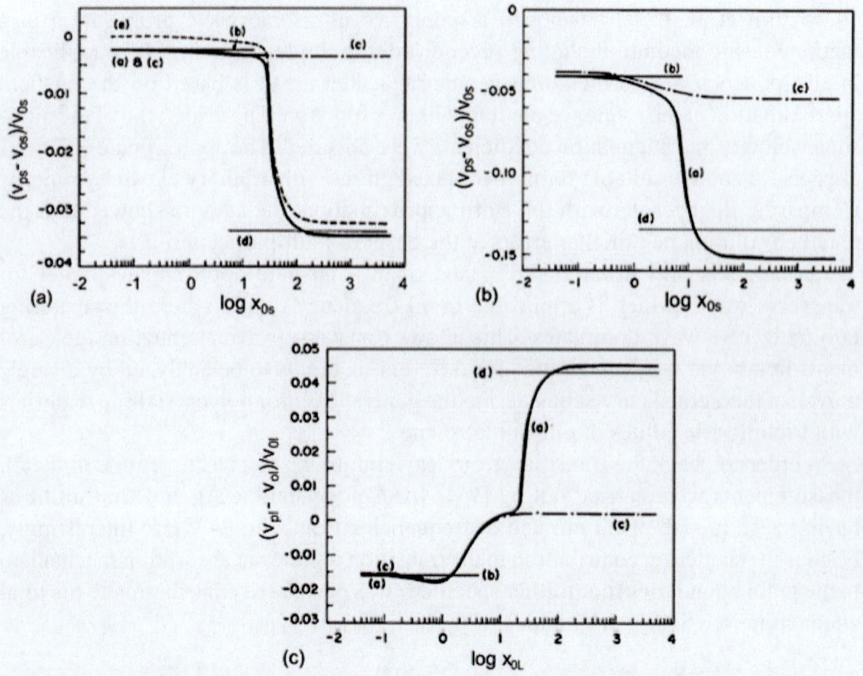

Fig. 1.48 Normalized variation of phase velocities $(v_p - v_0)/v_0$ as a function of normalized grain size $x_0 = k_0\overline{d}$. Here, v_0 is the Voigt dynamic average of the velocity calculated with the unified theory: **a** normalized shear wave velocity in polycrystalline aluminum; **b** normalized shear wave velocity in polycrystalline iron; **c** normalized variation of longitudinal-wave velocity in polycrystalline iron. The designations in the figures stand for: (a) unified theory, (b) Rayleigh limit, (c) Born approximation, and (d) lowest order of geometric asymptote. The high-frequency limit relative to the low-frequency limit is different in case of aluminum and iron. According to the authors, this difference is caused by the strong effect of multiple scattering in iron. Kube and Turner [137] obtained essentially the same behavior for iron, see also the discussion in this section and Fig. 1.73 (Figures from [279], with permission of the Acoustical Society of America)

of orthorhombic clusters in the microstructure was discussed [315], as well as the presence of lamellar duplex microstructures in pearlitic steels [54].

The theoretical models and their solutions for obtaining the attenuation coefficients are complex, in particular when applied to non-equiaxed grains and microstructures of lower crystallite symmetries. All theoretical models are approximate in accounting for the degree of material inhomogeneity and multiple scattering [33]. In these models, different approximations are employed. Numerical 2D and 3D simulations [65, 293, 294] of propagation and scattering in polycrystalline materials are free from those limitations, but are computationally very expensive. However, they serve to validate the applicability of analytical approximations and show good agreement with the measured ultrasonic attenuation coefficients and velocities in polycrystals even for relatively high grain anisotropy.

Rokhlin et al. [239] developed a model for ultrasonic wave propagation in a random elastic medium including second order multiple scattering. It is applicable in all frequency ranges including geometric scattering. It is based on the far-field approximation of the reference medium Green's function. The dispersion of the ultrasonic velocity and attenuation coefficients were obtained. The approximate solutions are general and are suitable for non-equiaxed grains with arbitrary elastic symmetry. Comparing their results with the Born approximation, the authors showed that the obtained solution has smaller errors at the onset of multiple scattering.

Further, Sha and Rokhlin [267] developed a far-field approximate model for transverse waves which is applicable in all frequency ranges where the scattering into transverse wave dominates. This allows transverse wave attenuation measurements in a broad frequency range for different materials to be analyzed by a single universal theoretical curve considering the general case for polycrystalline materials with triclinic crystallites of ellipsoidal shape.

In order to check the transition from Rayleigh to the stochastic regime in detail, measurements were carried out by [161] for nanocrystalline Ag and Cu thin films having a grain size of 60 nm and at frequencies from 1 to 24 GHz. Interestingly, besides the scattering contribution in the transition regime, an absorption mechanism in the grain boundaries (not further specified) was postulated contributing to the total attenuation, see Sect. 1.17.1 and (1.146).

It may be noted from Table 1.6 that for the same frequency, the attenuation due to scattering for shear waves is about 7 times and about 50 times higher than that for longitudinal waves in the Rayleigh and stochastic regions, respectively. However, it may also be deduced that for the same wavelength the attenuation due to scattering for shear waves is about 0.4 (7/16) times and 12 (50/4) times as large compared to longitudinal waves in the Rayleigh and stochastic regions, respectively. In contrast, in the geometric scattering region, the attenuation is independent of frequency or wavelength and similar for both shear and longitudinal waves. This is visualized in Fig. 1.47b. The above explains the reason for selecting ultrasonic longitudinal wave for the inspection of weld joints in materials with large scattering factors such as austenitic stainless steels and nickel base alloys.

1.17.5 Concepts of Backscattering

The scattering of ultrasound at grain and phase boundaries in polycrystalline multiphase and/or porous materials causes attenuation and dispersion, as discussed above. Scattering also occurs in backscattering direction, i.e. opposite to the direction in which the ultrasound was injected into the material. The signals can be received by the same transducers, amplified and either as rf-signal or as rectified signal displayed on an oscilloscope and analyzed. Typical backscattering signals are shown in Fig. 1.49 [269]. An ultrasonic immersion transducer with 16 MHz center frequency was used to obtain multiple backwall echoes in an austenitic stainless-steel specimen of 121 μm grain size immersed in water. In the time-domain signal, an increase in the backscatter noise can be observed after the water-steel interface echo. Furthermore, it can be seen in the spectrogram that the backwall echoes have lower frequency content in comparison to the backscattered signals. This is attributed to the fact that the higher frequencies in the beam are scattered more strongly than the lower frequency components, because scattering increases with increasing frequency, see (1.153). This part keeps returning to the transducer after multiple scattering events, as shown in Fig. 1.49a, whereas the low frequency content of the beam propagates to the back surface and gets reflected to form the backwall echoes. The backscattered wave amplitudes can be used for materials characterization and for the detection and evaluation of defects, see Figs. 1.50, 1.51, 1.52, 1.61, 1.66, and 1.67 as examples.

Rose derived in a series of three papers presented at QNDE conferences [240–242] the theoretical foundation for the definition of the backscattering coefficient η. The contrast for backscattering is given by the changes of the acoustic impedance for an ensemble of anisotropic crystallites in propagating direction of the ultrasonic wave. The impedance changes are averaged over the transducer pressure field, which

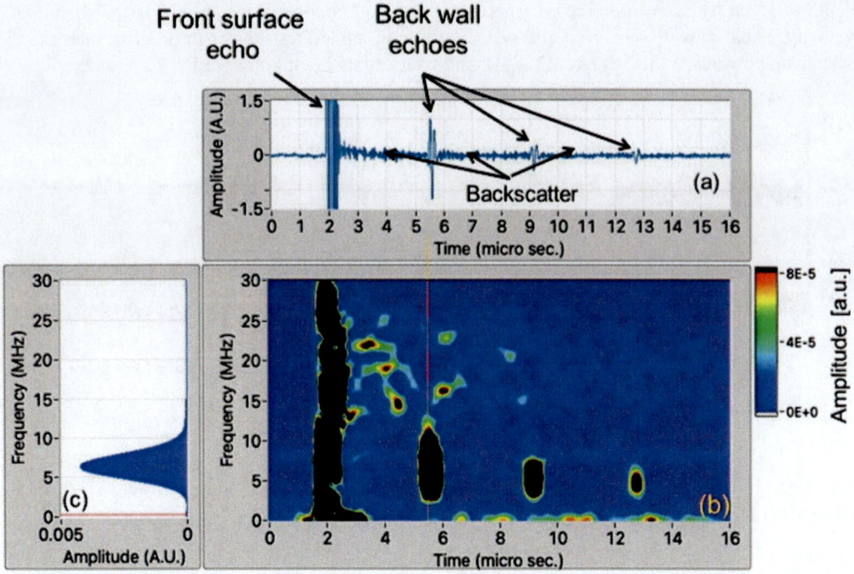

Fig. 1.49 **a** Time domain signal obtained in an austenitic stainless steel specimen of 121 μm grain size; **b** its spectrogram and **c** frequency content of the first back-wall [269]; with permission from Elsevier

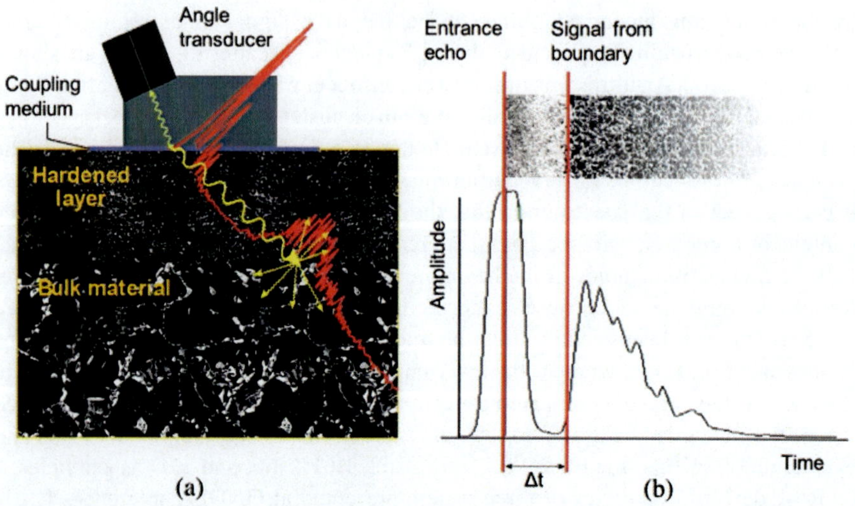

Fig. 1.50 **a** Backscattering signal obtained from grain scattering from a bulk ferritic-pearlite steel. The base material shows strong scattering due to the large grain size which causes the signal after the delay time Δt. The inductive hardening process of the surface material leads to a grain size refinement and martensitic transformation leading to a large reduction in the backscattering amplitude. **b** In the ultrasonic instrument with which the signals were measured, the backscattered signal was rectified and the time-of-flight Δt was measured in an automated way. The hardening depth is given by $\Delta s = \Delta t v_t \cos\alpha$, where α is the angle of incidence, and v_t is the shear wave velocity. Shear waves were used and generated by an angled transducer in order to increase the scattering power, see (1.156) (from [284]), and courtesy of Fraunhofer IZFP

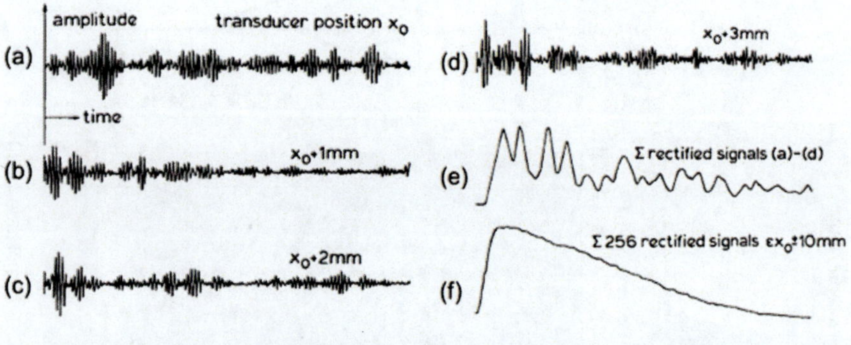

Fig. 1.51 Individual (**a–d**), rectified (**e**) and averaged backscattered signals from a steel with a grain size of 10 μm using 10 MHz longitudinal waves, see also [68]

is indicated by the tilde sign [240]:

$$\delta\tilde{Z} = \frac{\rho v_L}{2}\left\langle \frac{\delta\tilde{C}_{zzzz}}{(\lambda + 2\mu)} + \frac{\delta\tilde{\rho}}{\rho} \right\rangle \tag{1.166}$$

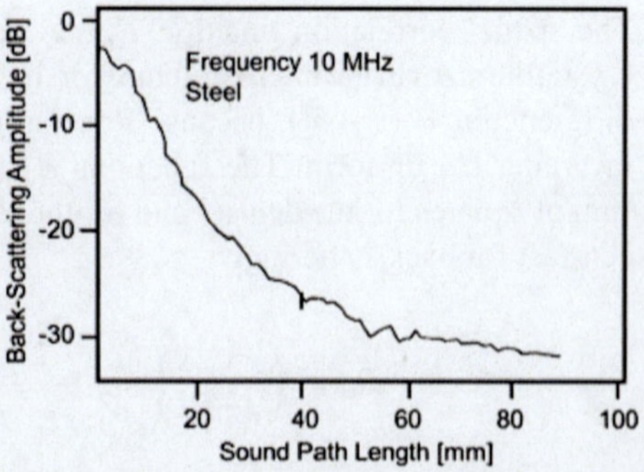

Fig. 1.52 Typical backscattered amplitude curve in a logarithmic scale, here in dB, obtained in a polycrystalline steel sample at a frequency of 10 MHz as a function of path length from the transducer

In the subsequent papers, Rose [241, 242] showed that the backscatter coefficient can be written as:

$$\eta = \sum_{\alpha=1}^{M} n_\alpha \langle |A_\alpha|^2 \rangle \tag{1.167}$$

Here, n_α denotes the number of grains per unit volume of the phase designated α. The scattering amplitude A_α, based on the Born approximation (see for example [128]), is the scattered amplitude at a single grain of the α-phase in an otherwise uniform effective medium, whose density and elastic constants are determined by the Voigt average over all the grains. Equation (1.167) means that the total backscattered power is equal to the sum of the power backscattered from each grain independently. As stated by [242], this is possible because the deviations in the material properties of each grain are assumed to be statistically independent. As a result, the amplitudes of the waves scattered by each grain are statistically independent, with zero mean in the Born approximation, which linearly relates the material property deviations and the scattered amplitudes. Consequently, the contributions to the total scattered power, that would arise due to the interference between scattering from different grains, average to zero. The total power is hence determined by summing the power scattered by each grain and for each phase. The total backscattering coefficient $\eta(\omega)$ from all phases with the fraction f_α is then given by [242]:

$$\eta(\omega) = \sum_{\alpha=1}^{M} f_\alpha \eta_\alpha(\omega) \tag{1.168}$$

where η_α was calculated to be:

$$\eta_\alpha(\omega) = \left(R_\alpha + Q_\alpha\right) k^4 \int d^3u \frac{\Gamma_\alpha(u)}{\Gamma_\alpha(0)} (exp(2ik(u))) \tag{1.169}$$

This equation does not take attenuation nor near-field diffraction into account. Hence, it is only valid for sufficiently short propagation distances from the transducer.

The spatial correlation function $\Gamma_\alpha\ (u)$ describes the size and distribution of the crystallite sizes in the microstructure for the phase α. It only depends on the coordinate difference $u = (r - r')$, because Rose assumed that the material is translationally invariant, i.e. uniform. The functions R_α and Q_α describe the deviation from the sum of squares for the density and for the elastic constants, and hence determine the contrast for backscattering:

$$R_\alpha = \left(\frac{1}{4\pi^2}\right)\left(\frac{\delta\rho_\alpha^2}{\rho_0^2}+\frac{2\delta\rho_\alpha\langle\delta C_{11,\alpha}\rangle}{\rho_0 C_{11,0}}+\frac{\langle\delta C_{11,\alpha}\rangle^2}{C_{11,0}^2}\right) \tag{1.170}$$

The factor R_α is due to the multiphase nature of a material and is zero in case it is single phase. It represents the contrast for scattering due to the different densities of the phases and the average elastic constants.

$$Q_\alpha = \left(\frac{1}{1575(4\pi C_{11,0})^2}\right)\left(\begin{array}{c}192C_{11,\alpha}^2-128C_{11,\alpha}C_{13,\alpha}+48C_{13,\alpha}^2-256C_{11,\alpha}C_{33,\alpha}\\+32C_{13,\alpha}C_{33,\alpha}+112C_{33,\alpha}^2-256C_{11,\alpha}C_{44,\alpha}\\+192C_{13,\alpha}C_{44,\alpha}+64C_{33,\alpha}C_{44,\alpha}+192C_{44,\alpha}^2\end{array}\right) \tag{1.171}$$

The parameter Q_α is due to the contrast that is caused by the anisotropy of the elastic constants.

In order to calculate the frequency dependence of the backscattering coefficient $\eta(\omega)$ in (1.169), one needs the correlation functions $\Gamma_\alpha(r)$. Based on experimental results, [242] used

$$\frac{\Gamma_\alpha(r)}{\Gamma_\alpha(0)} = \left(exp-\frac{|r|}{a_\alpha}\right) \tag{1.172}$$

Here, a_α represents the mean grain size of the phase α. This led to:

$$\eta(\omega) = 8\pi\sum_{\alpha=1}^{M} f_\alpha(R_\alpha+Q_\alpha)\frac{k^4a_\alpha^3}{(1+(2ka_\alpha)^2)^2} \tag{1.173}$$

In a multi-phase material, the parameter R_α stands for the contrast between the phases and Q_α stands for the scattering due to the anisotropy of the crystallites of the individual phases. The backscattering coefficient has the dimension $(\text{length})^{-1}$ because the backscattered signals are considered from a disc of thickness δz. For large ka_α, i.e. large frequencies, $\eta(\omega)$ saturates.

Experimentally, the backscattered signal amplitudes may be measured relative to the amplitudes of the backwall-echo sequence or to the front or entrance echoes [180]. The latter contains the transfer function of the transducer, and the diffraction of the ultrasonic beam. In this way the measured output voltages of the ultrasonic set-up can be related to the backscattering coefficient $\eta(\omega)$.

At the Fraunhofer IZFP, the backscattering amplitude was derived in a more heuristic way as discussed above. The approach was based on earlier work by [130]. Starting from Eqs. 1.8, 1.1, and 1.147 and squaring (1.8), one obtains for the attenuation of the intensity of an ultrasonic beam:

$$I(x) = I_0 exp(-2\alpha x) \tag{1.174}$$

The attenuated intensity dI_S at a point x in the material is $dI_s = -2\alpha_S I(x)dx$. Hence, the backscattered amplitude intensity I_{Sb} returning to the transducer from a section in the material at a distance $x/2$ from the transducer is given by:

$$I_{S,b}(x) = P(\pi)I_0 2\alpha_S \Delta x exp(-2\alpha x) \tag{1.175}$$

Here, only the scattering coefficient α_S enters and $dx \approx \Delta x$ is the ultrasonic pulse-length. In the Rayleigh regime, the directivity pattern of the scattered intensity varies differently according to different authors, see for example the references in [68]. Together with the directivity pattern of the transducer, this may be cast into a pre-factor $P(\pi)$ in $I_{S,b}(x)$, which stands for the received intensity, subtended by the opening angle of the transducer.

Based on (1.175), we may define a backscattered coefficient for the signals originating from a disc of width Δx:

$$\eta(\omega) = \frac{I_{S,b}}{I_0} = P(\pi) 2\alpha_S exp(-2\alpha x) \approx 2P(\pi)\alpha_S \Delta x \tag{1.176}$$

For small distances, we can neglect the exponential damping factor. However, the factor $P(\pi)$ is not known. Considering Eqs. 1.154, 1.155, and 1.156, one notices that (1.176) shows the same overall dependencies on the parameters which also enter in (1.173): $\eta(\omega) \propto A^2 a^3 f^4$, where A is the anisotropy factor (1.32), a is the grain size, and f is the frequency. This means that the same parameters which determine the attenuation due to scattering, also determine backscattered amplitudes. The dimension of $\eta(\omega)$ is again an inverse length. We refrain from discussing differences between Eqs. 1.173 and 1.176 in detail, but we will give further information in the applications which follow.

Let us return to the amplitude presentation. For the backscattered amplitude, (1.175) may be written as:

$$A_{S,b}(x) = G(\pi) A_0 \sqrt{2\alpha_S \Delta x} exp(-\alpha x) \tag{1.177}$$

where $G(\pi)$ has the same function for the amplitude as $P(\pi)$ for the intensity. Typical backscattered curves can be seen in Figs. 1.49a, 1.50b, 1.51, 1.66, and 1.67. Without averaging, one obtains individual backscattered amplitudes which decay as a function of distance, as can be seen in Fig. 1.51a. Taking again a backscattered curve at a different position on the same material, here steel, one obtains again slightly different decay curves (Fig. 1.51b–d). Rectifying these signals electronically (or digitally) and summing them up (Fig. 1.51f) leads to a smoothed backscattering curve which may be analyzed by Eqs. 1.176 and 1.177. Concerning absolute amplitude values, backscattered signals are typically about 30–40 dB below the back-wall echo sequence and they are usually measured with tone-burst excitation, because then the signal-to-noise ratio is considerably larger than with spike excitation [68, 287].

For a homogeneous microstructure the averaged and rectified amplitude first decays as $\exp(-\alpha x)$ (1.177). In the regime of multiple scattering, i.e. for $k\overline{d} \rightarrow 1$ or, because of large anisotropy of the polycrystals, the slope of the decay in a semi-log plot decreases continuously, see Fig. 1.52.

In [68] an expression was developed taking into account multiple scattering in the backscattered signals:

$$A_{SM,b}(x) = A_{S,b}(x)\sqrt{1 + 2\alpha_S \Delta exp(2\alpha_S x)} \quad (1.178)$$

where $A_{S,b}(x)$ is the backscattering amplitude without considering multiple scattering. Equation (1.178) can be written as:

$$ln\big((A^2_{SM,b}(x)/A^2_{S,b}(x)) - 1\big) = ln(2\Delta\alpha_S) + 2\alpha_S x \quad (1.179)$$

allowing one to determine the attenuation coefficient α_S due to scattering. Data based on this technique are shown in Fig. 1.53. One can clearly discern the Rayleigh scattering regime, where $\alpha_S \propto f^4$, and the transition to the stochastic or resonant regime with $\alpha_S \propto f^2$. In the latter, the pre-factor Σ for resonance scattering (1.160) can be determined by measurement in materials with known grain sizes.

Eventually, weak localization of elastic waves due to coherent backscattering of ultrasonic waves (Anderson localization) influences the amplitude of the backscattered signals in the vicinity of the source. Weak localization is a manifestation of interference of multiply scattered waves in disordered media and was first described in the context of electrical conductivity and scattering of electrons in disordered metals [148]. In acoustics [8, 112], [304] the effect is better known as coherent

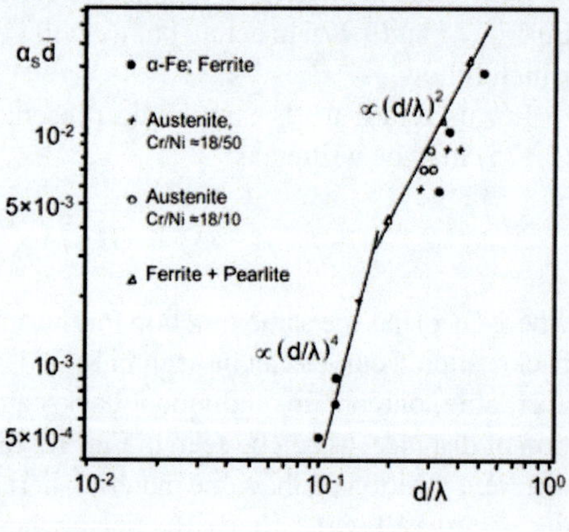

Fig. 1.53 Normalized attenuation coefficient $\alpha_S\overline{d}$ versus d/λ for different steels and α-iron, see also [68]

backscattering. It is caused by the constructive interference between reciprocal paths in wave scattering. This enhances the probability of the return of the wave to the source by a factor of two, which results in the local energy density enhancement by the same factor in the vicinity of a source. Hence, weak localization should lead to a modification of Eqs. (1.73 and 1.76). However, other effects contribute to backscattering, such as a local anisotropies and inhomogeneous grain size distribution in the materials, may be more important. In addition, absorption destroys the effect of weak localization.

1.17.6 Grain Size Measurements

Based on Eqs. (1.146 and 1.153), attenuation data due to scattering, for example obtained by a back-wall echo sequence (Fig. 1.49), allow one to determine the mean grain diameter $\overline{d}$ by carrying out measurements at two frequencies, f_1 and f_2. The frequency f_1 is selected so that absorption dominates ($\alpha_A \propto f$), and the second, larger frequency f_2, so that Rayleigh scattering dominates ($d \ll \lambda$; $\alpha_S \propto f^4$). This needs some a-priori knowledge of the grain size or just practical experience. Then, the mean grain diameter can be calculated from

$$\overline{d} = \left(\frac{\alpha_2 - \alpha_1 f_2/f_1}{S f_2 \left(f_2^3 - f_1^3\right)}\right)^{1/3} \tag{1.180}$$

where S is the appropriate scattering factor. If single scattering dominates, the corresponding S parameter is given by (1.155), respectively (1.156). Figure 1.54 shows experimental results for grain size determination in comparison to metallographic evaluation for different steels using shear waves as well as Rayleigh waves [68, 310].

Thus, backscattering measurement in the Rayleigh regime as well in the resonance regime can be exploited to determine the grain size (Eqs. 1.177 and 1.179, Fig. 1.53).

Scattering and diffusion theories and the pertaining applications correlate ultrasonic scattering and propagation parameters with characteristic materials quantities. A considerable part of the ultrasonic scattering theory, applications and testing systems were developed at the Fraunhofer IZFP, for example in order to monitor hydrogen embrittlement in pipeline steels and the case-hardening depth of steels.

Based on Eqs. (11.53, 1.154, and 1.177), the backscattered amplitude depends on the grain size. Inverting these relations, one finds that

$$\overline{d} \propto \left(A_{S,b}^2\right)^{1/3} \tag{1.181}$$

This relation was exploited by Michel et al. [188] in order to continuously monitor the grain size of austenitic stainless steel cold-rolled strips in a production line. The rolling strip was insonified by the immersion technique with 25 MHz ultrasound under the critical angle for Rayleigh wave generation. Besides the backscattered

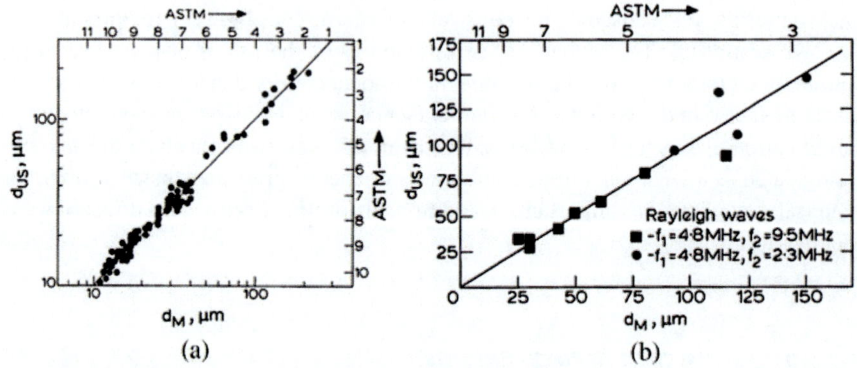

Fig. 1.54 Comparison of grain sizes obtained by backscattering (d_{US}) of **a** shear and **b** Rayleigh waves versus metallography data (d_M). The grain size scale according to the ASTM standard is also indicated (from [310], with permission from Francis & Taylor)

Rayleigh waves, volumes were generated by mode conversion as well. The backscattered Rayleigh waves were separated from the volume waves by their different time-of-flight. After calibration with metal sheets of well-defined grain size, which were put into the water path at the place of the rolling metal sheets, the volume waves were used for monitoring the grain size.

1.17.7 Detection of Porosity Agglomerations

For the detection and evaluation of defects (pores) close to the surface up to about 2 mm depth in die-casting components made of aluminum, magnesium, and zinc alloy components of complicated shape, the use of ultrasonic backscattering signals is promising. Pores and pore clusters must be detected, because they would cause leakage if cut open in cylinder bores. The complicated shapes required the use of a robot scanning system, which learnt the geometry of the component before testing it [94]. Then, in a second step, the component is scanned with a constant pre-determined distance to its surface, focusing the ultrasound to a predetermined depth in order to maximize the backscattering signal, see Fig. 1.55a and [67]. The signals are recorded and permit the one-, two-, and three-dimensional visualization and signal processing of the ultrasonic data. The contrast due to the pores can be masked by scattering from the microstructure of the metallic components.

Porosity or individual pores can only be detected when their scattering signals are large enough in comparison to the grain noise. As detection limit, it is assumed that the amplitude of the pore scattering signal has to be at least comparable to the grain noise. Using the ultrasonic scattering theories outlined above for grain as well as for scattering from pore agglomerations, this ratio can be calculated. The electronic noise in the signals can be eliminated by averaging A-scans recorded successively.

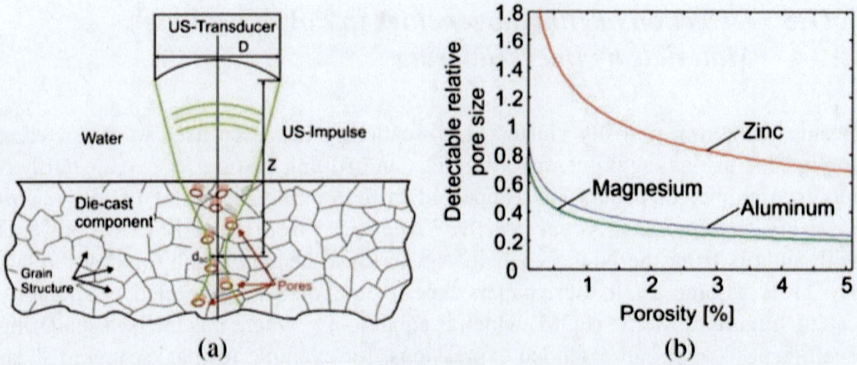

Fig. 1.55 **a** Focusing of the ultrasound to maximize the signal-to-noise ratio for pore detection; **b** smallest detectable relative pore size (ratio of pore to grain diameter) as a function of porosity in the range of 0.1–5%. In aluminum and magnesium die-castings pores can be detected that are smaller than the grains. The detection limit for pores in zinc is significantly larger because of its high single-crystal anisotropy, leading to a larger grain scattering (in modified form from [107]), with permission from de Gruyter

For the case of longitudinal waves, one obtains for the ratio of pore to grain scattering for $k\overline{d} \ll 1$ [107]:

$$\frac{\alpha_{S,pores,L}}{\alpha_{S,grains,L}} = 93.75 \frac{S_p}{S_g} V_p \left(\frac{d_p}{d_g} \right)^3 \tag{1.182}$$

Here, $\alpha_{Sgrain,L}$ and $\alpha_{Spores,L}$ are the ultrasonic attenuation coefficients for grain and pore scattering, respectively, S_g and d_p are the corresponding scattering factors, and d_g and d_p are the grain and the pore diameters, respectively. The porosity is V_p (volume fraction of the pores). The derivation of the scattering coefficients in (1.182) was based on an isotropic grain-orientation distribution. As said above, the criterion for the limit of the detectability of pores is

$$\frac{\alpha_{S,pores,L}}{\alpha_{S,grains,L}} \geq 1 \tag{1.183}$$

Equations (1.182 and 1.183) were evaluated numerically for aluminum, magnesium, and zinc die-castings. Figure 1.55b shows the smallest detectable relative pore size (ratio of pore to grain diameter) as a function of porosity in the range of 0.1–5.1%. In aluminum and magnesium die-castings pores can be detected that are smaller than the grains. The detection limit in zinc is, because of its high single-crystal anisotropy and the corresponding high grain-scattering in this material, at significantly larger pores. In all cases pores of diameters down to the μm range are detectable already at a porosity of only 0.1%. A comparison of ultrasonic measurements with metallographic micrographs showed that porosities of 0.1% and pores of 30 μm diameter were detectable.

1.17.8 Detection of Inhomogeneities in Polycrystalline Materials by Backscattering

Besides detecting porosity clusters, backscattering has been used to characterize foreign inclusions, agglomerates of cracks and rolling texture [68, 310]. Furthermore, a number of papers have appeared in the conference series of "*Review of Progress in Quantitative Nondestructive Evaluation*" from the early 1990's onwards with authors from the NDE center in Ames, Iowa, headed at the time by the late Prof. R.B. Thompson. In these papers experimental data are presented as a quantity called "Figure of Merit" (FOM) which is equal to $\eta^{0.5}$ where η is the backscattering coefficient (1.173) and extended expressions, for example for backscattered shear waves. The backscattering coefficient has the dimension of an inverse length. The backscattered noise observed in experiments is not simply the backscattering coefficient since the experimental observations are also influenced by the transduction efficiency and directivity of the transducer and the measurement frequency.

An overview of backscattering data is presented in [288] for polycrystalline copper, nickel base superalloys, and titanium alloys. For example, the backscattered FOM is of the order of 10^{-2} cm$^{0.5}$ for polycrystalline equiaxed Cu with a grain size of 15 μm at a frequency of 10 MHz. Further test samples examined were cut from billets of an IN718 alloy and from billets of Waspaloy. These materials each contain 50% Ni by weight and 50% other different alloying elements. The close relations between grain size, measured attenuation due to scattering and the theoretical and experimental backscattering factor $\eta(\omega)$ can be seen in Fig. 1.56a, b. Further experiments dealt with the detection of colonies, multiple backscattering using pitch-catch arrangements, micro- and macro-textures, grain elongations, and hard alpha inclusions in titanium alloys. We refrain from discussing further details, but refer to [85, 178–180, 285, 288, 323], which contain all pertinent references.

From Fig. 1.56a, we read that at a FOM = 0.018 cm$^{-0.5}$ or FOM^2 = 3.24×10^{-4} cm^{-1}, the measured attenuation due to scattering was $\alpha_S \approx 1$ dB/in = 0.394 dB/cm^{-1} $\hat{=}$ 4.42×10^{-2} cm^{-1} and hence $2\alpha_S = 0.884 \times 10^{-2}$ cm^{-1} which is a factor of 27 larger than FOM^2. This tells us that $P(\pi)$ is an important calibration factor and that (1.175) is only an approximation. $P(\pi)$ contains the characteristics of transducers as does the FOM.

The effect of texture, grain shape, and duplex microstructures on back-scattering in titanium was also modeled and measured for cubic, orthotropic and trigonal grains by [165, 172, 320]. Eventually, stresses also influence the shape and amplitude of backscattering curves, because they change the anisotropy of the grain structure [135].

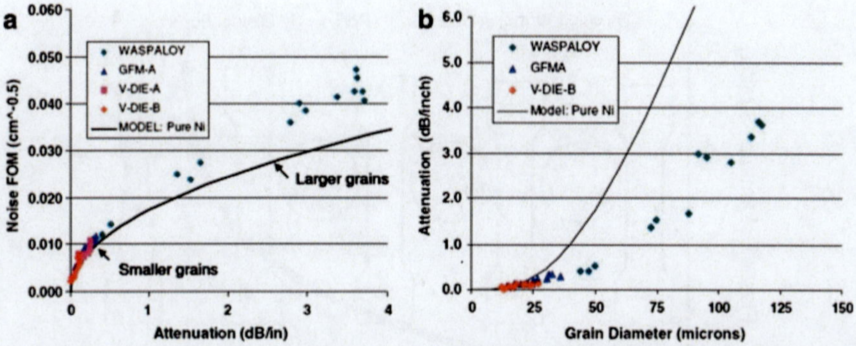

Fig. 1.56 Longitudinal-wave measurements at 7.5 MHz in nickel base superalloys; **a** correlation between the measured attenuation and grain noise FOM ("Figure of Merit" FOM $= \eta^{0.5}$); **b** dependence of the measured attenuation on grain size. The theoretical curve was calculated using scattering theory. The deviation from the experimental data was explained by noting that the billets contained only about 50% Ni. The remaining constituents were alloying elements. The unit 1 dB/in corresponds to ≈ 0.04 dB/mm (from [288], with permission from Elsevier). Further details are also discussed in [179]

1.18 Creep

During creep, materials deform continuously with time under a constant load at elevated temperatures. In the primary creep stage, the creep rate steadily decreases due to work hardening. In the secondary stage it stays constant due to the balance between strain hardening and recovery, and eventually in the tertiary stage the creep rate increases again until creep fracture occurs. There are a number of effects occurring in the microstructure which are summarized in [73]. The primary and secondary creep stages are characterized by dislocation reactions, and the development of micropores which coalesce to micro-cracks, and finally their coalescence to macro-cracks in the tertiary creep stage Fig. 1.57 [306] and Fig. 1.58 [309].

By monitoring changes in the absorption, in the ultrasonic velocity as well as in the scattering coefficient, it is possible to detect the materials damage before critical degradation occurs [308]. Small velocity changes of the order of 1–2% were found in the crack-growth region, where the density variation was even below 1%, Fig. 1.59a, b [309].

Comparable results were found for α-iron, with velocity changes up to 4%, while the longitudinal wave velocity reacted more sensitively than the shear wave velocity [17], Fig. 1.60. This finding recommends the ultrasonic survey of critical areas in order to detect creep-damage provided that there is an accurate knowledge of the sound velocity as a base-line for calibration.

An interesting approach was undertaken by [88] to evaluate creep damage for boiler piping of fossil-fuel combustion power plants. The concern was whether creep damage occurred in the heat-affected zone of a tube weldment of 9Cr–1Mo steel.

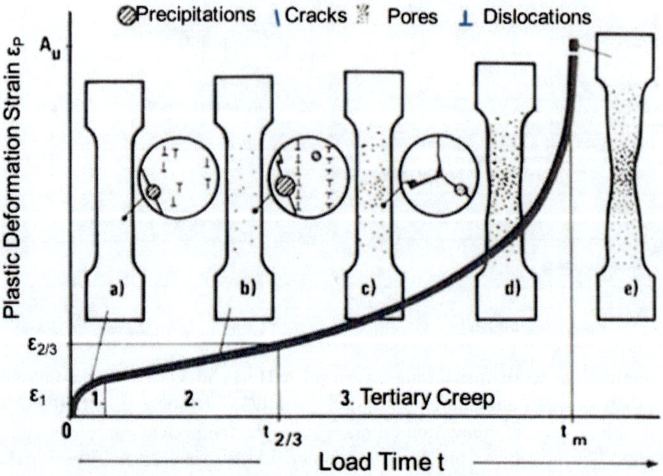

Fig. 1.57 Creep stages: **a** dislocation density increases and stabilizes in the primary and **b** secondary creep stage; **c** in the tertiary creep stage, material damage develops due to the formation of micro-pores and micro-cracks which lead to macro-crack formation (from [306], with permission from Springer Nature)

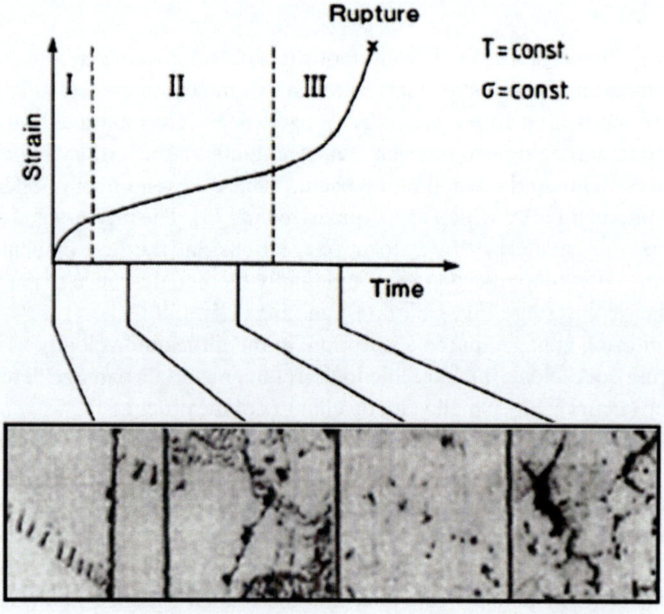

Fig. 1.58 Schematic creep curve $\varepsilon(t)$ and evolution of the microstructure during creep (from [309], with permission from Elsevier)

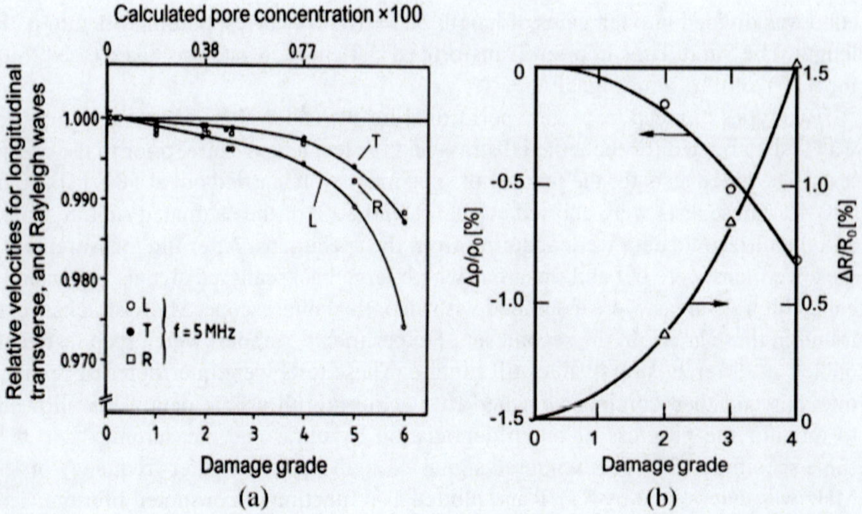

Fig. 1.59 **a** Relative change of sound velocity due to creep damage; **b** density variation as a function of damage grade, together with the change in electrical resistivity in creep damaged steel 14 MoV 6 3 (0.5 Cr, 0.5 Mo, 0.25 V) (in modified form from [309]), with permission from Elsevier

Fig. 1.60 Relative change of sound velocity as a function of creep-strain for the longitudinal (v_L) and shear wave (v_T) velocity (from[17], with permission from Springer Nature)

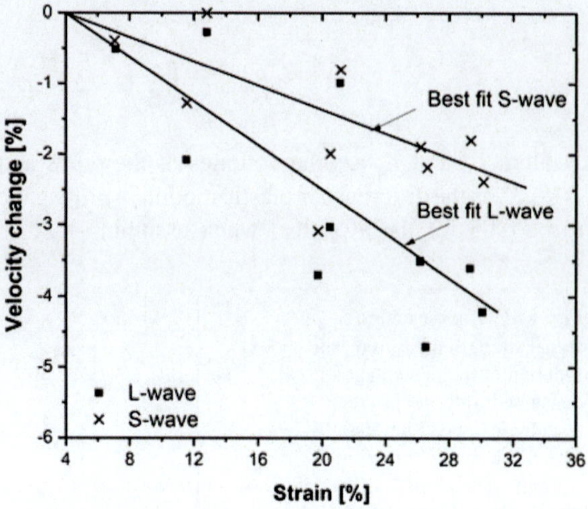

In order to develop the method, the rf-signals of a 15 MHz ultrasonic pulse (corresponding to $\lambda = 0.4$ mm) were evaluated in test specimens. A damage-free material zone showed very small backscattering signals between the electronic leakage signal and the first back-wall echo. For the damaged material, however, backscattered signals caused by creep voids and micro-cracks with amplitudes as large as 1/3 of the amplitude of the first back-wall echo were observed. This finding was used as a first indicator in B-scans. Then, the time span between entrance signal and back-wall echo was divided into ten gates of length $\Delta t = 167$ ns, corresponding to 1 mm path length. The Short-Time Fourier-Transform (STFT) in each gate was taken (see [269] for STFT of ultrasonic signals).

Two types of creep tests were performed for the calibration. The time to rupture was first estimated for the uniaxial stress and the test temperature, prior to the creep tests. The creep tests for the first set of specimens were carried out at 58.8 MPa and 650 °C. These tests were aborted at 40, 60, and 80% of the estimated rupture time, and the ultrasonic data were acquired from the specimens. After the measurement, the specimens were cut and the areas where large backscattered signals originated, called high P_s-areas, were examined with an optical microscope. Micro-cracks were found in these areas. In the second set of experiments, samples were creep tested at 650 °C at different stress values till rupture. These tests were interrupted at regular intervals and then continued further after acquiring ultrasonic data. This allowed to monitor the progress of creep degradation by ultrasonic measurements in the same specimens. The backscattered signal strength P_s at the center-frequency of 15 MHz was determined by STFT and plotted as a function of consumed lifetime, see Fig. 1.61. A linear correlation was found between P_s and the percentage of the level of life consumed [%]. For low consumed material life percentages, scattering from grains in the heat affected zone led to disturbing effects.

Using the theory of Rayleigh scattering [88, 281] calculated the ratio R_{amp} of the backscattered energies due to void and grain scattering:

$$R_{amp} = \left(\frac{2}{3}\right)^{0.5} \frac{T_p}{T_g} / \frac{\Delta K_g}{K_g} \tag{1.184}$$

Here T_p and T_g are the volumes of the pores and the grains, respectively, and $\Delta K_g / K_g$ is the difference in elastic moduli in propagation direction which is measured to be 0.05 for the 9Cr steel when assuming that the grain diameter is 5 μm. It

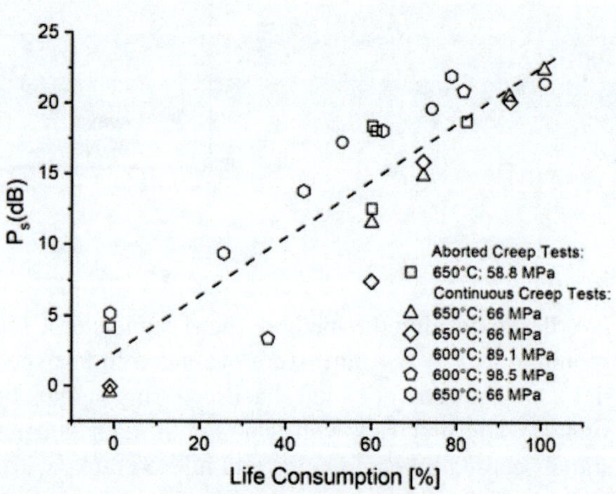

Fig. 1.61 Backscattered signal strength measured as a function of the percentage of consumed lifetime for creep tests (redesigned after data from [88], with permission from the American Society of Mechanical Engineers)

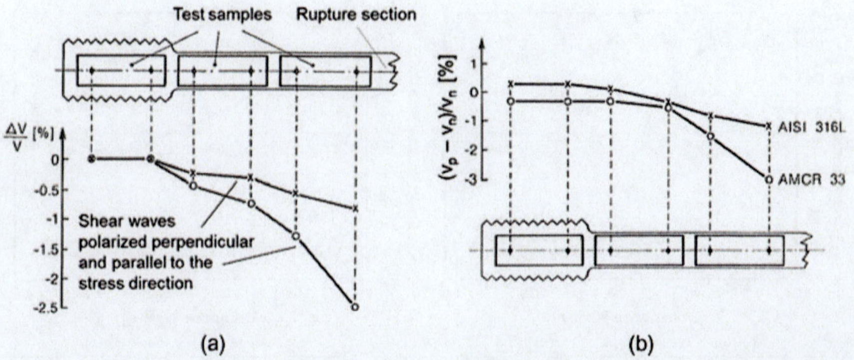

Fig. 1.62 **a** Relative velocity change of shear waves with polarization parallel and perpendicular to the creep-load direction; **b** Relative change in velocity of shear waves for creep-loaded test pieces of AISI 316 L and AMCR 33 steel. The closer the test piece to the rupture section, the higher the change in sound velocity (in modified form from [149], with permission from Springer Nature)

is interesting to note that (1.184) is similar to (1.182) obtained by [109] for the detection of pores in aluminum, magnesium, and zinc die-castings.

Shear wave ultrasonic velocities showed a clear correlation to the creep strain in austenitic steel, as demonstrated for the materials AISI 316 L and AMCR 33 [149]. Velocities were measured for test samples as a function of distance to the rupture section. In this case, the strongest effect was found when applying shear waves with polarization parallel to the creep-load direction (Fig. 1.62a). The velocity changes of max. 3% allow a survey of critical areas over time, with enough safety margin to the time of rupture, see Fig. 1.62b. Furthermore, it was suggested to use the velocity with polarization perpendicular to the stress direction as reference value eliminating the need to know the path-length.

For predicting the residual life-time of materials by ultrasonic methods [127] for components under high temperature creep conditions, one needs to consider the different changes of the microstructure during the life-time of the component. For two steels of the type 9Cr–2W, designated P92 and P122, with 20 MHz longitudinal waves, back-wall-echoes were evaluated in order to measure the attenuation coefficient α_L in the frequency range from 17 to 23 MHz. The steel samples were isothermally aged at T = 600 °C, with aging times from 0 to 10,000 h. The creep damage leads to an increase of ultrasonic attenuation α_{Lp} with increasing content of precipitates and cavities and to an increased α_{Lg}, due to the grain growth as a function of temperature and time. The two contributions were separated using single scattering theory. The total attenuation is then given by

$$\alpha_{L,total} = \alpha_{Lp} + \alpha_{Lg} \tag{1.185}$$

Studying the development of precipitates and grain growth under creep load led to a quantitative value for α_{Lp}, which could be correlated to the residual life time. The predicted and measured curves of creep rupture time (in h) and α_{Lp} are in

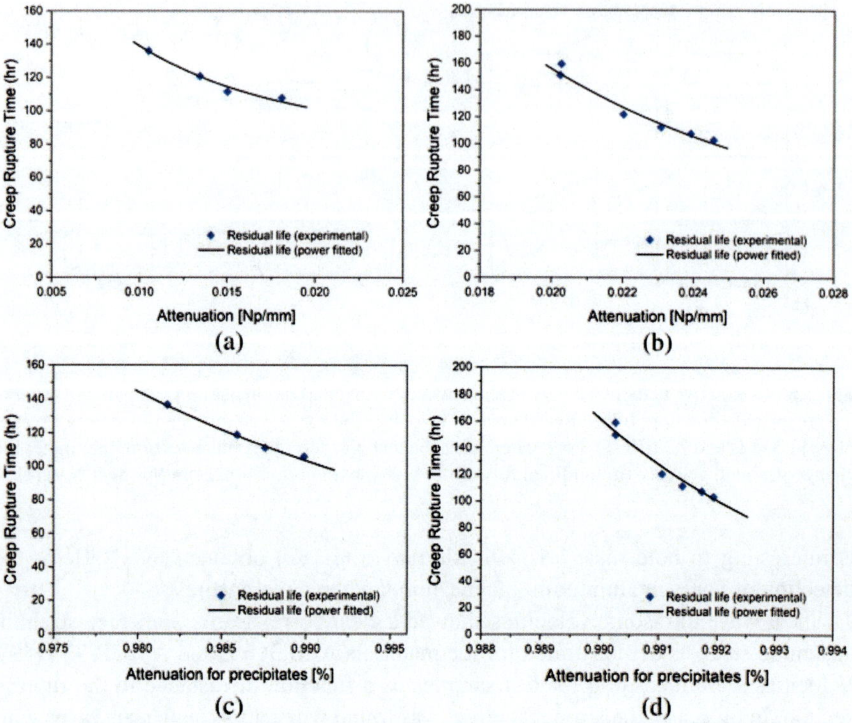

Fig. 1.63 Predicted life-time based on the total attenuation for the samples P92 **a** and P122 **b**; Predicted life-time based on the attenuation due to precipitates for P92 **c** and P129 **d** (in modified form from [127], with permission from American Institute of Physics Publishing)

good agreement, to be used for comparable conditions as a life-time-prediction tool. Figure 1.63a, b show the values of $\alpha_{L,total}$ versus creep rupture time of creep damaged specimens, and Fig. 1.63c, d the creep rupture time against relative α_{Lp} in %.

$$\Delta\alpha = \frac{\alpha_{L,total} - \alpha_{Lg}}{\alpha_{L,total}} \tag{1.186}$$

Finally, a critical review of the detection of creep damage by a number of NDT techniques, in particular using ultrasonics, can be found in [278].

1.19 Fatigue

Fatigue is responsible for most of the failures of components subjected to dynamic loading in practical applications. Therefore, the interest in NDMC is high to detect fatigue at an early stage. The majority of materials and components in practice are designed for a maximum dynamic load or for a specific average load with defined dynamic excursions above and/or below the average. The loads can be of cyclic, statistical or of stochastic nature. Therefore, the design of components and the choice of material strength parameters have to be specified, so that the components can bear the limits of the expected loads. Specific design rules exist for this case, for a limited number of load cycles as well as for unlimited durability. Unexpected loads at significant higher levels can lead to fatigue and finally failure. Low-cycle fatigue, as well as high-cycle fatigue, will start with initial degradation in microstructure and substructure as under creep conditions. Then fatigue gliding-structures followed by stable micro-cracks will develop parallel to the shear stress direction. Later, crack growth changes to an orientation perpendicular to the maximum longitudinal stress direction.

While the relative change of ultrasonic velocities in aluminum with the exhausted life-time is only of the order of 5×10^{-3} [9, 10], the ultrasonic attenuation coefficient strongly increases with the number of load cycles allowing to react timely to the incipient failure, as already shown in some early experiments [220, 225]. Figure 1.64 shows the increase of the attenuation until fracture for the frequencies 2.25, 10, and 25 MHz in an aluminum alloy.

The fatigue behavior of metal-matrix titanium-based composites was intensively studied by Chu et al. [13]. Assessment of damage was done by measuring ultrasonic phase velocities in the composites to determine the composite elastic moduli. Due to damage, the measured composite moduli change, one can assess the damage severity by ultrasonic velocity measurements. The composite had an 8-ply symmetry lay-up made by hot isostatic pressing of a foil/fiber/foil lay-up. The matrix was the metastable titanium alloy Ti–15V–3Cr–3AI–3Sn (Ti–15-3) with SiC fibers as reinforcement). The fiber volume fraction of the composite was 35% and the composite density was 4.18 g/cm^3. The heat-treated samples were first fatigued to failure under

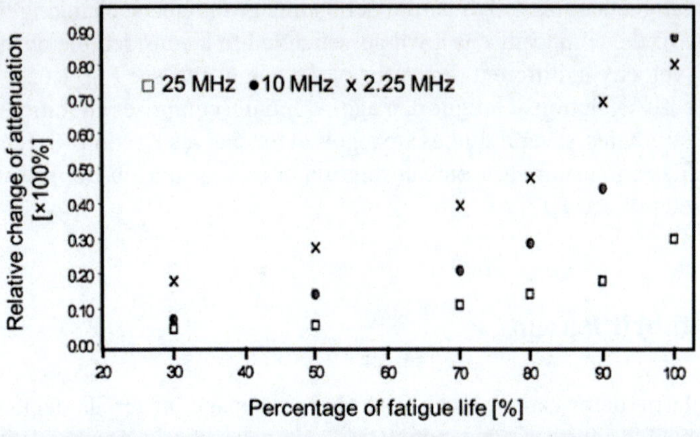

Fig. 1.64 Increase of the ultrasonic attenuation in an aluminum alloy for three different frequencies 2.25 MHz (x), 10 MHz (●), and 25 MHz (□) (from [10], with permission from American Institute of Physics Publishing)

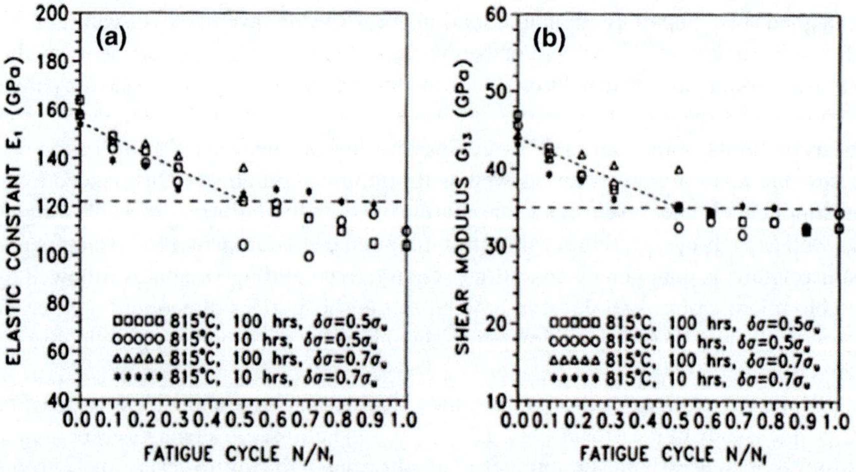

Fig. 1.65 Measured composite elastic constants obtained by sound velocity measurements: **a** E_1 and **b** G_{13} (shear modulus in the composite plane in 1-direction) versus number of fatigue cycles for samples heat treated at 815 °C. The measured composite moduli decrease rapidly (≈ 50%) with the development of debonding until the point when the fibers in the 90° plies no longer resist the load (at about 50% of the fatigue life) (from [43] with permission from Springer Nature)

different stress levels to obtain S–N curves (cycling stress versus logarithm of number of cycles until rupture) which showed that the composite fatigue properties were greatly affected by the heat treatments. Two stress-controlled fatigue tests (equal to 50–70% of the ultimate strength with $R = \sigma_{\min}/\sigma_{\max} = 0.1$), were selected for damage assessment using samples with different heat treatments. The fatigue cycling frequency was 10 Hz. Ultrasonic measurements were performed on samples prior to fatigue as well as at different stages of fatigue (with step size 0.1 of the fatigue life). Here, the fatigue loading led to partial debonding of the fibers, changing the elastic anisotropy of the composites as a whole which led to a considerable change in the ultrasonic velocity in different directions of the composite, see Fig. 1.65.

As the early evolution of fatigue damage essentially comprises alterations in dislocation substructures, several studies are reported for characterization of accumulation of fatigue damage using ultrasonic absorption or nonlinear acoustic measurements, as discussed in Sect. 1.13.

1.20 Embrittlement

Besides plastic deformation, creep, and fatigue damage, materials degradation by embrittlement is a further subject of NDMC. Here, we consider embrittlement originating from hydrogen attack, irradiation of plastics, elastomers, and composites by

UV-light, as well as from irradiation by gamma-rays and neutrons in the nuclear industry.

Materials parameters, which allow to obtain information on the grade of embrittlement, are hardness H, Charpy impact tests yielding the energy A_K needed for fracture (or when related to the impact area in energy per area, a_K [J/cm^2]), in conjunction with measurements of the ductile–brittle transition temperature, fracture toughness K [MPa m$^{1/2}$], fracture strain (%), and fracture cross-section reduction (%), see for example [268]. Out of these parameters, only the hardness can be measured in operational components.

1.21 Hydrogen Attack

Refinery steel vessels under high pressure and at elevated temperatures are exposed to hydrogen attack. Hydrogen atoms diffuse into the steel and react with carbides. The reaction entails the formation of methane gas and as a consequence leads to intergranular cracks and finally to the loss of strength and toughness.

Back-wall echo amplitudes or velocity measurements fail to serve as reliable indications of the level of a hydrogen attack due to uneven surfaces caused by corrosion. Using longitudinal waves at 10 MHz in hydrogen damaged C–0.5Mo steel, [302] observed backscattered signals from the damaged volume located inside a hydrogen-damaged heat exchanger due to the agglomeration of cracks. Figure 1.66 displays the backscattered ultrasonic signals measured from (a) the undamaged outside surface and (b) from the inside damaged surface.

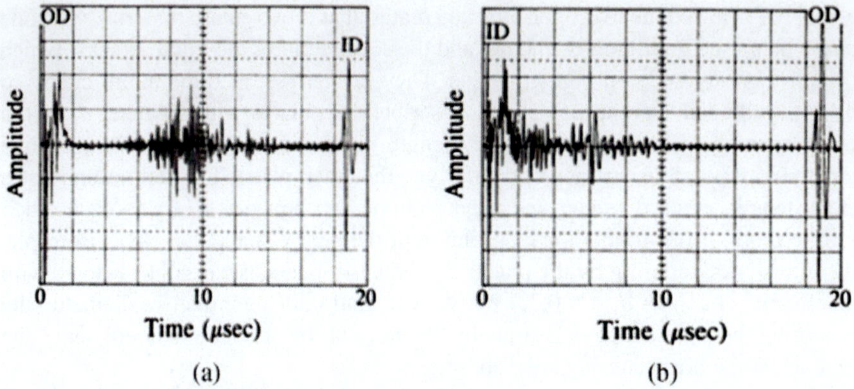

Fig. 1.66 Backscattered ultrasonics signals measured from **a** the outside surface, i.e. the undamaged side and **b** from the inside surface, i.e. the damaged side of a hydrogen-damaged heat exchanger. *OD* and *ID* stand for outer diameter and inner diameter, respectively. The backscattered signals stem from a hydrogen-damaged volume at a depth of about 2 cm from the outside surface, as can be easily estimated from the time-of-flight of the signals (from [302], with permission from Springer Nature)

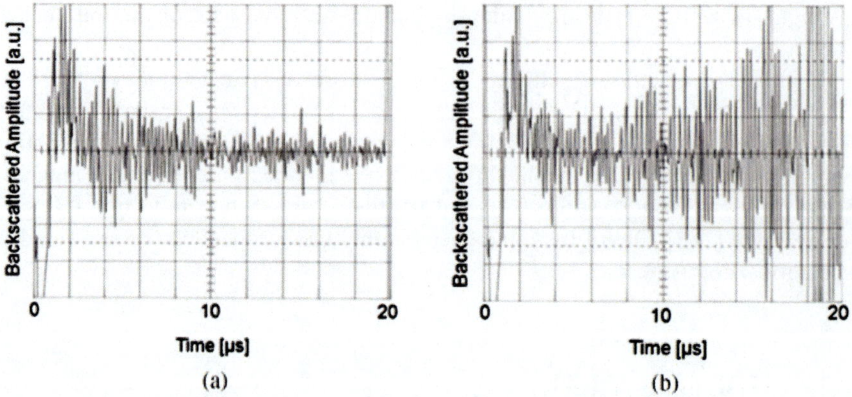

Fig. 1.67 **a** Backscattered signal from cracks caused by a hydrogen attack in a 0.5Mo steel tube from inside to outside and **b** from outside to inside ([109], with permission from Francis & Taylor)

A similar approach was undertaken to monitor the embrittlement due to different grades of hydrogen attack by studying backscattering theoretically as well as experimentally in a 0.5Mo steel-tube having a wall thickness of 76 mm. Figure 1.67 shows in analog way as Fig. 1.66 the backscattered signals between the damaged inside and the unaffected outside [109].

In hydrogen attacked steel, ultrasonic grain scattering and scattering at cracks are superposed. To detect and evaluate the damage from attenuation or velocity dispersion measurements, the part of the attenuation caused by scattering from cracks and its related velocity dispersion must be known separately. If the NDMC tests are aimed at backscattering measurements, attenuation is to be evaluated. Hirsekorn et al. [109] viewed the hydrogen attacked material as a two-phase material, with one phase being the grain microstructure and the second phase the micro-cracks, which is equivalent to a certain porosity value. They further assumed the micro-cracks to be penny-shaped, so that they can be described by oblates with diameter d_m and a crack opening of $\approx d_m/10$. The crack length L_c was defined as the averaged line length of all possible cuts perpendicularly to the crack plane. This definition relates crack length, crack diameter and crack volume and hence porosity to each other. Furthermore, if the orientation distribution of the penny-shaped cracks is isotropic, the ultrasonic scattering by the cracks can be described as scattering by pores with an effective diameter $d_{eff} \approx 0.3d$. For the calculation of the effective diameter, the scattering theory is adapted. Eventually, the ratio of the attenuation coefficients due to crack and grain scattering is given by

$$\frac{\alpha_{SL,cracks}}{\alpha_{SL,grains}} = 7.49V_p\left(\frac{L_c}{d_g}\right)^3 \tag{1.187}$$

in close analogy to (1.182). Here, d_g is again the grain diameter. For practical use, this equation can be modified in order to determine the minimum detectable crack length. Cracks can only be detected by ultrasonic backscattering measurements if they cause backscatter amplitudes at least 12 dB stronger than grain scattering due to the margin needed to avoid false calls during in-field inspections [109]. The resulting minimum detectable crack length L_m is given by

$$L_m = \left(\frac{P}{N}\right)^{0.2} d_g^{0.6} \tag{1.188}$$

Here, N is the number of cracks per area, and d_g is the grain diameter. The parameter P (≈ 2.11) takes into account the ratio between crack to grain scattering, detection criteria, crack and grain shape factors, and a correction for anisotropy due to preferred crack orientation.

For example, for a grain diameter of 45 μm and a crack density of N = 200/mm², the minimum detectable crack length is 63 μm. For further details see [109].

Furthermore, micro-cracks in a material will reduce the compressional and shear wave velocities with increasing crack density or porosity. This follows not only from static calculations of the effective elastic constants [246], but also from ultrasonic scattering theories which allow one to calculate the sound velocities in materials with microscopic inhomogeneities [102, 254]. For example, a porosity of 10% causes a decrease of nearly 3% for compressional and 2.5% for shear waves in the 0.5Mo steel-tube examined. Based on many experimental tests, the ratio v_T/v_L is sometimes used as an indicator of the level of hydrogen attack, for example if $v_T/v_L > 0.55$ [18]; [87, 138]. However, the usefulness of the ratio v_T/v_L depends on the velocities at zero porosity of the special material of interest and hence the exact knowledge of these values is required [109]. In summary, ultrasonic backscattering seems to be a more appropriate means to detect hydrogen attack.

1.22 Micro-cracks

Micro-crack detection and evaluation are of high interest for ceramic materials, where on the one side the grain sizes are some micrometers and on the other side micro-cracks of comparable sizes must be detected due to fracture mechanics considerations, for example in applications like implants [154]. Such small cracks are difficult to detect with standard ultrasonic NDT-techniques. If we take the criterion that the wavelength should be comparable to the defect size for a sufficiently large signal-to-noise ratio, frequencies of the order of 500 MHz are required. The attenuation coefficient at these frequencies gets very large in many materials, even in structural ceramics, except in single crystals [291], amorphous materials [114], and ultra-fine

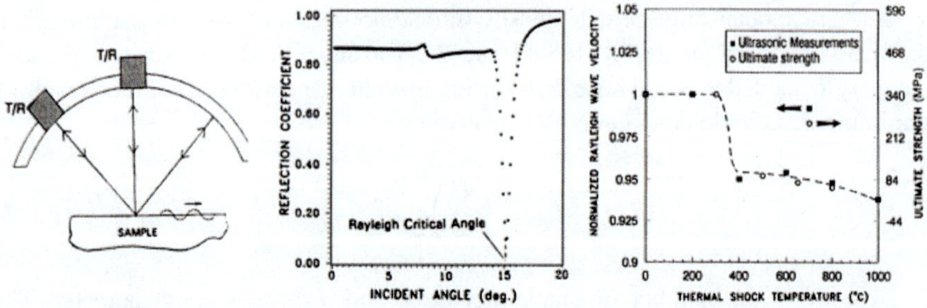

(a) (b) (c)

Fig. 1.68 **a** Principle of the measurement of the Rayleigh wave velocity by measuring the critical angle using a goniometer (**b**); and **c** effect of thermal shock on Rayleigh wave velocities and ultimate bending strength (from [42], with permission from Springer Nature)

grained materials like nano-crystalline solids [151]. In order to circumvent this situation, one operates in the Rayleigh regime, where the wavelength is much larger than the crack size. Whether a crack or another defect can then be detected depends on its contrast [211]. However, an ensemble of micro-cracks may be detectable at much lower frequency by their influence on the ultrasonic velocity or ultrasonic attenuation.

Such an approach was selected by [42] for alumina and reaction bonded silicon nitride (RBSN) in order to monitor the materials strength loss due to thermal shock. The use of the goniometer technique to measure the reflection coefficient allowed one to determine small changes of the Rayleigh wave velocity caused by micro-cracks, Fig. 1.68. It was found that for RBSN a good correlation exists between the ultimate bending strength and the Rayleigh-wave velocity, both measured as a function of the thermal shock temperature, where defects are created leading to failure at much lower loads than the material is specified for.

Furthermore, the influence of thermal shock on the longitudinal wave velocity was measured with a double transmission technique through test samples as a function of refraction angle θ_r. The velocity profile as a function of the refraction angle shows that thermal shock produces dominantly a near-surface crack population, see Fig. 1.69.

1.23 Grain Size in Polycrystalline Materials and Yield Strength

The yield strength σ_e in polycrystalline materials is related to the grain diameter by the so-called Hall–Petch relationship:

$$\sigma_e = \sigma_U + k_y/\sqrt{d} \tag{1.189}$$

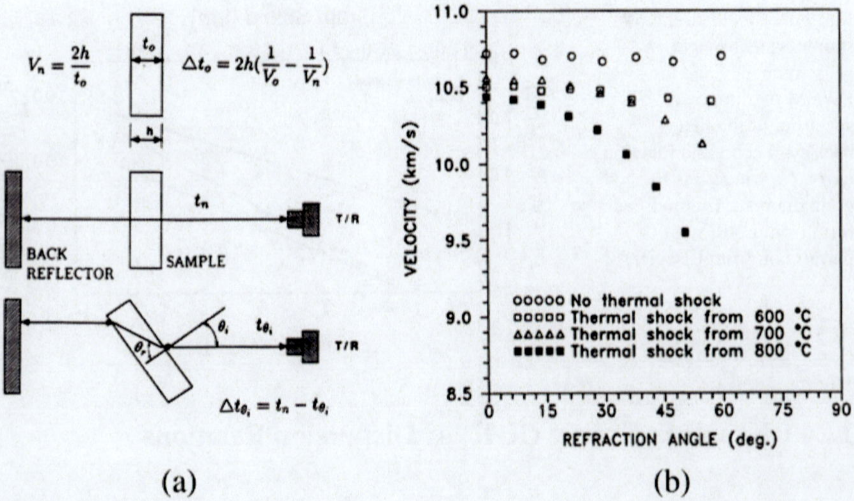

Fig. 1.69 **a** Longitudinal velocity v_L of reaction bonded silicon nitride (RBSN) measured with a double-ended ultrasonic system. By increasing the incidence angle θ_i, the refracted angle θ_r gets larger and the longitudinal waves probe the elastic properties of the material closer to the surface, resembling the use of surface skimming longitudinal waves, discussed in Sect. 1.11.5. **b** Because the reduction of the velocity v_L gets more pronounced with increasing θ_r, the damage caused by thermal shock apparently happens dominantly in the surface area (from [42], with permission from Springer Nature)

Here, σ_U is the yield strength of the single crystal, i.e. the strength for the ensemble of polycrystals for $d \to \infty$. The second part of the equation is caused by the fact that when deforming the ensemble, there is a pile-up of the dislocation at the grain boundaries. Their motion is hindered due to the elastic mismatch at the grain boundaries and the back-stress acting on the pile-up. For more details see for example [73, 84]. The Hall–Petch relation opens the possibility for the strengthening of a material by grain refinement.

In AISI type 316 stainless steel specimens of different grain sizes, Kumar et al. [140, 141] and Sharma et al. [269] measured parameters characterizing the variation of the ultrasonic attenuation as a function of frequency [140, 141, 269]. The Hall–Petch equation (1.189), predicts σ_e to vary linearly with $d^{-0.5}$. Indeed, there is a linear relation of the peak frequency (PF) and the full width at half maximum (FWHM) of the amplitude spectrum of the first backwall echo with $d^{-0.5}$ for samples of two different thicknesses. For the thicker specimens, the PF and the FWHM shift towards the lower values. These parameters can be fitted with a Hall–Petch type equation, as shown in Fig. 1.70. The decrease in the PF with increasing grain size is attributed to the increased scattering of higher frequency contents with increasing grain size.

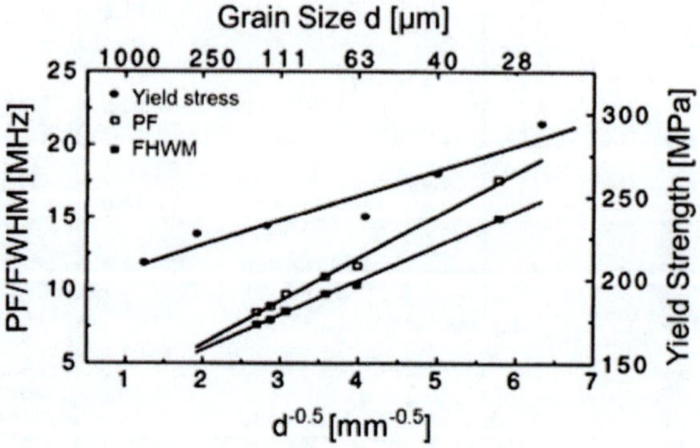

Fig. 1.70 Yield strength of stainless steel specimens, peak frequency and the FWHM of autopower spectrum of the first back-wall echo as a function of $d^{-0.5}$, where d is the grain diameter (in modified form from [140] with permission from Elsevier)

1.24 Kramers-Kronig (K-K) or Dispersion Relations

1.24.1 Theoretical Background

The phase velocity and the attenuation of ultrasonic waves are frequency dependent due to their interaction with internal degrees of freedom in the propagation media, for example by dislocation motions in metals which couple to mechanical stresses. In causal ("cause before effect") and linear ("result proportional to cause") systems $v(\omega)$ and $\alpha(\omega)$ are not independent of each other. The knowledge of one quantity for all frequencies enables one to calculate the other quantity for any frequency. This interdependence is known as the "Kramers–Kronig relation" [134]. These equations are rigorously derived in many textbooks [122, 187, 247, 312]. Here, we follow the work of [205] but leave out some of the more mathematical steps required to fully understand their derivation.

The Kramers–Kronig equations are based on the convolution integral

$$r(t) = \int_{-\infty}^{\infty} c(t')m(t - t')dt' \quad (1.190)$$

with the response $r(t)$ obtained by the cause or stimulus $c(t')$ and the material's susceptibility $m(t - t')$. The system response depends on the time difference $(t - t')$ because the response is delayed by the same amount as the stimulus. Here, the materials susceptibility stands for example for the compressibility $K(t)$. The causality demands that the response $r(t)$ is zero for all earlier times $t < t'$ before the stimulus occurred, which entails $m(t - t') = 0$ for $t - t' < 0$. If there exist Fourier transforms for r, c, and m, so that

$$R(\omega) = \int_{-\infty}^{\infty} r(t)e^{i\omega t} dt$$

$$C(\omega) = \int_{-\infty}^{\infty} c(t)e^{i\omega t}dt$$

$$M(\omega) = \int_{-\infty}^{\infty} m(t - t')e^{i\omega(t-t')}dt' \tag{1.191}$$

where the Fourier transform for $m(t)$ is defined as

$$m(t) = \int_{-\infty}^{\infty} \frac{M(\omega)}{2\pi} e^{-i\omega t} d\omega \tag{1.192}$$

we can rewrite (1.190) in the Fourier domain:

$$R(\omega) = C(\omega)M(\omega) \tag{1.193}$$

In general, the frequency dependent susceptibility is complex, i.e. $M(\omega) = M_1(\omega) + iM_2(\omega)$, with M_1 the real part and M_2 the imaginary part. If the system is excited by a stimulus $c(t)$ in the form of a delta pressure pulse, i.e. $c(t') = \delta(t')$, the response becomes

$$r(t) = K(t) = \int_{-\infty}^{\infty} \frac{d\omega'}{2\pi} K(\omega')e^{-i\omega' t} \tag{1.194}$$

where the response is now identified by the compressibility with $K(\omega) = K_1(\omega) + iK_2(\omega)$. Because the compressibility is physically measurable, it must be a real function, which entails $K(-\omega) = K^*(\omega)$, where the star denotes the complex conjugate. This means that the real part of the transform must be a symmetric function and the imaginary part an anti-symmetric function of frequency. Furthermore, the causality requires that the compressibility should not depend on the future pressure, i.e. $K(t) = 0$ for $t < 0$ as stated above. Based on the theory of the integration of complex functions, both requirements lead to

$$K_1(\omega) = \frac{2}{\pi} P \int_{0}^{\infty} \frac{\omega' K_2(\omega')}{\omega'^2 - \omega^2} d\omega' \tag{1.195}$$

$$K_2(\omega) = -\frac{2}{\pi} P \int_{0}^{\infty} \frac{\omega' K_1(\omega')}{\omega'^2 - \omega^2} d\omega', \tag{1.196}$$

where P stands for the principal part of the integral. Equations (1.195 and 1.196) are the Kramers–Kronig relations. Furthermore, the compressibility $K(\omega)$ is related to the wave vector k by the dispersion relation

$$k^2 = \omega^2 \rho K(\omega) \tag{1.197}$$

where ρ is the density of the material. Because $K(\omega)$ is complex, the wave vector $k = k_1 + ik_2$ must be complex as well. It is evident to write $k_I = \omega/v(\omega)$ and $k_2 = i\alpha$, where $v(\omega) = \omega/k$ (1.147) is the phase velocity and α is the attenuation. Equation (1.197) satisfies the standard expression for an attenuated plane of amplitude A_0 wave propagating in the x-direction (1.8):

$$A = A_0 e^{i(kx-\omega t)} = A_0 e^{i((k_1+ik_2)x-\omega t)} = A_0 e^{-\alpha x} e^{i(\omega x/v(\omega)-\omega t)} \tag{1.198}$$

It is straightforward to relate the complex wave vector to the complex compressibility:

$$\frac{\omega^2}{v^2(\omega)} - \alpha^2(\omega) + \frac{2i\omega\alpha(\omega)}{v(\omega)} = \omega^2 \rho (K_1(\omega) + iK_2(\omega)) \tag{1.199}$$

$$\frac{\omega^2}{v^2(\omega)} - \alpha^2(\omega) = \omega^2 \rho K_1(\omega) \tag{1.200}$$

$$\alpha(\omega) = v(\omega)\omega\rho K_2(\omega)/2 \tag{1.201}$$

Equations (1.195, 1.196, 1.199, 1.200, and 1.201) allow calculating the phase velocity $v(\omega)$ as a function of frequency when the attenuation $\alpha(\omega)$ is known *for all frequencies* and vice versa. Usually, this requirement is not fulfilled. However, the inertia of acoustical systems entails that their response falls off rapidly at high frequencies. This is equivalent to a limited bandwidth of integration. Furthermore, the magnitude of the imaginary part of the k-vector is much smaller than the real part, i.e. $\alpha v(\omega)/\omega \ll 1$. Then, in (1.200) the velocity decouples from the attenuation and it can be simplified to

$$v(\omega) = (\rho K_1(\omega))^{-0.5} \tag{1.202}$$

Besides the application for L-, and T-waves, the Kramers–Kronig relations are as well applicable for guided modes like Rayleigh, shear-vertical, and shear-horizontal waves [13]. Likewise, in the case of anisotropic materials, for example composites, wave propagation depends on the direction of the k-vector relative to the symmetries of the materials and this holds for the viscoelastic absorption as well and is covered by the Kramers-Kronig-relations [111]. The Kramers–Kronig relations hold also for optical phenomena [312] and many other dispersion phenomena in physics.

The examples presented in the following sections show that the Kramers–Kronig relations can be applied to attenuation due to scattering as well as due to internal friction or absorption, or to a combination for both loss mechanism, as they occur in composite materials [272]. The Kramers–Kronig relations help to understand the detection and quantification of changes in the microstructure of materials.

1.24.2 Applications of K-K Relations to Ultrasonic Relaxation Phenomena

It has been shown by O'Donell et al. [205] that if the dispersion occurs only locally, i.e. in a limited frequency band around a central frequency $\omega_0/2\pi$, and if it is sufficiently weak, the Kramers-Kronig-relations can be simplified to:

$$\alpha(\omega) = \frac{\pi\omega^2}{2v^2(\omega)}\frac{dv(\omega)}{d\omega} \tag{1.203}$$

$$\Delta v(\omega) = v(\omega) - v_0 = +\frac{2v_0^2}{\pi}\int_{\omega_0}^{\omega}\frac{\alpha(\omega')}{\omega'^2}d\omega' \tag{1.204}$$

If the dispersion is caused by a single relaxation phenomenon, $\Delta v(\omega)$ represents the dispersive part originating from the single relaxation absorption process as shown in Fig. 1.71, lower part.

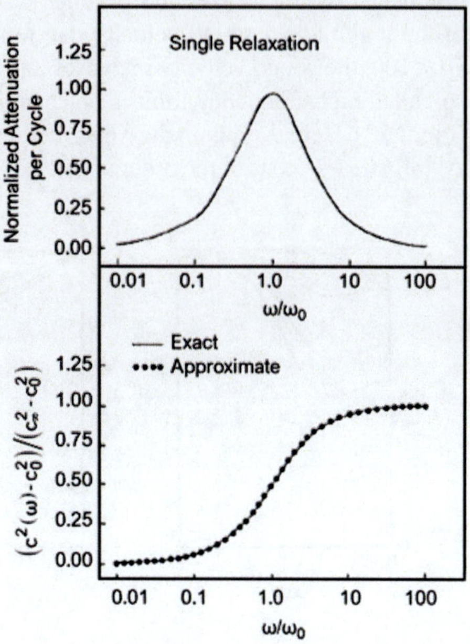

Fig. 1.71 Attenuation per cycle of a relaxation process with a unit strength with its maximum value at $\omega = \omega_0$ (top) and calculated dispersion of the phase velocity $c(\omega)$ using the Kramers–Kronig relations (Eqs. (1.195 and 1.196), respectively (1.204) (from [205], with the permission of the Acoustical Society of America)

1.24.3 Applications of K-K Relations to Ultrasonic Scattering

The Kramers–Kronig relations allow also to relate the velocity and the attenuation for damaged microstructures and hence material degradation of single phase or multiphase materials. Angel and Achenbach [5] considered a solid containing planar cracks with a density of $\varepsilon = na^2$, where a is the crack length. The cracks can be regarded as a thinly distributed second phase. In a first step, Angel and Achenbach calculated the attenuation due to scattering including multiple scattering. In a second step, they used the Kramers–Kronig relations to derive the accompanying velocity dispersion. The results are shown in Fig. 1.72a–c. The velocity shows a minimum at $ka \approx 1$ and the attenuation a maximum at $ka \approx 1.7$ (k is the wave-vector). For very high frequencies, the waves propagate unimpeded through the spaces between the thinly distributed cracks. Therefore, the limiting velocity is the same as for $\varepsilon \to 0$. This underscores the physical picture that cracks reduce the overall restoring force of a solid, in particular for frequencies where $ka \approx 1$ holds.

Rokhlin [238] developed an analytical expression for the ultrasonic attenuation as a function of frequency, covering the Rayleigh-, resonance- and diffuse scattering regime in polycrystalline materials. These expressions and the Kramers–Kronig relation were used by Beltzer and Brauner [12] in order to calculate the dispersive part of the longitudinal sound velocity due to scattering for three regimes: (1) For $1 < ka < 100$, the sound velocity increases and becomes dispersive until for $k \to \infty$ the so-called high-frequency limit is reached, and the velocity becomes constant again (Fig. 1.73). Here, k is again the wave vector and a is the grain size. The dispersive part $\Delta v(\omega)/v(\omega \to \infty)$ is of the order of 5×10^{-3}. This value does not change appreciably

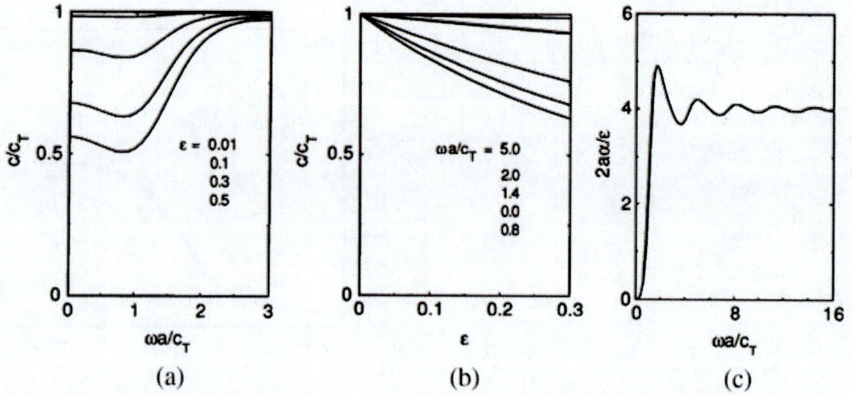

Fig. 1.72 **a** Relative shear wave velocity for a solid containing an ensemble of diluted cracks, with the density ε as a function of $ka = \omega a/c_T$. The velocity exhibits a minimum at $ka \approx 1$; **b** relative velocity as a function of the crack density parameter $\varepsilon = na^2$; **c** reduced and normalized attenuation due to scattering as a function of $ka = \omega a/c_T$. The attenuation has a maximum at $ka \approx 1.7$ (from [5], with the permission of the Acoustical Society of America)

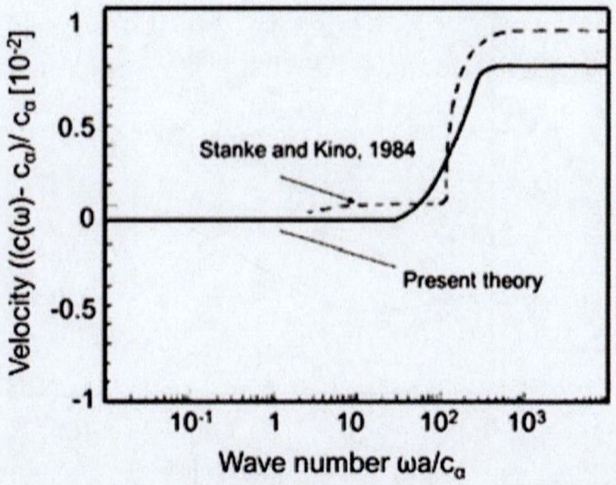

Fig. 1.73 Dispersion of the sound velocity due to scattering based on the theory of Rokhlin [238] and calculated by Beltzer and Brauner [12] using the Kramers–Kronig relations. Note the similarity of the dashed line to Fig. 1.48c for the shape $v(\omega)$ versus frequency as well as the absolute value of the dispersion according to the theory of Stanke and Kino [279] (from[12], with the permission of American Institute of Physics Publishing)

if the scattering theory of [279] is used as "input" for the Kramers-Kronig-relations to calculate $v(\omega)$, see Fig. 1.47.

A number of similar studies have been carried out for porous materials and materials containing lead particles [79, 201, 255]. As soon as the number of particles gets high enough, either pores or inclusions, multiple scattering becomes noticeable, which influences both the velocity and the attenuation for the regime $ka \approx 1$. In case of lead inclusions in an epoxy material, the dispersive part $\Delta v(\omega)/v(\omega \to \infty)$ is of the order of 0.2 [255], which is physically evident because of the large acoustic impedance mismatch between epoxy and lead, see Sect. 1.17.

These results are comparable to the theoretical calculation of the expected dispersion by second order perturbation theory [102]. Experimental data confirmed this behavior, see Fig. 1.74. The dispersive part for $ka < 1$ was also measured [68] and the data agreed with theory. Finally, the dispersive part due to scattering for $ka < 1$ was used by [207] to estimate the grain size.

1.24.4 Applications of K-K Relations to Ultrasonic Absorption

In highly absorbing materials, like polymers, composites or coarse-grained metals, when only a few back-wall echoes can be detected above noise, it may be difficult to determine the absorption coefficient $\alpha(\omega)$. However, it is often feasible to measure the velocity dispersion by evaluating a broad-band pulse and use these data to calculate or at least estimate $\alpha(\omega)$ (or inversely $v(\omega)$) using the Kramers–Kronig relations. In general, it is not necessary to measure or to evaluate the analyzed region of frequency from zero to infinity. So-called "nearly local solutions" around a center frequency

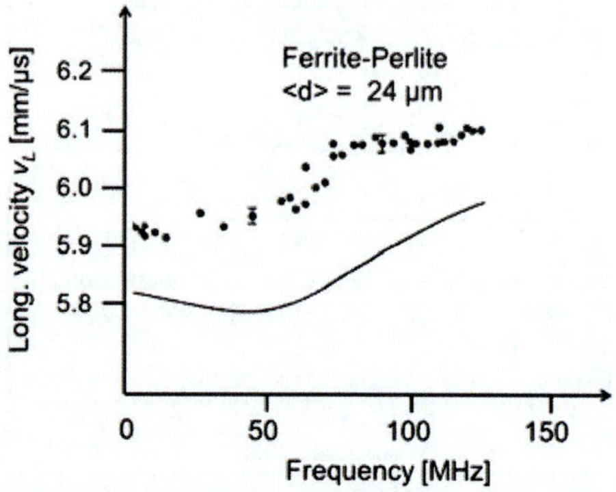

Fig. 1.74 Measurement of the ultrasonic dispersion due to scattering for a pearlite-ferrite steel in comparison to theory (solid line). Here, the grain size was $a = d/2 = 12$ µm and the frequencies employed extended from 10 to 120 MHz in order to cover the regimes $ka \ll 1$, $ka \approx 1$, and $ka > 1$. To accommodate for the strongly increasing attenuation, samples with decreasing thicknesses were used. For the measurements a tone-burst generator was used to ensure a well-defined excitation frequency for the ultrasonic transducer employed (from [224])

lead to useful results. This is the case when changes in dispersion and absorption occur over a limited frequency range, such as relaxation or resonance absorption peaks, as discussed above (Sects. 1.24.2 and 1.24.3). For example, amorphous and partially crystalline polymers do not show scattering but absorption due to their viscoelastic behavior. The α-, β-, γ-, and δ-relaxation peaks of the absorption versus frequency, $\alpha(\omega)$, are often quite narrow in frequency [185], and hence $\alpha(\omega)$ can be inverted to obtain the dispersive contribution of the relaxation peaks to $v(\omega)$ and vice versa, see Fig. 1.71. This is analog to the classical theory of dispersion in optics ("sum rules") [312].

Figure 1.75 shows the measured ultrasonic attenuation in polyethylene which varies almost linearly with frequency. A Kramers–Kronig inversion of these data from 1 to 10 MHz reproduces the measured velocity dispersion. Similar findings have been reported by [158] for polyurethane with and without tungsten powder as filler material so that the material became a composite, and by Zellouf et al. [326] for fully amorphous (PMMA) and for partially crystalline polymers, which show scattering phenomena as well. The measurements were carried out with a pulsed ultrasonic system in transmission mode. The signals were Fourier transformed in order to obtain $\alpha(\omega)$ and $v(\omega)$.

1.24.5 *Applications of K-K Relations to Ultrasonic Resonance Spectroscopy*

In resonance ultrasonic spectroscopy (RUS) measurements, the imaginary part of the elastic compressibility $K_2(\omega)$ was measured over the ferro-elastic phase transition

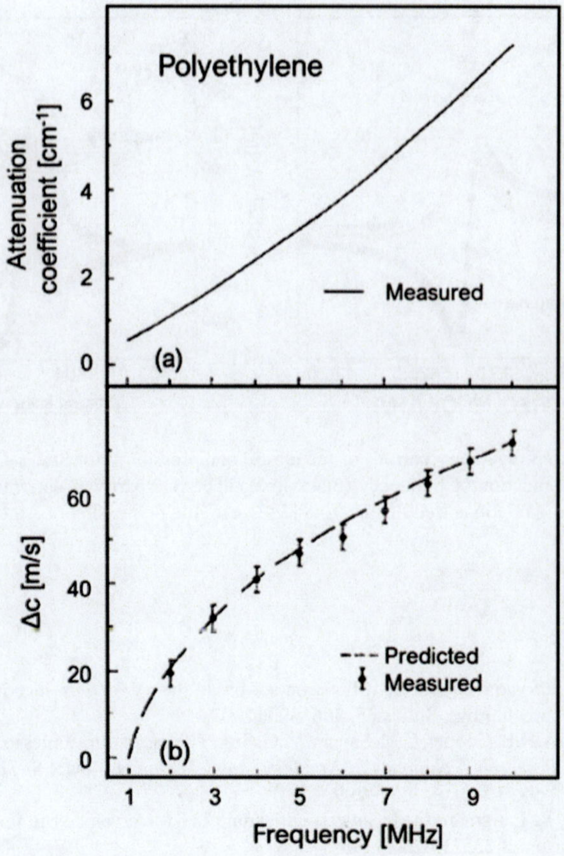

Fig. 1.75 **a** Measured ultrasonic attenuation $\alpha(\omega)$ for polyethylene as a function of frequency between 1 and 10 MHz and **b** contribution of this absorption band to the velocity dispersion (from [205], with the permission of the Acoustical Society of America)

of SnCrTe compounds by [248]. The imaginary part of the compressibility $K_2(\omega)$ is directly linked to the attenuation, see (1.201). Using the Kramers-Kronig-relations as formulated by O'Donnell et al. [205], the real part of the elastic susceptibility K_1 leading to the sound velocity was calculated and thus $v(\omega) = (\rho\, K_1(\omega))^{-0.5}$ (1.202). The result is shown in Fig. 1.76.

Simplified expressions of the Kramer-Kronig relations were derived for attenuation power laws from 0 to 2 [232, 282, 283, 303]. Finally, it should be mentioned that the Kramers–Kronig relations may become k-vector dependent as well, if there is spatial dispersion. For an example see [313]. Spatial dispersion is dominantly discussed and applied for optical phenomena, less in ultrasound. There, spatial dispersion is normally dealt with as effects of anisotropy and inhomogeneity.

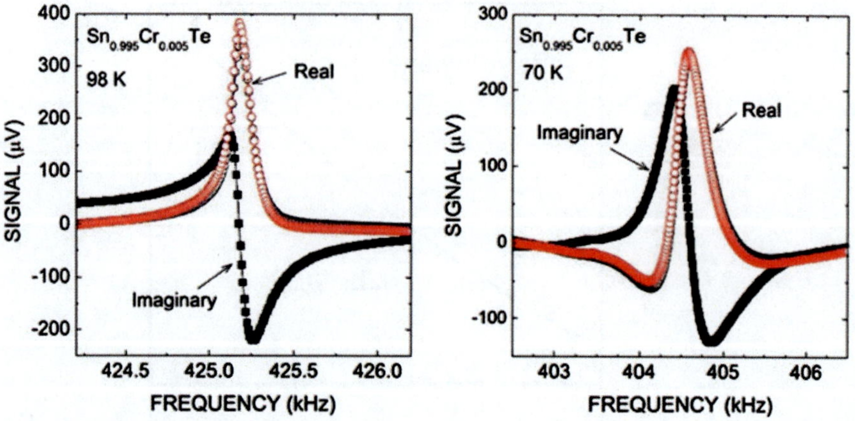

Fig. 1.76 Measured imaginary part K_2 of the elastic compressibility of $Sn_{0.995}Cr_{0.005}Te$ for two temperatures as a function of frequency (black lines). The real part was determined by applying the Kramers-Kronig relations (red lines) (from [248] with the permission of American Institute of Physics Publishing)

References

1. V. Aleshin, K. Van Den Abeele, Microcontact-based theory for acoustics in microdamaged materials. J. Mech. Phys. Solids **55**, 366–390 (2007)
2. D.R. Allen, W.H.B. Cooper, C.M. Sayers, M.G. Silk, The use of ultrasonics to measure residual stresses, in *Research Techniques in Nondestructive Testing*, ed. by R.S. Sharpe (Academic Press, New York, 1982), pp. 151–209
3. D.R. Allen, R. Langman, C.M. Sayers, Ultrasonic SH-wave velocity in textured aluminum plates. Ultrasonics **23**, 215–222 (1985)
4. O.L. Anderson, Determination and some uses of isotropic elastic constants of polycrystalline aggregates using single-crystal data, in *Physical Acoustics*. ed. by W.P. Mason (Academic Press, New York, 1965), pp. 43–76
5. Y.C. Angel, J.D. Achenbach, Attenuation and speed of antiplane waves in a cracked solid using the Kramers–Kronig relations. J. Acoust. Soc. Am. **90**, 2757–2762 (1991)
6. Anonymous, *Standard Test Method for Dynamic Young's Modulus, Shear Modulus, and Poisson's Ratio by Sonic Resonance (E 1875)* (American Society for Testing and Materials (ASTM), West Conshohocken, PA 19428, USA, 1998), p 1
7. W. Arnold, Nondestructive testing of electronic components and structural ceramics, in *Reference Module in Materials Science and Materials Engineering*. ed. by S. Hashmi (Elsevier, Oxford, 2019), p.1
8. G. Bayer, T. Niederdränk, Weak localization of acoustic waves in strongly scattering media. Phys. Rev. Lett. **70**, 3884–3887 (1993)
9. P. Bayer, L. Singher, A. Notea, Early detection of fatigue damage through ultrasonic non-destructive evaluation—Part II: experimental. J. Test. Eval. **27**, 432–439 (1999)
10. P. Bayer, L. Singher, A. Notea, Ultrasonic evaluation of fatigue damage, in *Review of Progress in Quantitative Nondestructive Evaluation*, vol. 23, ed. by D.O. Thompson, D.E. Chimenti (American Institute of Physics, 2004), pp.1240–1247
11. F.L. Becker, R.L. Richardson, Ultrasonic critical angle reflectivity, in *Research Techniques in Non-Destructive Testing*, ed. by R.S. Sharpe (Academic Press, London, 1970), pp. 91–131
12. A.I. Beltzer, N. Brauner, Elastic-waves propagation in polycrystalline media—a causal response. Appl. Phys. Lett. **47**, 1054–1055 (1985)

13. A.I. Beltzer, N. Brauner, SH-waves of an arbitrary frequency in random fibrous composites via the K–K relations. J. Mech. Phys. Solids **33**(471), 87 (1985)
14. R.T. Beyer, S.V. Letcher, *Physical Ultrasonics* (Academic, New York, 1969)
15. A.B. Bhatia, *Ultrasonic Absorption* (Oxford University Press, Oxford, 1967)
16. A.B. Bhatia, R.A. Moore, Scattering of high-frequency sound waves in polycrystalline materials. J. Acoust. Soc. Am. **31**, 1140–1141 (1959)
17. A.S. Birring, D.G. Alcazar, J.J. Hanley, Detection of creep damage by ultrasonics, in *Review of Progress in Quantitative Nondestructive Evaluation*, vol. 8B, eds. by D.O. Thompson, D.E. Chimenti (Springer, 1989), pp. 1833–1840
18. A.S. Birring, D.G. Alcazar, J.J. Hanley, S. Gehl, Ultrasonic detection of hydrogen damage. Mater. Eval. **47**, 345–369 (1989)
19. M. Borsutzki, C. Thoma, W. Bleck, W. Theiner, Online determination of the material properties of cold-rolled steel strip. Stahl Eisen **113**, 93–99 (1993)
20. D.E. Bray, R.K. Stanley, *Nondestructive Evaluation: A Tool for Design, Manufacturing, and Service* (McGraw-Hill Inc., New York, NY, 1989)
21. D.E. Bray, W. Tang, Subsurface stress evaluation in steel plates and bars using the L-CR ultrasonic wave. Nucl. Eng. Des. **207**(231), 40 (2001)
22. M.A. Breazeale, J. Ford, Ultrasonic studies of nonlinear behavior of solids. J. Appl. Phys. **36**, 3486–3490 (1965)
23. M.A. Breazeale, M. McPherson, Physical acoustics, in *Springer Handbook of Acoustics*, 2nd edn., ed. by T. Rossing (Springer, New York, 2014), pp.215–246
24. M.A. Breazeale, J. Philip, Determination of 3rd-order elastic constants from ultrasonic harmonic generation measurements, in *Physical Acoustics*, eds. by W.P. Mason, R.N. Thurston (Academic Press, New York, 1984), pp. 1–10.
25. L.M. Brekhovskikh, *Waves in Layered Media* (Academic Press, New York, 1960)
26. A. Briggs, O. Kolosov, *Acoustic Microscopy* (Oxford University Press, Oxford, UK, 2009)
27. G.A.D. Briggs, *An Introduction to Scanning Acoustic Microscopy* (Oxford University Press, Oxford, 1985)
28. E. Brinksmeier, E. Schneider, W.A. Theiner, H.K. Tönshoff, Nondestructive testing for evaluating surface integrity. CIRP Ann. Manuf. Technol. **33**, 489–509 (1984)
29. K.E. Bullen, *An Introduction to the Theory of Seismology* (Cambridge University Press, UK, 1980)
30. H.J. Bunge, Über die elastischen Konstanten kubischer Materialien mit beliebiger Textur. Krist. Tech. **3**(431), 38 (1968)
31. H.J. Bunge, *Mathematische Methoden der Texturanalyse* (Akademie-Verlag, Berlin, 1969)
32. H.J. Bunge, *Texture Analysis in Materials Science* (Butterworth-Heinemann, London, 1982)
33. M. Calvet, L. Margerin, Velocity and attenuation of scalar and elastic waves in random media: a spectral function approach. J. Acoust. Soc. Am. **131**, 1843–1862 (2012)
34. C.K. Campbell, *Surface Acoustic Wave Devices* (Academic Press, New York, 1998)
35. J.H. Cantrell, Crystalline structure and symmentry dependence of acoustic nonlinearity parameters. J. Appl. Phys. **76**, 3372–3380 (1994)
36. J.H. Cantrell, K. Salama, Acoustoelastic characterisation of materials. Int. Mater. Rev. **36**, 125–145 (1991)
37. O. Cassier, C. Donadille, B. Bacroix, Lankford coefficient evaluation in steel sheets by an ultrasonic method, in *Nondestructive Characterization of Materials*, eds. by P. Höller, V. Hauk, G. Dobmann, C.O. Ruud, R.E. Green (Springer, 1989), pp. 303–311
38. S. Chaki, W. Ke, H. Demouveau, Numerical and experimental analysis of the critically refracted longitudinal beam. Ultrasonics **53**, 65–69 (2013)
39. N. Chandrasekaran, K. Salama, Relationship between stress and temperature dependence of ultrasonic shear velocity, in *Nondestructive Methods for Material Property Determination*. ed. by C.O. Ruud, R.E. Green (Springer, Boston, MA, 1984), pp.393–403
40. J.D.N. Cheeke, *Fundamentals and Application of Ultrasonic Waves* (CRC Press, Boca Raton, FL, 2012)

41. S.L. Chu, E. Schneider, K. Salama, *Influence of Texture on the Variations of Temperature Dependence of Ultrasonic Wave Velocities with Stress* (Ultrasonics International, Butterworth, 1985), pp. 133–138.
42. Y.C. Chu, M. Hefetz, S.I. Rokhlin, Ultrasonic assessment of microcrack damage in ceramics, in *Review of Progress in Quantitative Nondestructive Evaluation*, vol. 12B, eds. by D.O. Thompson, D.E. Chimenti (Springer, 1993), pp. 2175–2182
43. Y.C. Chu, A. Lavrentyev, S.I. Rokhlin, 1995. Ultrasonic characterisation of interfacial fatige damage in metal matrix, in *Review of Progress in Quantitative Nondestructive Evaluation*, vol. 14B, eds. by D.O. Thompson, D.E. Chimenti (Springer, 1995), pp. 2011–2018.
44. R.C. Colwell, H.C. Hardy, LXXXIX. The frequencies and nodal systems of circular plates. London Edinburgh Dublin Philosoph. Mag. J. Sci. **24**, 1041–1055 (1937)
45. C. Comte, J. von Stebut, Microprobe-type measurement of Young's modulus and poisson coefficient by means of depth sensing indentation and acoustic microscopy. Surf. Coat. Technol. **154**, 42–48 (2002)
46. D.I. Crecraft, Measurement of applied and residual stresses in metals using ultrasonic waves. J. Sound Vib. **5**, 173–192 (1967)
47. R. De Batist, *Internal Friction of Structural Defects in Crystalline Solids* (North-Holland, Amsterdam, 1972)
48. I. De Wolf, Micro-Raman spectroscopy to study local mechanical stress in silicon integrated circuits. Semicond. Sci. Technol. **11**, 139–154 (1996)
49. P.P. Delsanto, R.B. Mignogna, A.V. Clark, Ultrasonic texture analysis for polycrystalline aggregrates of cubic materials displaying orthotropic symmetry, in *Nondestructive Characterization of Materials II*. ed. by J.F. Bussière, J.-P. Monchalin, C.O. Ruud, R.E. Green (Springer, Boston, MA, 1987), pp.535–543
50. R.J. Dewhurst, Q. Shan, Optical remote measurement of ultrasound. Meas. Sci. Technol. **10**, R139–R168 (1999)
51. D. Dorner, S. Zaefferer, L. Lahn, D. Raabe, Overview of microstructure and microtexture development in grain-oriented silicon steel. J. Mag. Mag. Mater. **304**, 183–186 (2006)
52. E. Drescher-Krasicka, Scanning acoustic imaging of stress in the interior of solid materials. J. Acoust. Soc. Am. **94**, 453–464 (1993)
53. E. Drescher-Krasicka, J.R. Willis, Mapping stress with ultrasound. Nature **384**, 52–55 (1996)
54. H.L. Du, J.A. Turner, Ultrasonic attenuation in pearlitic steel. Ultrasonics **54**, 882–887 (2014)
55. D.M. Egle, D.E. Bray, Measurement of acoustoelastic and third-order elastic constants for rail steel. J. Acoust. Soc. Am. **60**(741), 44 (1976)
56. T. Englert, G. Abstreiter, J. Pontcharra, Determination of existing stress in silicon films on sapphire substrate using Raman-spectroscopy. Solid-State Electron. **23**(31), 3 (1980)
57. V.I. Erofeev, A. Malkhanov, Nonlinear acoustic waves in solids with dislocations. Proc. IUTAM **23**, 228–235 (2017)
58. A.G. Evans, B.R. Tittmann, G.S. Kino, L. Ahlberg, B.T. Khuri-Yakub, Ultrasonic attenuation in ceramics. J. Appl. Phys. **49**, 2669–2679 (1978)
59. J.A. Evans et al., Determining elastic anisotropy of textured polycrystals using resonant ultrasound spectroscopy. J. Mater. Sci. **56**, 10053–10073 (2021)
60. G. Fantozzi, Phenomenology and definitions, in *Q^{-1}: Mechanical Spectroscopy Q^{-1} 2001 with Applications to Materials Science*, eds. by R. Schaller, G. Fantozzi, G. Gremaud (Trans Tech Publications, Switzerland, 2001)
61. D.G. Fertis, *Mechanical and Structural Vibrations* (John Wiley & Sons Inc., New York, 1995)
62. M.J. Fisher, Acoustoelastic measurements of elastic-plastic and residual stresses, in *Review of Progress in Quantitative Nondestructive Evaluation*, vol. 4, ed. by D.O. Thompson, D.E. Chimenti (Plenum Press, 1985), pp.1051–1059
63. D. Gallitelli et al., Simulation of shot peening: from process parameters to residual stress fields in a structure. C.R. Mec. **344**, 355–374 (2016)
64. G.R. Gerhart, Rayleigh wave velocity for a stress induced slightly anisotropic solid. J. Acoust. Soc. Am. **60**, 1085–1088 (1976)

65. G. Ghoshal, J.A. Turner, Numerical model of longitudinal wave scattering in polycrystals. IEEE Trans. Ultrason. Ferroelectr. Freq. Control **56**, 1419–1428 (2009)
66. C. Gilardoni, M. Gherbin, M. Carboni, A. Gianneo, High-performance methodology for residual stress measurement in railway wheels, in *11th European Conference on Non-Destructive Testing (ECNDT 2014)*, ed. by P. Mazal (Brno University of Technology (BUT), Prague, Czech-Republic, 2014)
67. R.S. Gilmore, K.C. Tam, J.D. Young, D.R. Howard, Acoustic microscopy from 10 MHz to 100 MHz for industrial applications. Philosop. Trans. R. Soc. London **A320**, 215–235 (1986)
68. K. Goebbels, Structure analysis by ultrasonic radiation, in *Research Techniques in NDT*. ed. by R.S. Sharpe (Academic Press, London, 1980), pp.87–157
69. K. Goebbels, Ultraschall-Spannungsmeßtechnik, in *Handbuch der Spannungsmeßpraxis*. ed. by A. Peiter (Vieweg, Braunschweig, 1992), pp.183–195
70. K. Goebbels, S. Hirsekorn, A new ultrasonic method for stress determination in textured materials. NDT Int. **17**, 337–341 (1984)
71. K. Goebbels, H.-J. Salzburger, Zerstörungsfreies Messen der plastischen Anisotropie von Blechen mit Ultraschall. Materialprüfung. **25**, 279–281 (1983)
72. N.P. Goss, New development in electrical strip steels characterized by fine grain structure approaching the properties of a single crystal. Trans. Am. Soc. Metals **23**(511), 31 (1935)
73. G. Gottstein, *Physical Foundation of Materials Science* (Springer, Berlin, 2004)
74. K. Graff, *Wave Motion in Elastic Solids* (Dover Publications, New York, 1991)
75. A. Granato, K. Lücke, Application of dislocation theory to internal friction pheonmena at high frequencies. J. Appl. Phys. **27**, 789–805 (1956)
76. R.E. Green, Nonlinear effects of high-power ultrasonics in crystalline solids. Ultrasonics **13**, 117–127 (1975)
77. G. Gremaud, Overview on dislocation-point defect interaction: the brownian picture of dislocation motion. Mater. Sci. Eng. **A370**(191), 98 (2004)
78. G. Grimvall, *Thermophysical Properties of Materials* (Elsevier, Amsterdam, 1999)
79. D. Gross, C. Zhang, Wave propagation in damaged solids. Int. J. Solids Struct. **29**, 1763–1779 (1992)
80. C.B. Guo, P. Höller, K. Goebbels, Scattering of ultrasonic waves in anisotropic polycrystalline metals. Acustica **59**, 112–120 (1985)
81. B. Haberer, M. Paul, H. Willems, W. Arnold, Material characterization by contactless measurement of ultrasonic absorption using laser ultrasound, in *Nondestructive Characterization of Materials VI*, eds. by R.E. Green, K.J. Kozaczek, C. Ruud (Springer, 1994), pp. 317–324
82. B. Haberer, M. Paul, H. Willems, W. Arnold, Measurement of internal friction in polycrystalline materials using laser-generated ultrasound. J. Alloy. Compd. **212**, 636–639 (1994)
83. P. Haldipur, F.J. Margetan, R.B. Thompson, Estimation of single crystal elastic constants from ultrasonic measurements on polycrystalline specimens, in *Review of Progress in Quantitative Nondestructive Evaluation*, vol. 23, eds. by D.O. Thompson, D.E. Chimenti (American Institut of Physics, 2004), pp. 1061–1068.
84. E.D. Hall, *Yield Point Phenomena in Metals and Alloys* (Springer, New York NY, 1970)
85. Y.K. Han, R.B. Thompson, Ultrasonic backscattering in duplex microstructures: theory and application to titanium alloys. Metall. Mater. Trans. A Phys. Metall. Mater. Sci. **28**(91), 04 (1997)
86. B. Hartmann, J. Jarzynski, Ultrasonic hysteresis absorption in polymers. J. Appl. Phys. **43**, 4304–4312 (1972)
87. Y. Hasegawa, Failures from hydrogen attack and their methods of detection. Weld. Int. **2**, 514–521 (1988)
88. H. Hatanaka, N. Ido, T. Ito, R. Uernichi, M. Tagami, H. Nakagawa, Ultrasonic creep damage detection by frequency analysis for boiler piping. J. Press. Vessel Technol. Trans. ASME **129**, 713–718 (2007)
89. M. Hayes, R.S. Rivlin, Surface waves in deformed elastic materials. Arch. Ration. Mech. Anal. **8**, 358–380 (1961)

90. M. Hefetz, S.I. Rokhlin, Thermal shock damage assessment in ceramics using ultrasonic waves. J. Am. Ceram. Soc. **75**, 1839–1845 (1992)
91. M. Heinrich, U. Rabe, S. Hirsekorn, B. Grabowski, Estimation of the geometry of randomly dimensioned components from their eigenfrequencies via a multiple linear regression analysis, in *Proceedings of International Conference on Noise and Vibration Engineering (ISMA 2014) and International Conference on Uncertainty in Structural Dynamics (USD 2014)*, eds. by P. Sas, H. Denayer, D. Moens (Katholieke Universiteit Leuven, Department of Mechanical Engineering, 2014), pp. 4467–4467
92. E.G. Henneke, J.C. Duke, Analytical ultrasonics for evaluation of composite materials response; part I: physical interpretation, in Materials Analysis by *Ultrasonics: Metals, Ceramics, Composites*, ed. by A. Vary (Noyes Data Corporation, Park Ridge, NJ, 1987), pp. 141–152
93. J. Herrmann, J.Y. Kim, L.J. Jacobs, J.M. Qu, J.W. Littles, M.F. Savage, Assessment of material damage in a nickel-base superalloy using nonlinear Rayleigh surface waves. J. Appl. Phys. **99**, 124913–124921 (2006)
94. R. Herzer, O. Frick, C.P. Keferstein, W. Arnold, Self-tracking ultrasonic inspection. Ind. Robot. **22**, 25–27 (1995)
95. R.W. Hill, The elastic behaviour of a crystalline aggregate. Proc. Phys. Soc. **A65**, 349–354 (1952)
96. M. Hirao, H. Fukuoka, K. Hori, Acoustoelastic effect of Rayleigh surface wave in isotropic material. J. Appl. Mech. Trans. ASME **48**, 119–124 (1981)
97. M. Hirao, H. Ogi, *EMATS for Science and Industry: Noncontacting Ultrasonic Measurements* (Kluwer Academic Publishers, Dordrecht, The Netherlands, 2003)
98. M. Hirao, H. Ogi, N. Suzuki, T. Ohtani, Ultrasonic attenuation peak during fatigue of polycrystalline copper. Acta Mater. **48**, 517–524 (2000)
99. S. Hirsekorn, The scattering of ultrasonic waves by polycrystals. J. Acoust. Soc. Am. **72**(1021), 031 (1982)
100. S. Hirsekorn, The scattering of ultrasonic waves by polycrystals, 2. shear waves. J. Acoust. Soc. Am. **73**, 1160–1163 (1983)
101. S. Hirsekorn, The scattering of ultrasonic waves in polycrystalline materials with texture. J. Acoust. Soc. Am. **77**, 832–843 (1985)
102. S. Hirsekorn, The scattering of ultrasonic waves by multiphase polycrystals. J. Acoust. Soc. Am. **83**(1231), 242 (1988)
103. S. Hirsekorn, Ultrasonic propagation in metal-matrix-composites, in *Nondestructive Characterization of Materials*, eds. by P. Höller, V. Hauk, G. Dobmann, C.O. Ruud, R.E. Green (Springer, 1989), pp. 184–189.
104. S. Hirsekorn, W. Arnold, High-resolution materials characterization by conventional and near-field acoustic microscopy. Ultrasonics **36**(491), 98 (1998)
105. S. Hirsekorn, S. Pangraz, Materials characterization with the acoustic microscope. Appl. Phys. Lett. **64**, 1632–1634 (1994)
106. S. Hirsekorn, S. Pangraz, G. Weides, W. Arnold, Measurement of elastic impedance with high-spatial-resolution using acoustic microscopy. Appl. Phys. Lett. **67**, 745–747 (1995)
107. S. Hirsekorn et al., Non-destructive testing of non-ferrous die casting components detection and evaluation of surface near porosity using high-frequency ultrasound. Materialprüfung. **50**, 206–215 (2008)
108. S. Hirsekorn, E. Schneider, Characterisation of rolling texture by ultrasonic dispersion measurement, in *Nondestructive Characterization of Materials*, eds. by P. Höller, V. Hauk, G. Dobmann, C.O. Ruud, R.E. Green (Springer, 1989), pp. 289–295
109. S. Hirsekorn, P.W. van Andel, U. Netzelmann, Ultrasonic methods to detect and evaluate damage in steel. Non-Destr. Test. Eval. **15**, 373–393 (2000)
110. A. Hoffmann, W. Arnold, Calculation and measurement of the ultrasonic signals generated by ablating material with a Q-switched pulse laser. Appl. Surf. Sci. **96–8**(71), 5 (1996)
111. B. Hosten, Heterogeneous structure of modes and Kramers-Kronig relationship in anisotropic viscoelastic materials. J. Acoust. Soc. Am. **104**, 1382–1388 (1998)

112. H.F. Hu, A. Strybulevych, J.H. Page, S.E. Skipetrov, B.A. Van Tiggelen, Localization of ultrasound in a three-dimensional elastic network. Nat. Phys. **4**, 945–948 (2008)
113. D.S. Hughes, J.L. Kelly, 2nd-order elastic deformation of solids. Phys. Rev. **92**, 1145–1149 (1953)
114. S. Hunklinger, W. Arnold, Ultrasonic properties of glasses at low temperatures, in *Physical Acoustics*. ed. by W.P. Mason, R.N. Thurston (Academic Press, New York, 1976), pp.155–215
115. D.C. Hurley, M. Kopycinska-Muller, A.B. Kos, R.H. Geiss, Nanoscale elastic-property measurements and mapping using atomic force acoustic microscopy methods. Meas. Sci. Technol. **16**, 2167–2172 (2005)
116. D.H. Hurley, S.J. Reese, F. Farzbod, Application of laser-based resonant ultrasound spectroscopy to study texture in copper. J. Appl. Phys. **111,** 053527 (2012)
117. Y. Iwashimizu, O. Kobori, Rayleigh wave in a finitely deformed isotropic elastic material. J. Acoust. Soc. Am. **64**, 910–916 (1978)
118. Y. Javadi, S. Hloch, *Employing the L-CR Waves to Measure Longitudinal Residual Stresses in Different Depths of a Stainless Steel Welded Plate* (Advances in Materials Science and Engineering, 2013)
119. K.Y. Jhang, Nonlinear ultrasonic techniques for non-destructive assessment of micro damage in material: a review. Int. J. Precis. Eng. Manuf. **10**, 123–135 (2009)
120. D.C. Jiles, *Introduction to the Principles of Materials Evaluation* (CRC Press, Boca Raton, FL, 2008)
121. A. Joseph, P. Palanichamy, T. Jayakumar, Evaluation of residual stresses in carbon steel weld joints by ultrasonic L-cr wave technique. J. Nondestr. Eval. **34** (2015)
122. J. Jäckle, *Einführung in die Transporttheorie* (Vieweg & Teubner, Wiesbaden, Germany, 1978)
123. S. Kalpakjian, R.S. Schmid, S. Werner, *Werkstofftechnik (Manufacturing Processes for Engineering Materials)* (Pearson Education Germany, Munich, 2011)
124. T. Kang, H.H. Kim, S.J. Song, H.J. Kim, Characterization of fatigue damage of Al6061-T6 with ultrasound. NDT E Int. **52**(51), 6 (2012)
125. S. Kenderian, T.P. Berndt, R.E. Green, B.B. Djordjevic, Ultrasonic monitoring of dislocations during fatigue of pearlitic rail steel. Mater. Sci. Eng. **A348**, 90–99 (2003)
126. A. Khedher, *L'Acoustoélasticité dans Les Solides Isotropes et son Expérimentation* (Docteur en Sciences Appliquées. Université Catholique de Louvain, Louvain-la-Neuve, Belgique, Département de Mécanique, Faculté des Sciences Appliquées, Vol, 1994)
127. D.Y. Kim, H.J. Kim, S.J. Song, B.J. Kim, B.S. Lim, Ultrasonic method for prediction of residual life of creep damaged materials, in *Review of Progress in Quantitative Nondestructive Evaluation*, vol. 28B, ed. by D.O. Thompson, D.E. Chimenti (American Institute of Physics, 2009), pp.1402–1409
128. G.S. Kino, *Acoustic Waves: Devices, Imaging, and Analog Signal Processing* (Prentice-Hall Inc., Englewood Cliffs, New Jersey, 1987)
129. C. Kittel, *Introduction to Solid State Physics*, 8th edn. (John Wiley & Sons Inc., Hoboken, NJ, 2005)
130. J. Koppelmann, B. Fay, Quantitative evaluation of ultrasonic back-scatter in metals. Acustica **29**, 297–302 (1973)
131. V.V. Kozhushko, G. Paltauf, H. Krenn, S. Scheriau, R. Pippan, Broadband optoacoustic measurements of ultrasound attenuation in severely plastically deformed nickel. J. Appl. Phys. **107** (2010)
132. V.V. Kozhushko, G. Paltauf, H. Krenn, S. Scheriau, R. Pippan, Attenuation of ultrasound in severely plastically deformed nickel. NDT E Int. **44**(261), 66 (2011)
133. J. Krautkrämer, H. Krautkrämer, *Ultrasonic Testing of Materials* (Springer, Berlin, 1990)
134. R.L. Kronig, H.A. Kramers, Absorption and dispersion in x-ray spectra. Z. Phys. **48**, 174–179 (1928)
135. C.M. Kube, H.L. Du, G. Ghoshal, J.A. Turner, Stress-dependent changes in the diffuse ultrasonic backscatter coefficient in steel: experimental results. J. Acoust. Soc. Am. **132,** EL43–EL48 (2012)

136. C.M. Kube, J.A. Turner, Acoustic attenuation coefficients for polycrystalline materials containing crystallites of any symmetry class. J. Acoust. Soc. Am. **137**, EL476–EL482 (2015)
137. C.M. Kube, J.A. Turner, Ultrasonic attenuation in polycrystals using a self-consistent approach. Wave Motion **57**, 182–193 (2015)
138. A. Kumar, Ultrasonic velocity ratio: an effective tool for in-situ materials characterization. Acta Acust. Acust. **104**(251), 58 (2018)
139. A. Kumar, R.R. Adharapurapu, J.W. Jones, T.M. Pollock, In situ damage assessment in a cast magnesium alloy during very high cycle fatigue. Scripta Mater. **64**, 65–68 (2011)
140. A. Kumar, T. Jayakumar, P. Palanichamy, B. Raj, Influence of grain size on ultrasonic spectral parameters in AISI type 316 stainless steel. Scripta Mater. **40**, 333–340 (1999)
141. A. Kumar, T. Jayakumar, B. Raj, Ultrasonic spectral analysis for microstructural characterization of austenitic and ferritic steels. Philosop. Mag. A Phys. Condens. Matter Struct. Defects Mech. Properties **80**, 2469–2487 (2000)
142. A. Kumar, T. Jayakumar, B. Raj, D. Banerjee, A new methodology for identification of beta-transus temperature in alpha+beta and beta titanium alloys using ultrasonic velocity measurement. Phil. Mag. **88**, 327–338 (2008)
143. A. Kumar, T. Jayakumar, B. Raj, K.K. Ray, Characterization of solutionizing behavior in VT14 titanium alloy using ultrasonic velocity and attenuation measurements. Mater. Sci. Eng. A Struct. Mater. Proper. Microstruct. Process. **360**, 58–64 (2003)
144. A. Kumar, K. Laha, T. Jayakumar, K.B.S. Rao, B. Raj, Comprehensive microstructural characterization in modified 9Cr–1Mo ferritic steel by ultrasonic measurements. Metall. Mater. Trans. A Phys. Metall. Mater. Sci. **33**, 1617–1626 (2002)
145. A. Kumar, V. Shankar, T. Jayakumar, K.B.S. Rao, B. Raj, Correlation of microstructure and mechanical properties with ultrasonic velocity in the Ni-based superalloy Inconel 625. Philos. Mag. A Phys. Condens. Matter Struct. Defects Mech Proper. **82**, 2529–2545 (2002)
146. A. Kumar, C. Torbet, J. Jones, T. Pollock, Nonlinear ultrasonics for in situ damage detection during high frequency fatigue. J. Appl. Phys. **106** (2009)
147. J. Kushibiki, N. Chubachi, Material characterization by line focus beam acoustic microscope. IEEE Trans. Sonics Ultrasonics **32**, 189–212 (1985)
148. A. Lagendijk, B. van Tiggelen, D.S. Diersma, Fifty years of Anderson localization. Phys. Today **62**, 24–29 (2009)
149. F. Lakestani, P. Rimoldi, Variation of the ultrasonic propagation velocity due to creep strain in austenitic steel, in *Nondestructive Characterization of Materials*, eds. by P. Höller, V. Hauk, G. Dobmann, C. Ruud, R.E. Green (Springer, 1989), pp. 391–397
150. M. Lang, *Die Bestimmung elastischer und anelastischer Eigenschaften nanokristalliner Materialien.* Ph.D. Thesis, Naturwissenschaftlich-Technische Fakultät III (Saarland University, Saarbrücken, Germany, 2001)
151. M.J. Lang, M. Duarte-Dominguez, R. Birringer, R. Hempelmann, H. Natter, W. Arnold, Measurement of elastic and anelastic properties of nanocrystalline metals. Nanostruct. Mater. **12**(811), 16 (1999)
152. K.J. Langenberg, P. Fellinger, R. Marklein, On the nature of the so-called subsurface longitudinal wave and/or the surface longitudinal "creeping" wave. Res. Nondestr. Eval. **2**, 59–81 (1990)
153. W.T. Lankford, S.C. Snyder, J.A. Bausher, New criteria for predicting the press performance of deep drawing sheets. Trans. Am. Soc. Metals **42**, 1197–1205 (1950)
154. B. Lawn, *Fracture of Brittle Solids* (Cambridge University Press, Cambridge, UK, 1993)
155. H.M. Ledbetter, Elastic properties of materials at low temperatures, in *Materials at Low Temperatures*, eds. by R.P. Reed, A.F. Clark (American Society for Metals, Metals Park, OH 44073, 1983), pp 1–5
156. H.M. Ledbetter, Monocrystal polycrystal elastic-constants of a stainless steel. Physica Status Solidi A Appl Res. **85**, 89–96 (1984)
157. H.M. Ledbetter, Monocrystal-polycrystal elastic-constant models, in *Dynamic Elastic Modulus Measurements in Materials*, ed. by A. Wolfenden (ASTM, Philadelphia, PA 19103, 1990), pp. 135–148

158. C.C. Lee, M. Lahham, B.G. Martin, Experimental verification of the Kramers-Kronig relationship for acoustic waves. IEEE Trans. Ultrason. Ferroelectr. Freq. Control **37**, 286–294 (1990)
159. S.S. Lee, J.F. Smith, R.B. Thompson, Inference of crystallite orientation distribution function from the velocity of ultrasonic plate modes, in *Nondestructive Characterization of Materials II*, eds. by J.F. Bussière, J.-P. Monchalin, C.O. Ruud, R.E. Green (Springer, 1987), pp. 555–562
160. Y.C. Lee, J.O. Kim, J.D. Achenbach, Measurement of stresses by line-focus acoustic microscopy. Ultrasonics **32**, 359–365 (1994)
161. B. Lehr, H. Ulrich, O. Weis, Hypersound and attenuation and phase velocities in evaporated polycrystalline films of silver, copper, and constantan between 1 GHz and 24 GHz. Zeitschrift Fur Physik B-Condensed Matter. **48**, 23–30 (1982)
162. J.E. Lennard-Jones, On the determination of molecular fields-I: from the variation of the viscosity of a gas with temperature. Proc. R. Soc. Lond. A **106**, 441–462 (1924)
163. J. Li, S.I. Rokhlin, Propagation and scattering of ultrasonic waves in polycrystals with arbitrary crystallite and macroscopic texture symmetries. Wave Motion **58**, 145–164 (2015)
164. J. Li, S.I. Rokhlin, Elastic wave scattering in random anisotropic solids. Int. J. Solids Struct. **78–79**, 110–124 (2016)
165. J. Li, L. Yang, S.I. Rokhlin, Effect of texture and grain shape on ultrasonic backscattering in polycrystals. Ultrasonics **54**, 1789–1803 (2014)
166. Y. Li, J.F. Smith, R.B. Thompson, Characterization of textures in plates by ultrasonic plate wave velocities, in *Nondestructive Characterization of Materials*, eds. by P. Höller, V. Hauk, G. Dobmann, C.O. Ruud, R.R. Green (Springer, 1989), pp. 312–319
167. E.A. Lindgren, T.S. Jones, H. Berger, M. Rosen, Determining residual stress in plates with Rayleigh waves, in *Review of Progress in Quantitative Nondestructive Evaluation*, eds. by D.O. Thompson, D.E. Chimenti, vol. 13B (Springer, 1994), pp. 2039–2046
168. C.J. Lissenden, Nonlinear ultrasonic guided waves—principles for nondestructive evaluation. J. Appl. Phys. **129,** 021101–0211018 (2021)
169. M. Liu, J.-Y. Kim, L. Jacobs, J. Qu, Experimental study of nonlinear Rayleigh wave propagation in shot-peened aluminum plates-feasibility of measuring residual stress. NDT E Int. **44**, 67–74 (2011)
170. M. Liu, J.-Y. Kim, J. Qu, L.J. Jacobs, Measuring residual stress using nonlinear ultrasound, in *Review of Progress in Quantitative Nondestructive Evaluation*, vol. 29B, ed. by D.O. Thompson, D.E. Chimenti (American Institute of Physics, 2010), pp.1365–1372
171. S. Liu, S. Best, S.A. Neild, A.J. Croxford, Z. Zhou, Measuring bulk material nonlinearity using harmonic generation. NDT E Int. **48**, 46–53 (2012)
172. O.I. Lobkis, L. Yang, J. Li, S.I. Rokhlin, Ultrasonic backscattering in polycrystals with elongated single phase and duplex microstructures. Ultrasonics **52**, 694–705 (2012)
173. B. Lüthi, *Physical Acoustics in the Solid State* (Springer, Berlin, Heidelberg, 2007)
174. C.F. Ma et al., Stick-to-sliding transition in contact-resonance atomic force microscopy. Appl. Phys. Lett. **113**, 0831021 (2018)
175. E. Macherauch, *Praktikum in der Werkstoffkunde. Friedr* (Vieweg & Sohn, Braunschweig, Germany, 1992)
176. E. Macherauch, H.W. Zoch, *Röntgenographische Eigenspannungsbestimmung - Praktikum in Werkstoffkunde* (Vieweg+Teubner - Springer Fachmedien, 2011)
177. P. Mahadevan, Effect of frequency on texture induced ultrasonic wave birefringence in metals. Nature **211**(621), 22 (1966)
178. F.J. Margetan, T.A. Gray, D.O. Thompson, A technique for quantitatively measuring microstructurally induced ultrasonic noise, in *Review of Progress in Quantitative Nondestructive Evaluation*, eds. by D.O. Thompson, D.E. Chimenti, vol. 10B (Springer, 1991), pp. 1721–1728
179. F.J. Margetan, et al., Fundamental studies of nickel billet materials-engine titanium consortium phase II. Report No. DOT/FAA/AR-05/17 (2005). http://www.tc.faa.gov/its/worldpac/techrpt/ar05-17.pdf

180. F.J. Margetan, R.B. Thompson, I. Yalda-Mooshabad, Backscattered microstructural noise in ultrasonic toneburst inspections. J. Nondestr. Eval. **13**(111), 36 (1994)
181. W.P. Mason, H.J. McSkimin, Energy losses of sound waves in metals due to scattering and diffusion. J. Appl. Phys. **19**, 940–946 (1948)
182. K.H. Matlack, J.Y. Kim, L.J. Jacobs, J. Qu, Review of second harmonic generation measurement techniques for material state determination in metals. J. Nondestr. Eval. **34** (2015)
183. O.E. Mattiat (ed.), *Ultrasonic Transducer Materials* (Springer, New York, New York, NY, 1971)
184. A. Maurel, V. Pagneux, F. Barra, F. Lund, Wave propagation through a random array of pinned dislocations: velocity change and attenuation in a generalized Granato and Lucke theory. Phys. Rev. B. **72** (2005)
185. N.G. McCrum, B.E. Read, G. Williams, *Anelastic and Dielectric Effects in Polymeric Solids* (Dover Publications, New York, 1991)
186. A.D.W. McKie, J.W. Wagner, Optical sensing of in-plane ultrasonic transients. Appl. Phys. Lett. **53**, 1043–1044 (1988)
187. E. Meyer, D. Guicking, *Schwingungslehre* (Vieweg + Teubner Verlag, Wiesbaden, 1974)
188. K.H. Michel, H. Mülders, A. Hecht, Kontinuierliche Bestimmung der Korngröße in austenitischen nichtrostenden Kaltbreitbändern. Stahl Eisen **106**, 377–382 (1986)
189. R.B. Mignogna, P.P. Delsanto, A.V. Clark, B.B. Rath, C.L. Vold, Ultrasonic measurements on textured materials, in *Nondestructive Characterization of Materials II* (Springer Boston MA, 1987), pp. 545–554
190. Mohrbacher, H., Lee, D., Schneider, E., Salama, K., 1991. Acoustic nonlinearity in metal-matrix composites, in *Review of Progress in Quantitative Nondestructive Evaluation*, eds. by D.O. Thompson, D.E. Chimenti, vol. 10B (Springer, 1991), pp. 1821–1828
191. H. Mohrbacher, K. Salama, Elastic nonlinearity in metal-matrix composites. Res. Nondestr. Eval. **3**, 159–170 (1991)
192. J.-P. Monchalin, Optical detection of ultrasound. IEEE Trans. Ultrason. Ferroelectr. Freq. Control **33**, 485–499 (1986)
193. J.-P Monchalin, Progress towards the application of laser-ultrasonics in industry, in *Review of Progress in Quantitative Nondestructive Evaluation*, vol. 12, eds. by D.O. Thompson, D.E. Chimenti (Springer, 1993), pp. 495–506
194. J.-P. Monchalin, J.F. Bussière, J. F., 1985. Infrared detection of ultrasonic absorption and application to the determination of absorption in steel, in *Review of Progress in Quantitative Nondestructive Evaluation*, vol. 4B, eds. by D.O. Thompson, D.E. Chimenti (Springer, 1985), pp. 965–973
195. P. Mora, M. Spies, On the validity of several previously published perturbation formulas for the acoustoelastic effect on Rayleigh waves. Ultrasonics **91**, 114–120 (2019)
196. A. Moreau et al., On-line measurement of texture, thickness and plastic strain ratio using laser-ultrasound resonance spectroscopy. Ultrasonics **40**, 1047–1056 (2002)
197. P.M. Morse, K.U. Ingard, *Theoretical Acoustics* (Princeton University Press, Princeton, 1968)
198. G. Mott, M.C. Tsao (eds.), *Acoustoelastic Effects in Two Structural Steels* (Springer, Boston, MA, 1983)
199. R. Murayama, K. Fujisawa, H. Fukuoka, M. Hirao, Development of an on-line evaluation system of formability in cold-rolled steel sheets using electromagnetic acoustic transducers (EMATs). NDT E Int. **29**(141), 46 (1996)
200. F.D. Murnaghan, *Finite Deformation of an Elastic Solid* (John Wiley and Sons, New York, 1951)
201. A.A. Naumenko, S.A. Shcherbinin, D.I. Makariev, A.N. Rybyanets, Experimental study of relationships between ultrasonic attenuation and dispersion for ceramic matrix composite. Phys. Procedia **70**(171), 74 (2015)
202. D. Nicoletti, D. Kasper, Ultrasonic attenuation based on the roney generalized theory and multiple power-law grain-size distributions. IEEE Trans. Ultrason. Ferroelectr. Freq. Control **41**, 144–149 (1994)

203. A.N. Norris, Symmetry conditions for third-order elastic-moduli and implications in nonlinear-wave theory. J. Elast. **25**, 247–257 (1991)
204. A. Nowick, B.S. Berry, *Anelastic Relaxation of Solids* (Academic Press, Ney York, 1972)
205. M. O'Donnell, E.T. Jaynes, J.G. Miller, Kramers-Kronig relationship between ultrasonic attenuation and phase velocity. J. Acoust. Soc. Am. **69**, 696–701 (1981)
206. H. Ogi, H. Ledbetter, S. Kim, M. Hirao, Contactless mode-selective resonance ultrasound spectroscopy: electromagnetic acoustic resonance. J. Acoust. Soc. Am. **106**, 660–665 (1999)
207. P. Palanichamy, A. Joseph, T. Jayakumar, B. Raj, Ultrasonic velocity measurements for estimation of grain-size in austenitic stainless-steel. NDT E Int. **28**, 179–185 (1995)
208. S. Palm, W. Arnold (eds.), *Ultrasonic Absorption Measurement* (Springer, Heidelberg, 2004)
209. R. Pamnani, G.K. Sharma, S. Mahadevan, T. Jayakumar, M. Vasudevan, B.P.C. Rao, Residual stress studies on arc welding joints of naval steel (DMR-249A). J. Manuf. Process. **20**, 104–111 (2015)
210. S. Pangraz, E. Babilon, W. Arnold, Observation of microcrack damage in Al_2O_3-ceramics by scanning acoustic microscopy and small angle X-ray scattering, in *19th International Symposium on Acoustical Imaging*, vol. 19, eds. by H. Ermert, H. P. Harjes (Springer, 1992), pp. 691–696
211. S. Pangraz, H. Simon, R. Herzer, W. Arnold, Non-destructive evaluation of engineering ceramics by high-frequency acoustic techniques, in *18th International Symposium Acoustical Imaging*, vol. 18, eds. by H. Lee, G. Wade (Springer, 1990), pp. 189–195
212. Y.H. Pao, W. Sachse, H. Fukuoka, Acoustoelasticity and ultrasonic measurements of residual stresses, in *Physical Acoustics*, eds. by W.P. Mason, R.N. Thurston (Elsevier, New York, 1984), pp. 61–143
213. E.P. Papadakis, Ultrasonic attenuation caused by scattering in polycrystalline media, in *Physical Acoustics*. ed. by W.P. Mason (Academic Press, New York, 1968), pp.269–328
214. E.P. Papadakis, Effects of input amplitude profile upon diffraction loss and phase change in a pulse-echo system. J. Am. Soc. Acoust. **49**, 166–168 (1970)
215. E.P. Papadakis, Scattering in polycrystalline media, in *Methods of Experimental Physics*, ed. by P.D. Edmons (1981), pp. 237–298.
216. S. Parthasarathi, T. Prucher, C.J. Yu, J. Jo, R.J. Henry, Determination of Dynamic Elastic Properties of Powder Metallurgy Components, in *Review of Progress in Quantitative Nondestructive Evaluation*, vol. 12A, eds. by D.O. Thompson, D.E. Chimenti (Springer, 1993), pp. 1631–1638
217. M. Paul, *Wandstärkemessung im Rohrwalzwerk mittels Laserultraschall*. Technical Faculty, Ph.D. Saarland University, Saarbrücken (1998)
218. M. Paul, B. Haberer, W. Arnold, Materials characterization at high-temperatures using laser ultrasound. Mater. Sci. Eng. A **168**, 87–92 (1993)
219. M. Paul, A. Hoffmann, L. Oesterlein, G.J. Deppe, Laser ultrasonics in industry, in *7th European Conference on Non-Destructive Testing*, vol. 2 (Copenhagen, 1998), pp. 2134–2140
220. Z. Pawlowski, Ultrasonic attenuation during cyclic straining, in *4th International Conference on Non-Destructive Testing* (Butterworth, London, 1964), pp. 192–195
221. H. Peukert, E. Schneider, Zerstörungsfreier Nachweis und Analyse von Eigenspannungszuständen in geschweißten Platten mit Ultraschallverfahren, in *DACH Tagung der Deutschen Gesellschaft für zerstörungsfreie Prüfung 1987*, vol. 10, ed. by DGZFP (Deutsche Gesellschaft für zerstörungsfreie Prüfung, 1987), pp. 356–363
222. M.K. Phani, A. Kumar, W. Arnold, K. Samwer, Elastic stiffness and damping measurements in titanium alloys using atomic force acoustic microscopy. J. Alloy. Compd. **676**, 397–406 (2016)
223. P. Polukhin, S. Gorelik, V. Vorontsov, *Physical Principles of Plastic Deformation* (Mir Publishers, Moscow, 1983)
224. H.J. Pongratz, Interpretation und Auswertung von Ultraschall-Rückstreusignalen. Fachbereich Werkstoffwissenschaften und Fraunhofer IZFP, Vol. Diplom-Arbeit (Master Thesis). Saarland University, Saarbrücken (1983)

225. J. Pouliquen, A. Defebvre, Use of surface acoustic waves for the detection of modifications produced by deformation fatigue or corrosion on metallic surfaces, in *Ultrasonic International 1975*. ed. by Z. Novak (IPC Science and Technology Press, London, 1975), pp.102–106
226. M. Prikhodko, M.S. Miao, W.R.L. Lambrecht, Pressure dependence of sound velocities in 3C-SiC and their relation to the high-pressure phase transition. Phys. Rev. B **66** (2002)
227. C.F. Quate, A. Atalar, H.K. Wickramasinghe, Acoustic microscopy with mechanical scanning—a review. Proc. IEEE **67**, 1092–1114 (1979)
228. U. Rabe, Atomic force acoustic microscopy, in *Applied Scanning Probe Microscopy*. ed. by B. Bhushan, H. Fuchs (Springer, Berlin, 2006), pp.37–90
229. U. Rabe, K. Janser, W. Arnold, Vibrations of free and surface-coupled atomic force microscope cantilevers: theory and experiment. Rev. Sci. Instrum. **67**(3281), 293 (1996)
230. R. Raghavan, V.V. Shastry, A. Kumar, T. Jayakumar, U. Ramamurty, Toughness of as-cast and partially crystallized composites of a bulk metallic glass. Intermetallics **17**, 835–839 (2009)
231. M. Rahman, J.R. Barber, Exact expressions for the roots of the secular equation for Rayleigh waves. Trans. ASME **62**, 250–252 (1995)
232. R. Raisutis, R. Kazys, L. Mazeika, Application of the ultrasonic pulse-echo technique for quality control of the multi-layered plastic materials. NDT E Int. **41**, 300–311 (2008)
233. K.V. Rajkumar, A. Kumar, T. Jayakumar, B. Raj, K.K. Ray, Characterization of aging behavior in M250 grade maraging steel using ultrasonic measurements. Metall. Mater. Trans. A Phys. Metall. Mater. Sci. **38A**, 236–243 (2007)
234. S. Razvi, P. Li, K. Salama, J.H. Cantrell Jr., W.T. Yost, Nondestructive characterization of aluminum alloys, in *Review of Progress in Quantitative Nondestructive Evaluation*, vol. 6A, eds. by D.O. Thompson, D.E. Chimenti (Springer, 1987), pp. 1403–1409
235. A. Reuss, Berechnung der Fließgrenze von Mischkristallen auf Grund der Plastizitätsbedingung für Einkristalle. Z. Angew. Math. Mech. **9**, 49–58 (1929)
236. R.J. Roe, Description of crystallite orientation in polycrystalline materials. 3. general solution to pole figure inversion. J. Appl. Phys. **36**, 2024–2031 (1965)
237. G. Roebben, B. Bollen, A. Brebels, J. Van Humbeeck, O. Van der Biest, Impulse excitation apparatus to measure resonant frequencies, elastic moduli, and internal friction at room and high temperature. Rev. Sci. Instrum. **68**(4511), 515 (1997)
238. L.L. Rokhlin, Scattering of ultrasound in polycrystalline materials. Soviet Phys. Acoust. USSR **18**, 71–000 (1972)
239. S.I. Rokhlin, J. Li, G. Sha, Far-field scattering model for wave propagation in random media. J. Acoust. Soc. Am. **137**, 2655–2669 (2015)
240. J.H. Rose, Ultrasonic backscattering from polycrystalline aggregates using time-domain linear response theory, in *Review of Progress in Quantitative Nondestructive Evaluation*, vol. 10B, eds. by D.O. Thompson, D.E. Chimenti (Springer, 1991), pp. 1715–1720
241. J.H. Rose, Ultrasonic backscatter from microstructure, in *Review of Progress in Quantitative Nondestructive Evaluation*, vol. 11, ed. by D.O. Thompson, D.E. Chimenti (Plenum Press, 1992), pp.1677–1684
242. J.H. Rose, Theory of ultrasound backscatter from multiphase polycrystalline solids, in *Review of Progress in Quantitative Nondestructive Evaluation*, vol. 12, eds. by D.O. Thompson, D.E. Chimenti (1993), pp. 1719–1726
243. J.L. Rose, *Ultrasonic Waves in Solid Media* (Cambridge University Press, Cambridge, UK, 1999)
244. W.R. Rose, S.I. Rokhlin, P.B. Nagy, L. Adler, Polar characteristics of the group and phase velocities as well as the frequency dependence of lamb waves in graphite/epoxy composites, in *Nondestructive Characterization of Materials II*, eds. by J.F. Bussière, J. Monchalin, C.O. Ruud, R.E. Green (Springer, 1987), pp. 61-68
245. F. Roters, P. Eisenlohr, L. Hantcherli, D.D. Tjahjanto, T.R. Bieler, D. Raabe, Overview of constitutive laws, kinematics, homogenization and multiscale methods in crystal plasticity finite-element modeling: theory, experiments, applications. Acta Mater. **58**, 1152–1211 (2010)
246. D.J. Roth, D.B. Stang, S.M. Swickard, M.R. Deguire, L.E. Dolhert, Review, modeling, and statistical analysis of ultrasonic velocity pore fraction relations in polycrystalline materials. Mater. Eval. **49**, 883–888 (1991)

247. W. Sachse, Y.H. Pao, Determination of phase and group velocities of dispersive waves in solids. J. Appl. Phys. **49**, 4320–4327 (1978)
248. D.J. Safarik, E.K.H. Salje, J.C. Lashley, Spectral analysis of resonance ultrasonic spectroscopy: Kramers-Kronig analysis, fano profiles, and the case of precursor softening in SnTe:Cr. Appl. Phys. Lett. **97** (2010)
249. K. Sakata, D. Daniel, J.J. Jonas, J.F. Bussière, Acoustoelastic determination of the higher-order orientation distribution function coefficients up to I = 12 and their use for the online prediction of r-value. Metall. Trans. A Phys. Metall. Mater. Sci. **21**, 697–706 (1990)
250. V. Salinas, C. Aguilar, R. Espinoza-Gonzalez, F. Lund, N. Mujica, In situ monitoring of dislocation proliferation during plastic deformation using ultrasound. Int. J. Plast **97**, 178–193 (2017)
251. S. Sathish, R.W. Martin, T.J. Moran, Local surface skimming longitudinal wave velocity and residual stress mapping. J. Acoust. Soc. Am. **115**, 165–171 (2004)
252. J.C. Savage, Thermoelastic attenuation of elastic waves by Cracks. J. Geophys. Res. **71**, 3929–3938 (1966)
253. C.M. Sayers, Ultrasonic velocities in anisotropic polycrystalline aggregates. J. Phys. D Appl. Phys. **15**, 2157–2167 (1982)
254. C.M. Sayers, R.L. Smith, The propagation of ultrasound in porous media. Ultrasonics **20**(201), 05 (1982)
255. C.M. Sayers, R.L. Smith, Ultrasonic velocity and attenuation in an epoxy matrix containing lead inclusions. J. Phys. D Appl. Phys. **16**, 1189–1194 (1983)
256. V. Scherer, M. Reinstädtler, W. Arnold, Atomic force microscopy with lateral modulation, in *Applied Scanning Probe Methods*. ed. by H. Fuchs, B. Bhushan, S. Hosaka (Springer, Berlin, 2003), pp.76–115
257. U. Schleichert, K.J. Langenberg, W. Arnold, S. Faßbender, A quantitative theory of laser generated ultrasound, in *Review of Progress in Quantitative Nondestructive Evaluation*, vol. 8A, eds. by O. Thompson, D.E. Chimenti (Springer, 1989), pp. 489–496
258. E. Schneider, Ultrasonic birefringence effect—its application for materials characterisations. Opt. Lasers Eng. **22**, 305–323 (1995)
259. E. Schneider, Ultrasonic techniques, in *Structural and Residual Stress Analysis by Nondestructive Methods*. ed. by V. Hauk (Elsevier, Amsterdam, 1997), pp.522–563
260. E. Schneider, Untersuchung der materialspezifischen Einflüsse und verfahrungstechnische Entwicklungen der Ultraschallverfahren zur Spannungsanalyse an Bauteilen. Fakultät für Maschinenwesen, Ph.D. Rheinisch-Westfälische Technische Hochschule, Fraunhofer IRB Verlag, Stuttgart (2000)
261. E. Schneider, S.L. Chu, K. Salama, Influence of texture on the temperature dependence of ultrasonic velocities, in *IEEE Ultrasonic Symposium*. ed. by B.R. McAvoy (IEEE, New York, 1984), pp.944–949
262. E. Schneider, S.L. Chu, K. Salama, Nondestructive determination of mechanical properties, in *Review of Progress in Quantitative Nondestructive Evaluation*, vol. 4, ed. by D.O. Thompson, D.E. Chimenti (Plenum Press, 1984), pp.867–876
263. B. Scholtes, Röntgenographisches Verfahren, in *Handbuch für experimentelle Spannungsanalyse*. ed. by C. Rohrbach (VDI-Verlag, Düsseldorf, Germany, 1989), pp.435–464
264. C.B. Scruby, L.E. Drain, *Laser ultrasonics* (Taylor & Francis, New York, NY, 1990)
265. V.N. Serebryany, Plastic anisotropy prediction by ultrasonic texture data. Textures and Microstructures. **25**, 223–228 (1996)
266. M. Sgalla, D. Vangi, A device for measuring the velocity of ultrasonic waves: an application to stress analysis. Exp. Mech. **44**, 85–90 (2004)
267. G. Sha, S.I. Rokhlin, Universal scaling of transverse wave attenuation in polycrystals. Ultrasonics **88**, 84–96 (2018)
268. J.F. Shackelford, *Introduction to Material Science for Engineers* (Pearson Education, Upper Saddle River, NJ, USA, 2005)
269. G.K. Sharma, A. Kumar, C.B. Rao, T. Jayakumar, B. Raj, Short time Fourier transform analysis for understanding frequency dependent attenuation in austenitic stainless steel. NDT E Int. **53**, 1 (2013)

270. J.F. Smith, R.B. Thompson, Simultaneous ultrasonic evaluation with differentiation of stress and texture. J. Mater. Eng. Perform. **3**, 273–281 (1994)
271. R.T. Smith, R. Stern, R.W. Stephens, Third order elastic moduli of polycrystalline metals from ultrasonic velocity measurements. J. Acoust. Soc. Am. **40**, 1002–1008 (1966)
272. Y.G. Sokolovskaya, N.B. Podymova, A.A. Karabutov, Verification of the Kramers–Kronig relations between ultrasonic attenuation and phase velocity in a finite spectral range for CFRP composites. Ultrasonics **95**, 37–44 (2019)
273. I. Solodov, K. Pfleiderer, H. Gerhard, G. Busse, Nonlinear acoustic approach to material characterisation of polymers and composites in tensile tests. Ultrasonics **42**(1011), 015 (2004)
274. M. Spies, K. Salama, Texture of metal-matrix composites by ultrasonic velocity measurements. Res. Nondestr. Eval. **1**, 99–109 (1989)
275. M. Spies, E. Schneider, Non-destructive analysis of textures in rolled sheets by ultrasonic techniques. Textures Microstruct. **12**, 219–231 (1990)
276. M. Spies, M. Schneider, Nondestructive analysis of the deep-drawing behaviour of rolling sheets with ultrasonic techniques, in *Nondestructive Characterization of Materials*, eds. by P. Höller, V. Hauk, G. Dobmann, C. Ruud, R. Green (Springer, 1989), pp. 296–302
277. S. Spinner, W.E. Tefft, A method for determining mechanical resonance frequencies and for calculating elastic moduli from these frequencies. Proc. Am. Soc. Testing Mater. **61**(1221), 238 (1961)
278. G. Sposito, C. Ward, P. Cawley, P.B. Nagy, C. Scruby, A review of non-destructive techniques for the detection of creep damage in power plant steels. NDT E Int. **43**, 555–567 (2010)
279. F.E. Stanke, G.S. Kino, A unified theory for elastic wave propagation in polycrystalline materials. J. Acoust. Soc. Am. **75**, 665–681 (1984)
280. S. Suresh, *Fatigue of Materials* (Cambridge University Press, Cambridge UK, 1991)
281. V.A. Sutilov, *Physik des Ultraschalls* (Springer, Vienna, 1984)
282. T.L. Szabo, Time-domain wave-equations for lossy media obeying a frequency power-law. J. Acoust. Soc. Am. **96**(491), 00 (1994)
283. T.L. Szabo, Causal theories and data for acoustic attenuation obeying a frequency power-law. J. Acoust. Soc. Am. **97**, 14–24 (1995)
284. W. Theiner, R. Kern, M. Stroh, Process-integrated nondestructive testing of ground and case hardened parts, in *8th European Conference on Nondestructive Testing. Spanish Society for NDT (AEND)*, ed. by AEND (Barcelona, Spain, 2002), pp. Code MC-4.1
285. B.R. Thompson, Elastic-wave propagation in random polycrystals: fundamentals and application to nondestructive evaluation, in *Imaging of Complex Media with Acoustic and Seismic Waves*. ed. by M. Fink, W.A. Kuperman, J.-P. Montagner, A. Tourin (Springer, Berlin, 2002), pp.233–256
286. R.B. Thompson, S.S. Lee, J.F. Smith, Relative anisotropies of plane waves and guided modes in thin orthorhombic plates-implication for texture characterisation. Ultrasonics **25**, 133–137 (1987)
287. R.B. Thompson, F.J. Margetan, Microstructural noise in titanium alloys and its influence on the detectability of hard-alpha inclusion, in *Review of Progress in Quantitative Nondestructive Evaluation*, vol. 11, ed. by D.O. Thompson, D.E. Chimenti (Plenum Press, 1992), pp.1717–1724
288. R.B. Thompson et al., Scattering of elastic waves in simple and complex polycrystals. Wave Motion **45**, 655–674 (2008)
289. R.B. Thompson, J.F. Smith, S.S. Lee, Microstructure independent acoustoelastic measurement of stress. Appl. Phys. Lett. **44**, 296–298 (1984)
290. R. Truell, C. Elbaum, B.B. Chick, *Ultrasonic Methods in Solid State Physics* (Academic Press, New York, 1969)
291. J.W. Tucker, V.W. Rampton, *Microwave Ultrasonics in Solid State Physics* (North-Holland Publishing Company, Amsterdam, 1972)
292. K.E.-A. Van Den Abeele, P.A. Johnson, A. Sutin, Nonlinear ElasticWave spectroscopy (NEWS) techniques to discern material damage, part I: nonlinear wave modulation spectroscopy (NWMS). Res. Nondestr. Eval. **12**, 17–30 (2000)

293. A. Van Pamel, G. Sha, M.J.S. Lowe, S.I. Rokhlin, Numerical and analytic modelling of elastodynamic scattering within polycrystalline materials. J. Acoust. Soc. Am. **143**, 2394–2408 (2018)
294. A. Van Pamel, G. Sha, S.I. Rokhlin, M.J.S. Lowe, Finite-element modelling of elastic wave propagation and scattering within heterogeneous media, in *Proceedings of the Royal Society A—Mathematical Physical and Engineering Sciences*, 473 (2017)
295. I.A. Victorov, *Rayleigh and Lamb Waves* (Springer, New York, NY, 1967)
296. A. Viswanath, A. Kumar, T. Jayakumar, B.P.C. Rao, Ultrasonic characterization of microstructural changes in Ti-10V-4.5Fe-1.5Al beta-titanium alloy. Metall. Mater. Trans. A Phys. Metall. Mater. Sci. **46A**, 3393–3405 (2015)
297. C.M. Vlad, Verfahren zur Ermittlung der Anisotropie-Kennzahlen für das Tiefziehverhaltens kohlenstoffarmer Feinbleche. Materialprüfung. **14**, 179–182 (1972)
298. M. Voigt, *Lehrbuch der Kristallphysik* (Springer Fachmedien, Wiesbaden, 1966)
299. J.W. Wagner, Optical detection of ultrasound, in *Physical Acoustics*, eds. by R.N. Thurston, A.D. Pierce (Academic Press Inc., New York, NY, 1990), pp. 201–264
300. A.J. Walton, *Three Phases of Matter* (Oxford University Press, New York, NY, 1983)
301. L.M. Wang et al., Ultrasonic investigation of $Pd_{39}Ni_{10}Cu_{30}P_{21}$ bulk metallic glass upon crystallization. Appl. Phys. Lett. **77**, 1147–1149 (2000)
302. W.D. Wang, Inspection of refinery vessels for hydrogen attack using ultrasonic techniques, in *Review of Progress in Quantitative Nondestructive Evaluation*, vol. 12B, eds. by D.O. Thompson, D.E. Chimenti (Springer, 1993), pp. 1645–1652
303. K.R. Waters, M.S. Hughes, G.H. Brandenburger, J.G. Miller, On a time-domain representation of the Kramers–Kronig dispersion relations. J. Acoust. Soc. Am. **108**, 2114–2119 (2000)
304. R.L. Weaver, Anderson localization of ultrasound. Wave Motion **12**, 129–142 (1990)
305. R.L. Weaver, Diffusivity of ultrasound in polcrystals. J. Mech. Phys. Solids **38**, 55–86 (1990)
306. H. Weber, Development of creep damages on heatresistant ferritic steel, in *Nondestructive Characterization of Materials*. ed. by P. Höller, V. Hauk, G. Dobmann, C. Ruud, R.E. Green (Springer, Saarbrücken, Germany, 1989), pp.474–485
307. H. Willems, A new method for the measurement of ultrasonic absorption, in *Review of Progress in Quantitative Nondestructive Evaluation*, vol. 6A, eds. by D.O. Thompson, D.E. Chimenti (Springer, 1987), pp. 473–481
308. H. Willems, Assessment of creep damage by NDT, in *Materials Research Society Symposium*, vol. 142, ed. by M. R. Society (Boston, 1989), pp. 165–170
309. H. Willems, G. Dobmann, Early detection of creep damage by ultrasonic and electromagnetic techniques. Nucl. Eng. Des. **128**, 139–149 (1991)
310. H. Willems, K. Goebbels, Characterisation of microstructure by backscattered ultrasonic waves. Metal Sci. **15**, 549–553 (1981)
311. A. Wolfenden, *Dynamic Elastic Modulus Measurements in Materials* (American Society for Testing and Materials, Philadelphia, PA 19103, USA, 1990)
312. F. Wooten, *Optical Properties of Solids* (Academic Press, New York, 1972)
313. B.Q. Xu, J. Feng, G.D. Xu, J.J. Wang, H.X. Sun, G.R. Cao, Laser-generated thermoelastic acoustic sources and Lamb waves in anisotropic plates. Appl. Phys. A Mater. Sci. Process. **91**, 173–179 (2008)
314. K. Yamanaka, T. Tsuji, A. Noguchi, T. Koike, T. Mihara, Nanoscale elasticity measurement with in situ tip shape estimation in atomic force microscopy. Rev. Sci. Instrum. **71**, 2403–2408 (2000)
315. L. Yang, J. Li, S.I. Rokhlin, Ultrasonic scattering in polycrystals with orientation clusters of orthorhombic crystallites. Wave Motion **50**, 1283–1302 (2013)
316. L. Yang, O.I. Lobkis, S.I. Rokhlin, Explicit model for ultrasonic attenuation in equiaxial hexagonal polycrystalline materials. Ultrasonics **51**, 303–309 (2011)
317. L. Yang, O.I. Lobkis, S.I. Rokhlin, Shape effect of elongated grains on ultrasonic attenuation in polycrystalline materials. Ultrasonics **51**, 697–708 (2011)
318. L. Yang, O.I. Lobkis, S.I. Rokhlin, An integrated model for ultrasonic wave propagation and scattering in a polycrystalline medium with elongated hexagonal grains. Wave Motion **49**, 544–560 (2012)

319. L. Yang, S.I. Rokhlin, On comparison of experiment and theory for ultrasonic attenuation in polycrystalline niobium. J. Nondestr. Eval. **31**, 77–79 (2012)
320. L. Yang, S.I. Rokhlin, Ultrasonic backscattering in cubic polycrystals with ellipsoidal grains and texture. J. Nondestr. Eval. **32**, 142–155 (2013)
321. C.F. Ying, Photoelastic visualization and theoretical analyses of ultrasonic pulses in solids, in *Physical Acoustics*, eds. by R.N. Thurston, A.D. Pierce (Academic Press, Inc., Elsevier B.V., 1990), pp. 291–342
322. W.T. Yost, J.H. Cantrell, M.A. Breazeale, Ultrasonic nonlinearity parameters and 3rd order elastic constants of copper between 300 K and 3 K. J. Appl. Phys. **52**, 126–128 (1981)
323. L. Yu, R.B. Thompson, F.J. Margetan, The spatial correlation of backscattered ultrasonic grain noise: theory and experimental validation. IEEE Trans. Ultrason. Ferroelectr. Freq. Control **57**, 363–378 (2010)
324. P.A. Yuya, D.C. Hurley, J.A. Turner, Relationship between Q-factor and sample damping for contact resonance atomic force microscope measurement of viscoelastic properties. J. Appl. Phys. **109**, 113528–113531 (2011)
325. V. Zaitsev, V. Gusev, B. Castagnede, Luxemburg-Gorky effect retooled for elastic waves: a mechanism and experimental evidence. Phys. Rev. Lett. **89** (2002)
326. D. Zellouf, Y. Jayet, N. SaintPierre, J. Tatibouet, J.C. Baboux, Ultrasonic spectroscopy in polymeric materials. Application of the Kramers–Kronig relations. J. Appl. Phys. **80**, 2728–2732 (1996)
327. A. Zinck, S. Krishnaswamy, Ultrasonic nonlinearity measurements on rolled polycrystalline copper, in *Review of Progress in Quantitative Nondestructive Evaluation*, vol. 29, ed. by D.O. Thompson, D.E. Chimenti (American Institute of Physics, 2009), pp.1404–1409
328. T. Lüthi, Determination of Biaxial and Triaxial Stress Distributions using Ultrasonics. NDT International. **23**, 351–356 (1990). https://doi.org/10.1016/0308-9126(90)90861-h
329. C. Esling, E. Bechler-Ferry, H.J. Bunge, Three-dimensional texture analysis after Bunge and Roe: Correspondence between the respective mathematical techniques. Textur. Microstruct. **5**, 95–125 (1982)

Chapter 2
Non-destructive Materials Characterization using Ionizing Radiation

Abstract This chapter discusses the physical principles and the applications of non-destructive materials characterization (NDMC) techniques using X-rays, neutrons, electrons, and positrons. Various means of generation and detection of X-rays are presented followed by the basics of their interaction with materials. These interactions lead to attenuation and diffraction, which are elucidated by examples from different metallurgical and material science applications of NDMC. The basics of X-ray diffraction and their applications for the measurement of elastic stresses at different length scales and of textures are discussed as well. The specific differences of the interaction of neutrons with materials as compared to X-rays in radiography and scattering studies are worked out. Applications of electron and positron annihilation for NDMC are also discussed.

2.1 NDMC using X-rays

X-ray radiography is the most commonly used NDT technique for the inspection of materials and components in order to detect volumetric defects. In this NDT method, the differential absorption property of X-rays is utilized. This chapter deals with this property of X-rays for characterizing textures, stresses, and microstructures. In addition, advanced X-ray imaging techniques based on the differential absorption properties of materials are also discussed.

2.2 Properties of X- and γ-rays

There are numerous textbooks explaining the basics of the generation of X- and γ-rays and their interaction with matter (Fig. 2.1) and therefore the reader may use the references given here or others for deeper studies. We only present the main facts required for understanding NDMC using X- and γ-rays.

W. Arnold et al., *Non-destructive Materials Characterization and Evaluation*, Springer Series in Materials Science 329,
https://doi.org/10.1007/978-3-662-66489-6_2

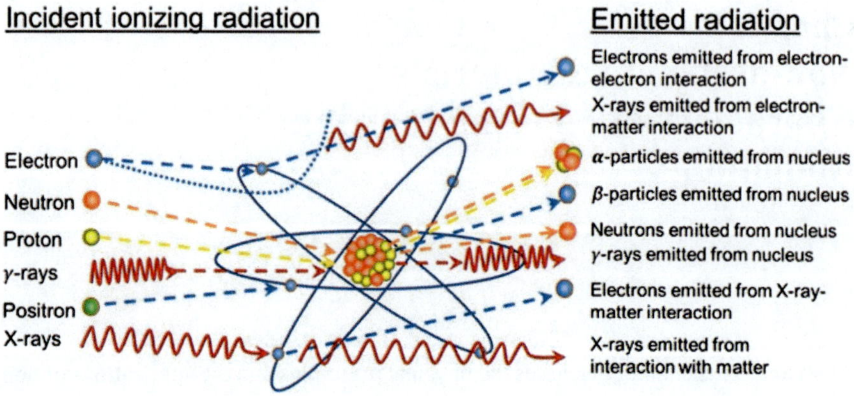

Fig. 2.1 Interaction of ionizing radiations with matter

2.2.1 *Electromagnetic Spectrum and Energy of X- and γ-rays*

X-rays are electromagnetic waves with wavelengths ranging from about 10 nm to 4 pm. Their energies span usually from about 100 eV to 300 keV. Like X-rays, gamma rays (also designated as γ-rays), are electromagnetic waves with energies beyond about 200 keV. The difference between X-rays and γ-rays is their source of generation. While the generation of X-rays involves interactions with electrons, γ-rays are generated by nuclear reaction. The relations between energy E and frequency f of photons, the quanta of electromagnetic waves, are given by

$$E_x = hf = \frac{hc}{\lambda} \tag{2.1}$$

with $h = 6.626 \times 10^{-34}$ [Js] as the Planck constant and the light velocity $c = 2.998 \times 10^8$ [m/s]. The relation between wavelength λ, frequency f, and velocity c is given by

$$c = \lambda f \tag{2.2}$$

An energy of 100 eV corresponds to a frequency of $f \approx 2.5 \times 10^{16}$ Hz and 300 keV to a frequency of 7.2×10^{19} Hz. The corresponding wavelengths are 10 nm and 4 pm, respectively.

2.2.2 *Sources of X- and γ-rays*

Until recently, for radiography and X-ray diffraction experiments, X-ray beams emanating from an X-ray tube were dominantly used. Since about two decades, various X-ray sources were invented and developed into a viable technology in order to obtain higher X-ray fluxes and brilliances. These are dominantly liquid metal targets, laser wake-field generators, linear accelerators, Betatrons, and synchrotrons. We describe them shortly in what follows, respectively in their applications.

2.2.2.1 X-ray Tubes

In X-ray tubes, electrons from a cathode material are accelerated by a voltage to energies in the kV range and hit an anode target material (Fig. 2.2a). The energy must be high enough to remove electrons from the inner shells of the target material atoms. The X-rays emanate from the target into all directions.

The shells are re-occupied by recombination with electrons from the outer shells producing X-ray quanta with an energy characteristic for the target material, see Fig. 2.2b [1, 5–7]. For example, the K_α -line is caused by a transition from the L-shell ($n = 2$) to the K-shell ($n = 1$), the K_β -line by a transition from the M-shell ($n = 3$) to the K-shell, and the K_γ-line by a transition from the N-shell ($n = 4$) to the K-shell. The lines have a fine structure due to spin–orbit coupling and other effects which we do not discuss (Fig. 2.2b).

The energy $E_x(\lambda)$ of the emitted X-rays is proportional to the square of the atomic number Z, known as Moseley's law [7, 8]:

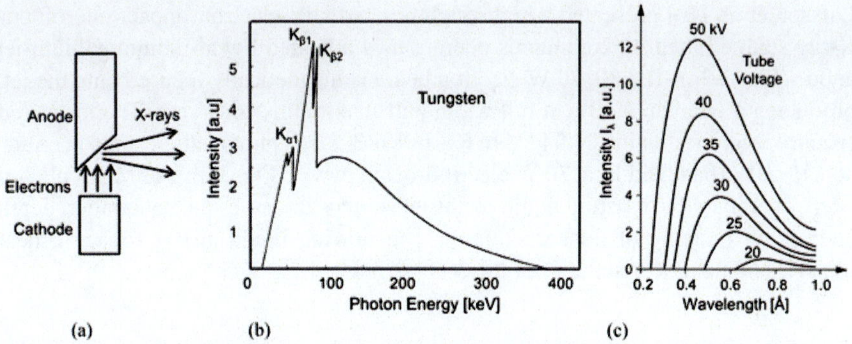

Fig. 2.2 **a** Operation principle of an X-ray tube; **b** Experimental spectrum for tungsten at U = 400 keV with 4 mm aluminum filtration, also showing the characteristic lines $K_{\alpha 1}$, $K_{\beta 1}$, and $K_{\beta 2}$ from the target material; **c** Bremsstrahlung spectrum from a tungsten target at various tube voltages. Graphs like in **b**, **c** can be found in many references and textbooks, see for example [1, 2] (**b c** in modified form from [3, 4], with permission from Springer Nature and the American Physical Society)

$$E_x \sim Z^2 \tag{2.3}$$

The electric field of the target nuclei leads to a negative acceleration of the penetrating electrons, which in turn produces the so-called bremsstrahlung radiation. The shape of this continuum or "white" spectrum (Fig. 2.1c) is approximately described by Kramer's law [9]. Furthermore, the bremsstrahlung exhibits an upper energy $E_{x,\,\max}$ and hence a minimum wavelength which is determined by (2.1):

$$E_{x,\max} = eU = hf_{\max} = \frac{hc}{\lambda_{\min}} \tag{2.4}$$

where λ_{min} is the minimum wavelength, and e is the electron charge (1.602×10^{-19} As). The minimal wavelength shifts to smaller values for larger tube voltages (Fig. 2.2c).

In regard to applications, the X-ray beam must possess sufficient high energy and intensity so that the component thicknesses can be penetrated. This depends primarily on the attenuation of the material under test which is a function of the X-ray energy, see Sect. 2.2.4, but also on the minimum intensity necessary to operate with a given signal-to-noise ratio when penetrating thick components. Another requirement is to attain high-resolution in imaging. This requires small emission spots for X-rays, which in turn leads to high temperatures of the target anode material of the X-ray tube and to instable operation. Therefore, there is a permanent effort to improve X-ray tubes and to develop technologies for other types of X-ray sources.

2.2.2.2 Liquid Metal-Jet X-ray Sources

Larsson et al. [10] presented a high-brightness 24 keV electron-impact microfocus X-ray source based on continuous operation of a heated liquid-indium/gallium-jet anode (Fig. 2.3a). The 30–70 W electron beam is magnetically focused onto the jet, producing a circular 7–13 μm full width half maximum X-ray spot. The measured spectral brightness at the 24.2 keV In $K\alpha$-line is 3×10^9 photons/(s × mm^2 × mrad^2 at 0.1% BW (bandwidth) at 30 W electron-beam power (Fig. 2.3b). The high photon energy compared to existing liquid metal-jet sources increases the penetration depth and allows imaging of thicker samples. The flowing liquid allows for rapid heat transfer and hence for stability of the X-ray source.

2.2.2.3 Synchrotron Radiation

Synchrotrons provide high brilliance and partially coherent X-rays for radiography and computed tomography, allowing one to image micro-structure and multicomponent samples with a resolution down to the sub-micrometer range. This is of high interest for solid state physics, material science, life science, and other fields. There are also efforts towards non-destructive materials characterization on actual

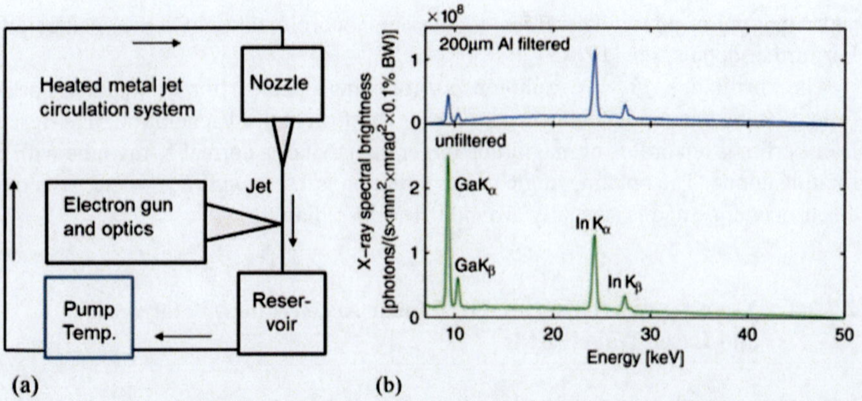

Fig. 2.3 **a** Experimental arrangement showing the heated In/Ga jet for X-ray generation; **b** Calibrated emission spectrum recorded at 30 W electron power with a 7 × 7 μm^2 X-ray spot. Lower curve: unfiltered spectrum. Upper curve: 200 μm thick Al hardened spectrum (according to [10] and with the permission of AIP Publishing)

components which are discussed in Sect. 2.3.5. There are several facilities worldwide at various locations for example at the ESRF in Grenoble, France, or at the BESSY in Berlin, Germany, at the Raja Ramanna Centre for Advanced Technology, Indore, India, and at the Argonne National Laboratory, USA.

Together with the availability of sensitive flat-panel X-ray detectors, the X-ray sources of synchrotrons contributed a lot to the development of X-ray imaging with a resolution down to some tens of nanometers. The high flux of X-rays in synchrotrons allows more projections to be taken in micro-computed tomography (μCT) in a given period with a significant increase in speed of 3D reconstruction. These imaging capabilities comprise X-ray microscopy (see Sect. 2.3.6 as well as 2.3.2). There are now commercially available instruments which may use either a conventional X-ray [11, 12] or a synchrotron X-ray beam as a source for imaging, compensating that synchrotron radiation sources will not become commonplace in the foreseeable future due to their size and cost.

In a synchrotron an electron beam is forced on a circular path by strong magnetic fields. The ensuing radial acceleration of the electron charges leads to electromagnetic radiation across a wide spectrum. There are two cases to distinguish: classical radiation when the velocity of the electrons is small compared to the light velocity and relativistic radiation when the electron velocity gets close to the light velocity. These cases are discussed in many textbooks of electromagnetism, see for example [13]. In the electron beam of a synchrotron, there are so-called insertion devices: wigglers and undulators. These devices are magnets with alternating magnetic fields along the beam path. They force the electron beam to an oscillatory motion and therefore to a transverse acceleration leading to additional radiation. In case of an undulator the distance between magnets is such that the radiation from each magnetic sector superimposes in phase. Therefore, a quasi-monochromatic radiation is generated. For further details, see [14].

The brilliance B_b = (photons/s)/((opening angle [mrad]^2)(source area [mm^2])(0.1% bandwidth)) of a beam of a synchrotron of third generation, like Bessy II in Berlin, is ten orders of magnitude larger than that of a normal X-ray tube with a rotating anode. The opening angle of the radiation is $1/\gamma = mc^2/E_e$, where E_e is the electron energy, and is typically around 0.1 milli-radian.

2.2.2.4 X- and γ-rays Generated by Linear Accelerators, Betatrons, and Laser Wake-Fields

Some devices, which were developed to study the properties of nuclei and nuclei interactions, are now also used for NDT and NDMC:

(i) A portable linear accelerator as a high energy X-ray source in the 1.5 to 9 MeV range was described by [15, 16]. Energy and power levels in the current generation of portable systems are suitable for radiography of very thick sections of steel, masonry, concrete and other construction materials, and hence is a tool for civil engineering applications. It allows to penetrate 75 mm to 400 mm steel in a maximum exposure time of 3 h. The radiation output is 3 to 300 R/min at 1 m distance.

(ii) Tomsk Polytechnic University developed a number of Betatron systems with Bremsstrahlung radiation having peak energies from 1–9 MeV. The radiation peak dose rate is respectively 2, 4, 7.5, or 20 cGy/min at 1 m distance. Betatrons produce narrow beams for small target sizes, which improves the resolution of the X-ray system [17]. The narrow-beam Betatron radiation is easy to collimate, so that spurious background scattered radiation is much lower than in systems based on other sources. These systems are used in inspection systems allowing to penetrate up to 250 mm steel at scanning speed of 3 m/s with a resolution of 3 mm [18]. For example, they are used to control cargo transported by road, by railway, and by river and in custom inspections.

(iii) In the last two decades, X-ray sources based on the so-called laser wake-field acceleration of electrons were developed by various groups, see for example [19–21]. A high-intensity short laser pulse with an intensity of 10^{19} W/cm^2 and larger is incident on a gas, which ionizes instantly and forms a plasma. The radiation pressure of the laser pulse creates an electron-depleted cavity. The plasma electrons trapped inside experience both longitudinal and radially focusing electric fields and oscillate transversely while accelerated in forward direction. These so-called Betatron oscillations cause the electrons to emit synchrotron-like radiation confined to a narrow cone in the same direction. The radiation is characterized by a peak E_{crit}, presently of the order of some 10 keV, in the spectrum. The Betatron oscillation radius determines the X-ray source size and can be as low as 1 μm, comparable to or even exceeding the state-of-the-art in conventional μ-CT sources, see Sect. 2.2.2.1.

Concerning the units [7]: 1 Röntgen (R) is the radiation dose which produces in 1 cm^3 of air ($m_{air} \approx 1.293 \times 10^{-6}$ kg) under normal conditions one electrostatic charge unit (1 esu $\approx 3.33564 \times 10^{-10}$ C) of electrons and ions. This means that 1 R produces 2.58×10^{-4} C/kg. Because the energy needed to produce one electron–ion pair is W = 34 eV $\triangleq 54.4 \times 10^{-19}$ J, and the charge on an electron is 4.8×10^{-10} esu, 1 R = N × W/m_{air} (where N = number of electron pairs in 1 esu charge = 2.1×10^9), corresponding to 8.8 mJ/kg or 0.88 cGy. One Gy is equal to an energy of 1 J deposited in 1 kg mass, hence 1 Gy = 1 J/kg.

2.2.2.5 Isotope γ-ray Sources

Gamma rays of discrete energies are emitted from isotope disintegration processes [1, 22] and serve a source in NDT and NDMC. Because isotopes have characteristic nuclear energy levels, the isotopes radiate characteristic lines and hence energies. This entails that γ-ray energies will remain constant for a particular isotope, but its intensity will decay with time. The decay process is governed by the equation

$$N = N_0 e^{-\lambda t} \tag{2.5}$$

where λ is the decay rate. If the number of atoms at the time $t = 0$ is equal to N_0, then the number N of atoms present at a later time t is given by 2.5. At the so-called half-life period $T_{1/2}$, one-half of the isotope material is disintegrated.

$$T_{1/2} = \frac{ln2}{\lambda} \tag{2.6}$$

Since the intensity $I(t)$ of the radiation is proportional to the number of atoms present, (2.5) may be written

$$I = I_0 e^{-\lambda t} \tag{2.7}$$

Gamma ray sources for radiography applications are assembled from pellets of a given source material, having some mm in diameter and some tens of a millimeter thickness. The activity is introduced by irradiating the source material to a high energy neutron source, for example a nuclear reactor or a linear accelerator. Table 2.1 lists the characteristics of four isotopes most commonly used in NDE.

Table 2.1 Common gamma ray isotope sources. Each of the isotopes possesses characteristics suitable for certain applications in NDMC. The dose rate ($=\Gamma A_s/d^2$) at distance d can be calculated from the Gamma factor (Γ) and the source activity (A_s). One Becquerel is the rate of one nuclear transmutation per second. A ^{60}Co source with a strength of 740 GBq has an intensity of 260 mGy/h at 1 m distance

Element	Cobalt	Iridium	Cesium	Thulium
Isotope	^{60}Co	^{192}Ir	^{137}Cs	^{170}Tm
Half-life $T_{1/2}$	5.27 y	74.3 d	30.1 y	129 d
Energy [MeV]	1.17; 1.33	0.31; 0.47; 0.60	0.66	0.052; 0.084
Optimum steel thickness [mm] for radiography inspection	50–150	10–70	50–100	2.5–12.5
Gamma factor (mGy/h from 1 GBq source activity at 1 m distance)	0.351	0.13	0.081	0.034
Activity [Ci] ([Bq])	20 (7.4×10^{11})	100 (3.7×10^{12})	75 (2.78×10^{12})	50 (1.8×10^{12})

2.2.3 X-ray Detectors

The processes which lead to the attenuation of X-rays, described in Sect. 2.2.4, are the excitation of electrons in their shells. These processes are used in detectors to monitor X-rays. The output signal of the detector can be produced by primary ionizations, thermal changes or scintillations that are converted into electric signals [23]. Usually, the amplitude of the output electric signal is close to a value proportional to the energy of the X-ray photon which hits the detector. The quantum processes of the interaction of X-rays with materials require the counting of events until a sufficient level of significance is obtained which manifests itself in a corresponding signal-to-noise ratio. For non-destructive materials characterization, the detector employed must enable an energy resolution sufficient for imaging, for materials sorting, for the determination of the composition of alloys, as well as for the detection of the diffraction spots due to the lattice structure.

Until recently, films were the standard recording medium for X-ray imaging and storage. Resolution, sensitivity, and artefacts for films are discussed in many textbooks of non-destructive testing and medical imaging, see for example [1]. The first available digital detectors were phosphor plates in which the X-ray photons generated quasi permanent charges proportional to their intensity. The charge distribution was read out by a laser coupled to photosensitive detectors like photomultipliers. Driven by X-ray computed tomography, films were eventually replaced by digital imaging detectors. They are now ubiquitously used, also in the applications of X-ray techniques for NDMC discussed in this book. They allow digital picture archiving and communication systems with anytime availability, hence increasing overall operator throughput and efficiency. An introduction to their working principles is given by [24].

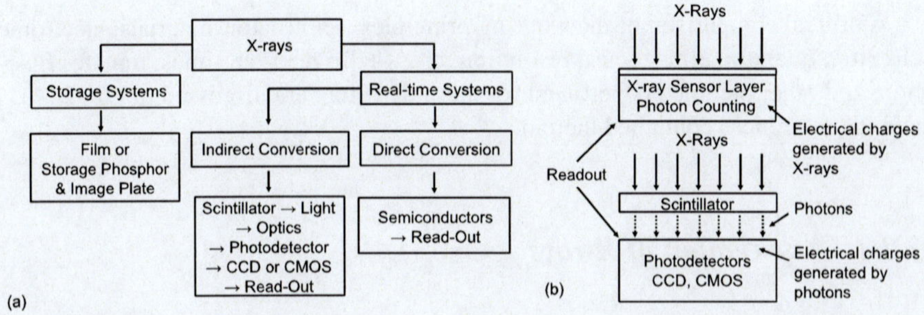

Fig. 2.4 **a** Schematic operational principle of X-ray detectors; **b** Schematic principle of direct and indirect conversion detectors

There are storage and real-time X-ray detection systems, as depicted schematically in Figs. 2.4a and b. The real-time systems can be subdivided into systems which directly convert the X-ray into electrical charges in semiconductors which are read out electronically, and indirect conversion systems using scintillators for example.

In the indirect detection systems, the X-rays first generate light photons in a scintillator (also called phosphor). The optical photons produce electron–hole pairs, whose charges are then read out by charged coupled detectors (CCD), or by an array of semiconductor detectors (CMOS technique).

In CCD detectors, the X-ray photons converted by the scintillator into optical photons are guided by fiber-optics to the CCD device. At the end of the image recording, charges stored in potential wells are transferred from well to well before being amplified and finally digitized by an analog–digital converter. The recorded image is a histogram of numerical values, one for each pixel, which are proportional to the number of X-ray photons arriving on each pixel.

In a CMOS detector, now the state-of-the art, the principle of photon detection and the charge accumulation in a potential well are the same as in the CCD detector with very short fiber-optics. The main difference is that the additional complementary metal oxide semiconductor (CMOS) electronics are attached at each potential well. They perform the conversion of the accumulated charges of each pixel into a voltage signal which is amplified accordingly. It also allows for each pixel to be read without the need for a charge transfer like in the CCD detector.

Photon counting detectors are based on the direct detection and individual counting of X-ray photons and can be considered as an assembly of thousands of independent point detectors for each pixel. The incident X-ray photon is absorbed in a sensor layer (Si, CdTe, GaAs) which generates electron–hole pairs, i.e. a current. It allows for the conversion of the accumulated charges into a voltage and the amplification of the signal at each pixel, which is hybridized to the electronics that count the photons one by one. These detectors are also called hybrid pixel detector, photon counting detector, or hybrid photon counting detector.

A critical comparison of the working principles, scintillator materials, electronic circuitry, quantum efficiencies, resolution, cross-talk, read-out times, transfer function, and other parameters pertinent for these detectors are discussed in [23, 25–27] and the references contained therein.

2.2.4 *Attenuation of X-rays*

The interaction of X-rays with matter is caused by various processes. These are Rayleigh-scattering (rs), photoelectric effect (pe), Compton scattering (cs), and pair-production (pp). They add up to a total attenuation coefficient μ. The intensity of X-rays decreases exponentially with the path x through the material [22]:

$$I(x) = I_0 e^{-\mu x} \tag{2.8}$$

where I_0 is the incident intensity, x is the X-ray path length in the material, and μ is the linear attenuation coefficient (usually given in cm^{-1}). If the attenuation is viewed as being caused by scattering processes at the individual atoms, the attenuation coefficient can be written as:

$$\mu = n\sigma \tag{2.9}$$

where σ is the cross-section of the scattering process concerned, and n is the number of atoms per volume. The dimension of σ is that of an area. It represents an assumed geometrical cross-section or collision area for the attenuation process which is not the geometrical cross-section of the atoms, see Fig. 2.5 (in Chapter 1 on ultrasound, Sect. 1.17.2 a similar definition was used for the scattering cross-section for ultrasonic waves at spherical discontinuities, however based on the true geometrical cross-section).

A quantity often used is the mass attenuation coefficient μ_m which is defined as:

$$\mu_m = \frac{\mu}{\rho} = \frac{N\sigma}{A} \tag{2.10}$$

where ρ is the density, A is the mass number, i.e. the number of protons and neutrons in the nuclei, and N is Avogadro's number. N is related to the number of atoms per unit volume n by $N = nA/\rho$. Thus, the linear attenuation coefficient can also be written as

$$\mu = \frac{N\sigma\rho}{A} \tag{2.11}$$

Photoelectric absorption (μ_{pe}) is the dominant attenuation process for lower energies (the limit depends on the atomic number of the element) and results from a

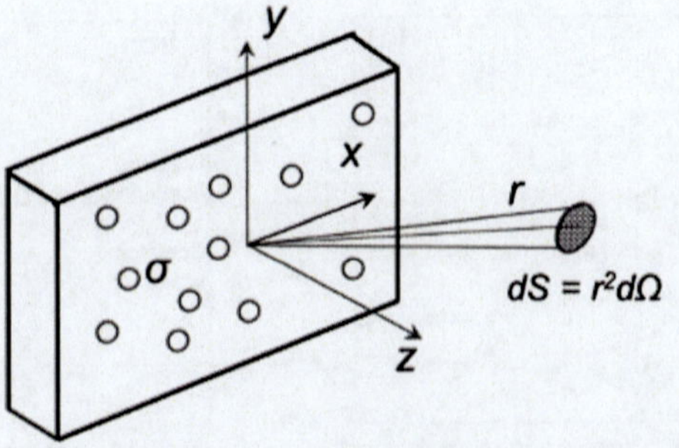

Fig. 2.5 Scattering of X-ray or gamma rays by the individual nuclei in a material. It is viewed as being caused by the nuclei, as having a geometrical cross-section σ with the dimension of an area. The absorption coefficient is then given by (2.9), where n is the number of scatterers/volume. The differential cross-section $dS/d\Omega$ is a measure of how many photons are scattered into the angle subtended by $d\Omega(\varphi,\theta)$, where φ is the azimuth and θ is the polar angle. For further details see for example [28]

process removing one electron from an inner atomic shell. This shell will be filled up again by electrons from outer shells accompanied by fluorescence radiation. The corresponding cross-section for this process is [29]:

$$\sigma_{pe} = \left(\frac{32}{\varepsilon_x^7}\right)^{0.5} \alpha^4 Z^5 \sigma_{Th}^e \tag{2.12}$$

where Z is the atomic number, $\varepsilon_x = E_x/m_e c^2$, is the reduced energy, m_e is the electron mass, $\alpha = e^2/(4\pi\varepsilon_0\hbar c)$ is the Sommerfeld fine structure constant, E_x is the X-ray energy, and σ_{Th}^e is the Thompson cross-section for elastic scattering:

$$\sigma_{Th}^e = \frac{8}{3}\pi r_e^2 = 6.65 \cdot 10^{-25}\ [\text{cm}^2/\text{atom}] \tag{2.13}$$

where $r_e = e^2/8\pi\varepsilon_0 m_e c^2$ is the classical electron radius.

Rayleigh scattering, also called classical scattering, is a process by which photons are scattered by bound electrons and in which the atoms are neither ionized nor excited. The scattering from different parts of the atom's charge distribution is thus "coherent", entailing interference effects. This process occurs mostly at low energies and for high Z materials, in the same region where electron binding effects influence the Compton-scattering cross-section. It is necessary to consider the charge distribution of all electrons. This can be done through the use of an "atomic form factor", $F(q, Z)$, based on the Thomas–Fermi, Hartree, or other models. The square of the form factor, $(F(q, Z))^2$, is the probability that the Z electrons of an atom take up the recoil momentum q, without absorbing any energy. The photons are dominantly scattered in forward direction. The differential cross-section is $\propto \left(1 + cos^2\theta\right)/2$, where θ is the scattering angle (Fig. 2.5). The total cross-section is given by [29, 30]:

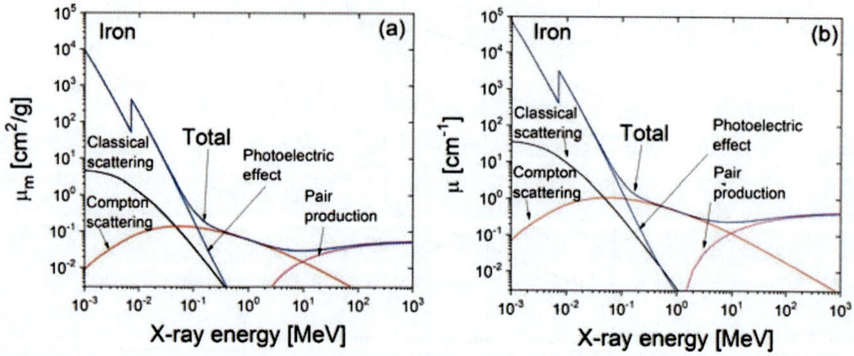

Fig. 2.6 **a** X-ray mass absorption coefficient μ_m as a function of X-ray energy (2.10) and **b** X-ray absorption coefficient μ (2.7) as a function of X-ray energy, both for iron. The characteristic absorption lines of iron are also seen (data shown in the figures are from [31]

$$\sigma_{rs} = nF(q,Z)^2\sigma_{Th}^e = \frac{NZ\rho}{A}F(q,Z)^2\sigma_{Th}^e \tag{2.14}$$

Rayleigh scattering represents only a minor part of the absorption. It decays rapidly with energy, see Fig. 2.6.

The Compton scattering loss, $\mu_{c,}$ sets an electron free from the outer shell, whose binding energy is small compared to the X-ray energy $\hbar\omega$. The cross-section per electron is given by [29]:

$$\sigma_{cs}^e = 2\pi r_e^2 \left(\begin{array}{l} \left(\frac{1+\varepsilon_x}{\varepsilon_x^2}\right)\left(\frac{2(1+\varepsilon_x)}{1+2\varepsilon_x} - \frac{1}{\varepsilon_x} ln(1+2\varepsilon_x)\right) \\ + \frac{ln(1+2\varepsilon_x)}{2\varepsilon_x} - \frac{1+3\varepsilon_x}{(1+2\varepsilon_x)^2} \end{array} \right) [\text{cm}^2/\text{electron}] \tag{2.15}$$

Like above, $\varepsilon_x = E_x/m_e c^2$ is the reduced energy. The cross-section per atom is obtained by multiplying (2.15) with the number of the Z electrons in an atom. Like in (2.14), the total cross-section is given by [29]:

$$\sigma_{cs} = n\sigma_{cs}^e = \frac{N\,Z\rho}{A}\sigma_{cs}^e \tag{2.16}$$

In X-ray imaging, Compton scattering contributes considerably to the background radiation reducing the images contrast.

Finally, σ_{pp} is the cross-section for pair production due to the creation of electron–positron pairs in the electric field of the nucleus:

$$\sigma_{pp} = 4\alpha r_e^2 Z^2 \left(\frac{7}{9}\ln 2\varepsilon_x - \frac{109}{54}\right) [\text{cm}^2/\text{atom}] \tag{2.17}$$

for ionized atoms. In case the nucleus charge is shielded, the value of the parenthesis is $((7/9)\ln(183/Z^{1/3}) - (1/54))$ [29]. Pair production absorption only occurs at energies above 1.02 MeV ($\triangleq 2\ m_ec^2$). The contribution becomes significant at energies $\geq$ 10 MeV. Due to the pair production, penetration beyond a particular thickness (~ 450 mm in steel) cannot be achieved even by increasing the X-ray energy further beyond about 10 MeV.

Because the attenuation is caused by processes in the individual isolated atoms, the total attenuation or linear attenuation coefficient μ can be written as the sum of four independent different absorption or scattering processes:

$$\mu = \mu_{pe} + \mu_{cs} + \mu_{rs} + \mu_{pp} \tag{2.18}$$

In Fig. 2.6, the absorption coefficient μ and the mass absorption coefficient μ_m are plotted as a function of X-ray energy E_x for iron. As one can see from the figures, the attenuation coefficient due to the photoelectric effect, $\mu_{pe,}$ and due to Rayleigh or classical scattering, μ_{rs}, are significant at lower energies, and become negligible above 0.3 MeV. Above this energy, the remaining attenuation processes is caused mainly by Compton scattering up to about 4 MeV and by pair production, μ_{pp}, beyond that. Due to the characteristic absorption edge at about 7 keV, as shown in Fig. 2.6, an iron foil can be effectively utilized to filter K_β (6.93 keV) radiations from the Co anode target to obtain a monochromatic K_α (7.649 keV) radiation beam to be used in X-ray diffraction studies.

The absorption coefficients μ and μ_m of X-rays for various metals, compounds and mixtures used in practice can be found in ([31] X-ray Mass Attenuation Coefficients | NIST (https://www.nist.gov/pml/X-ray-mass-attenuation-coefficients)). Figure 2.6 shows such data for iron. A related parameter is the penetration depth:

$$\delta = \frac{1}{2\mu} \tag{2.19}$$

For a vertical incident path, the intensity is reduced by 61% at a depth of δ. Two other definitions of penetration depths are in use: $\delta_{1/2} = \ln 2/\mu \approx 0.693/\mu$ and $\delta_{1/10} = \ln 10/\mu \approx 2.302/\mu$.

2.2.5 *Diffraction of X-rays*

Bragg's Law

In solids, X-rays are scattered elastically and inelastically by atoms, as discussed in the above section. The elastically scattered waves interfere with each other and due to the arrangement of the atoms in lattice planes, constructive interference occurs only in certain directions determined by the lattice distances, the X-ray wavelength, and the incident angle of the X-ray beam. The well-known Bragg relation yields the angle of constructive interference:

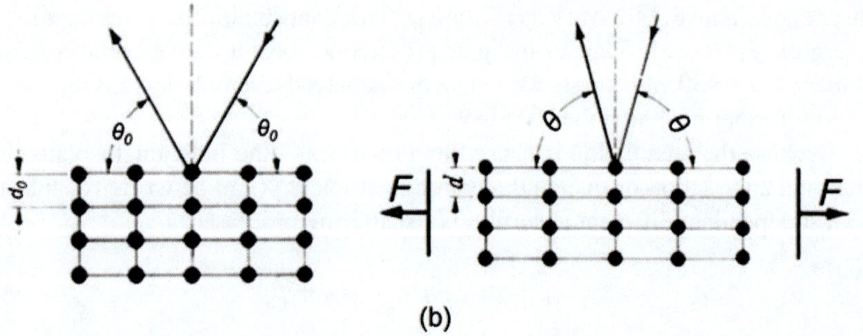

Fig. 2.7 **a** X-ray scattering at atoms in a crystalline lattice; **b** Change of the distance of lattice planes after application of a force *F*, which leads to strain in the lattice

$$n\lambda = 2d_0 sin(\theta_0) \tag{2.20}$$

Here, λ is the X-ray wavelength, d_0 is the distance between the interfering crystal planes, θ_0 is the incident angle, and *n* is an integer value, see Fig. 2.7a. More precisely, in order to maintain the constructive interference at the different lattice planes, the difference in path lengths must be an integer multiple of the X-ray wavelength. For a cubic crystal structure, this occurs when

$$sin\theta_{hkl} = \frac{n\lambda}{2d_{hkl}} = \frac{n\lambda}{2a_0}\sqrt{(h^2 + k^2 + l^2)} \tag{2.21}$$

Here, (*hkl*) are the Miller indices of the corresponding plane and a_0 is the lattice parameter. Equation (2.21) can be written in a more general form:

$$2d_{hkl} sin\theta_{hkl} = n\lambda \tag{2.22}$$

where d_{hkl} is the interplanar spacing of the ensemble of planes with the Miller indices (*hkl*), θ is the Bragg angle and *n* is the order of interference. For example, (222) is the second order ($n = 2$) of the (111) interference with 2θ the angle between the primary and the refracted beam. For a fixed wavelength λ, the 2θ positions are directly related to the interplanar spacings, see Fig. 2.8. Consequently, the measurement of the 2θ angles yield the *d*-spacings. If a force is applied to the lattice, strain develops which changes the lattice distances (Fig. 2.7b). They can be measured by X-ray diffraction, see Sect. 2.4.3.

The applications of Bragg's law is manyfold in NDMC, including the determination of the orientation of the single crystals in turbine blades, texture characterization, identification of microstructural constituents and their volume fractions and stress measurement. A few are discussed in detail in the following sections.

Laue Pattern and Debye–Scherrer Diagrams for Powder Diffraction

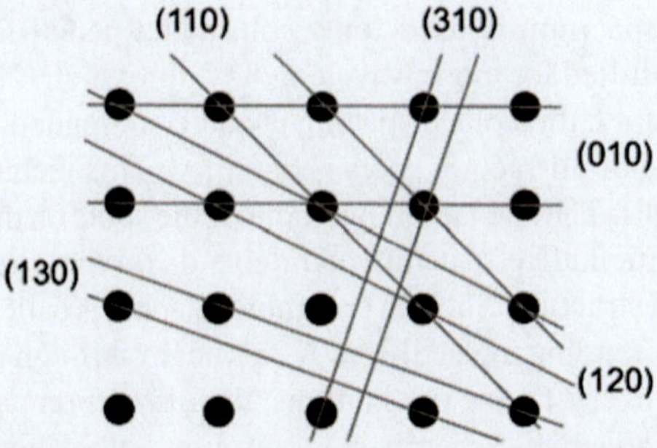

Fig. 2.8 Examples of crystalline planes of a cubic structure with their Miller indices in (001) plane cross-section

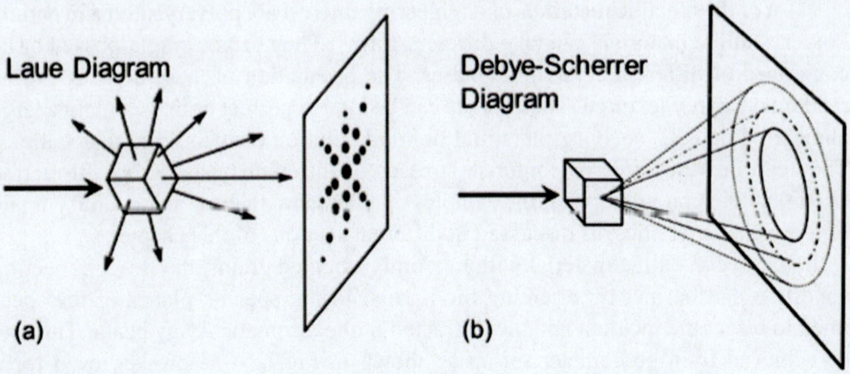

Fig. 2.9 **a** Laue diagram; **b** Debye–Scherrer diagram

Fig. 2.10 For X-ray diffraction measurements on a polycrystalline material, usually a monochromatic X-ray beam is used, with a goniometer arrangement to exploit Bragg reflection (2.20)

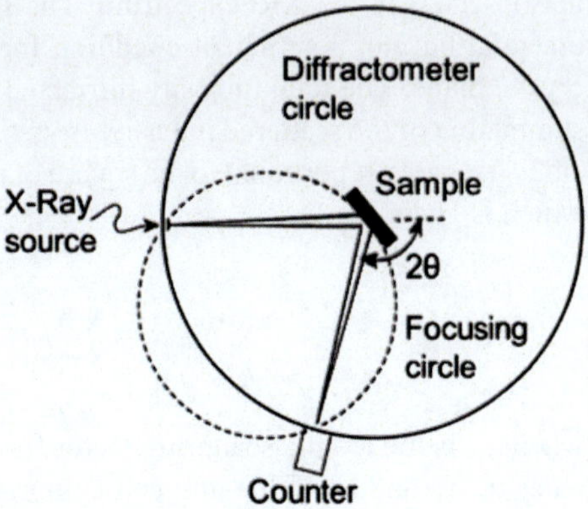

In case a white spectrum is used for a collimated pencil-like X-ray beam, the Bragg equation is fulfilled for those wavelengths λ for which (2.20) is obeyed by the right combination of a lattice plane spacing d and a subtended incident angle θ. As a result, an ensemble of diffraction spots is obtained. This technique is of interest to identify single crystals. The spatial arrangement of the spots on the recording medium is typical for a specific lattice structure and depends on the orientation of the beam relative to the crystal structure. Such an ensemble of spots is called a "Laue-diagram", see Fig. 2.9a. When sending a pencil-like X-ray beam through a powder containing particles with random crystalline orientations, the diffraction pattern shows circular rings. This is called a Debye–Scherrer-diagram, Fig. 2.9b. Knowing the wavelength of the X-ray beam, *d*-spacings can be calculated from the diameter of the rings.

X-ray Diffraction in Polycrystalline Materials

Most of the metallic materials of engineering interest are polycrystalline in nature. Polycrystalline materials can take different forms. They can be single-phased or be comprised of different crystalline phases. The orientation of these crystals can be random or highly textured. The crystals can be almost perfect or may contain a large number of defects, such as interstitial or solid solution elements and dislocations. They can be strain-free or comprise stresses of different types. X-ray diffraction (XRD) studies on polycrystalline samples enable us to analyze and quantify many of these characteristics, as discussed in different sections of this chapter.

In a polycrystalline material with randomly oriented grains, the Bragg reflection condition can be met by orienting the normal to the specific planes of the specimen to bisect the incident and the diffracted monochromatic X-ray beam. This can be achieved by a goniometer set-up as shown in Fig. 2.10. However, even for a polycrystalline material with grains much smaller than the X-ray beam, diffraction peaks of all possible crystallographic planes corresponding to all Bragg angles are not observed in the XRD spectrum. The Bragg condition defined by (2.22) is an essential but not a sufficient condition for diffraction peaks to appear for a given crystal plane. The total intensity diffracted by a crystal unit cell is described by the summation of the scattered intensity from the individual atoms. The diffracted intensities $I_{(hkl)}$ are proportional to the square of the crystallographic structure factor $F_{(hkl)}$ which is given by:

$$F_{(hkl)} = \sum_{1}^{N} f_n e^{2\pi i(hu_n + kv_n + lw_n)} \tag{2.23}$$

where f_n is the atomic scattering factor for an atom located at the fractional coordinates u_n, v_n and w_n in the unit cell comprising N atoms. The total scattering factor $F_{(hkl)}$ expresses both the amplitude and phase of the resultant wave scattered from an atom and its absolute value $|F|$ gives the amplitude of the resultant wave in terms of the amplitude of the wave scattered by a single atom. From (2.23), one can see that for a cubic structure with a single atom at the origin ($u = v = w = 0$), the structure factor is independent of the hkl values. Thus, reflections of all planes are expected for a simple cubic crystal. However, $F_{(hkl)}$ values for a few combinations of planes are zero for unit cells with additional atoms and thus reflections of these planes are not observed in the XRD spectrum. For example, $F_{(hkl)} = 0$ holds for BCC crystals with odd values of $h + k + l$ and for FCC crystals with mixed (odd and even) values of *hkl*. For an HCP crystal with lattice point coordinates such as (0 0 0) and (1/3 2/3 1/2), $F_{(hkl)} = 0$ when $h + 2\,k = 3n$ (n is an integer) and l is odd. This is visualized in Fig. 2.11 which shows typical XRD spectra of polycrystalline metals having the crystal structures FCC (Cu and Ni), BCC (Mo and Fe), and HCP (Zr and Ti). These data were simulated using the BRASS code provided by [32] (see also http://www.brass.uni-bremen.de). The CuK_α radiation with $\lambda = 0.154$ nm was used for the simulation. For obtaining the simulated spectra, an instrumental broadening and a background corresponding to an experimental system was incorporated. Reflections corresponding to the even values of $h + k + l$ and unmixed (only even or only odd) values of *hkl* can be observed for the BCC and the FCC crystals, respectively. In accordance with Bragg's law, an increase in the 2θ values for the different peaks, with decrease in the size of their unit cell (*a and c*) for the three considered crystal structures, can be clearly observed in Fig. 2.11. Further details can be found in textbooks for X-ray diffraction and a summary of the main facts in [33].

The total diffracted intensities at any 2θ position is given by:

$$I_{(hkl)} = K|F_{hkl}|^2 f_a e^{\frac{-B_t \sin^2\theta}{\lambda^2}} AL(\theta)P(\theta)m \tag{2.24}$$

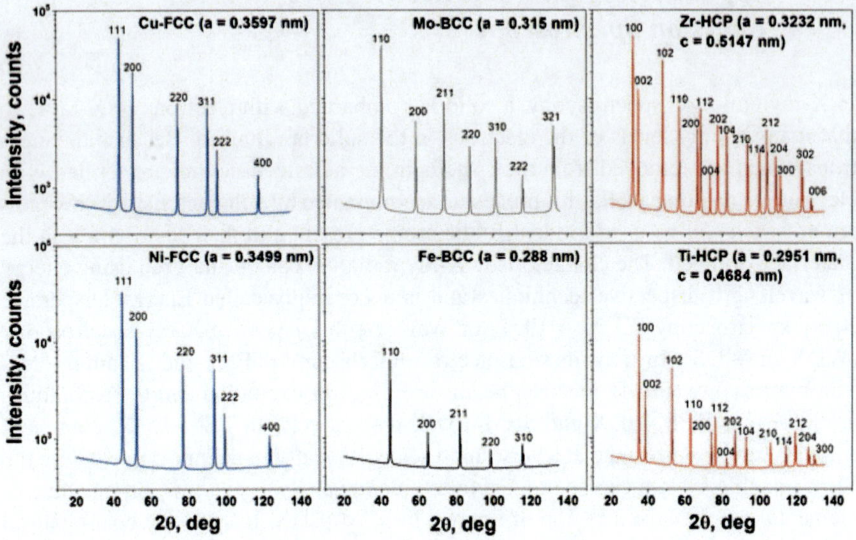

Fig. 2.11 Simulated XRD spectra for polycrystalline metals with BCC (Mo and Fe), FCC (Cu and Ni) and HCP (Zr and Ti) crystal structures

where, K is a constant independent of 2θ, $f_a \exp\left(-B_t \sin^2 \theta / \lambda^2\right)$ is the Debye–Waller factor describing the average displacement of atoms from their mean position due to temperature. The parameter B_t contains the average squared displacement of the atoms caused by the thermal agitation. Furthermore, A is the absorption factor, $L(\theta) = 1/\sin(2\theta)$ is the Lorentz factor, $P(\theta) = (1 + \cos^2(2\theta))/2$ is the polarization factor, and m is the multiplicity describing the number of equivalent planes that can diffract at a given Bragg angle [34].

A monochromatic X-ray beam for XRD applications is normally obtained by utilizing the characteristic radiations (K_α) of a specific anode target material (Fig. 2.2b) with a suitable excitation voltage just above that required for K_α excitation. Furthermore, filtering by foils of materials or crystal monochromators is performed in order to produce the monochromatic X-rays needed for sharp diffraction peaks. The lines $K_{\alpha 1}$ and $K_{\alpha 2}$ are sufficiently close in wavelength, such that a weighted average of the two is used. A list of several common target materials and corresponding wavelengths of their K_α and the K_β radiation, together with the minimum excitation potential and the appropriate filter material, is given in Table 2.2.

2.3 Applications of X-ray Interaction with Matter for NDMC

2.3.1 *Application of the Photoelectric Effect in X-ray Emission Spectroscopy*

In X-ray emission spectroscopy, a solid is bombarded with electrons or X-rays, so that many energy levels of the electrons in the solid are ionized. Because the electrons which are removed from their shells in the atomic structure, are refilled with electrons from other shells, this process is accompanied by a characteristic X-ray radiation, which can be used to identify the atoms present and their distribution in the material examined. The characteristic X-ray radiation can be analyzed using energy or wavelength dispersive techniques and is accordingly called Energy Dispersive X-ray spectroscopy (EDX or EDS) or Wavelength Dispersive X-ray spectroscopy (WDX or WDS). In transmission electron microscopy (TEM) and scanning electron microscopy (SEM), electron beams are used for excitation and the techniques are known as TEM–EDX and SEM–EDX, respectively. In TEM–EDX, elemental analysis can be performed at a very small scale, typically in the nm-range, while the SEM imaging system can be utilized for elemental analysis at a spot, along a line or elemental mapping of a region of interest in SEM–EDX. If an X-ray beam is used for excitation, the technique is called X-ray fluorescence (XRF). Unlike SEM–EDX and TEM–EDX, XRF instruments are portable and allow the *in-situ* analysis of large areas of samples or components. The photoelectric effect of X-ray-matter interaction is utilized in XRF.

Table 2.2 List of a few common target materials and corresponding wavelengths of K_α and K_β radiation, together with the minimum excitation potential in kV and the appropriate filter material [35]

Target	$K_{\alpha 1}$ [nm]	$K_{\alpha 2}$ [nm]	$K_{\alpha \text{ mean}}$ [nm]	K_β [nm]	Excitation potential [kV]	Filter	Corresponding K_α edgen for the filter [nm]
Cr	0.22897263	0.22936513	0.22910346	0.20848881	5.98	V	0.22687
Mn	0.21018543	0.21058223	0.21031770	0.19102164	6.54	Cr	0.20701
Fe	0.19360413	0.19399733	0.19373520	0.17566055	7.11	Mn	0.18961
Co	0.17889961	0.17928351	0.17902758	0.16208263	7.21	Fe	0.17430
Ni	0.16579301	0.16617561	0.16592054	0.15001523	8.33	Co	0.16083
Cu	0.15405929	0.15444274	0.15418711	0.13922346	8.98	Ni	0.14879
Mo	0.07093000	0.07135900	0.07107300	0.06322880	20.0	Zr	0.06889

In EDXRF spectrometers, all elements in the sample are excited simultaneously and an energy dispersive detector in combination with a multi-channel analyzer is used to collect the fluorescence radiation emitted from the sample. The different energies of the characteristic radiation allow to separate the different sample elements. The peak positions of this radiation are predicted by Mosley's law with an accuracy much better than experimental resolution of a typical EDX instrument [7]. The resolution of EDXRF systems depends on the detector, and typically ranges from 150 eV – 600 eV [https://www.xos.com/EDXRF]. The principal advantages of EDXRF systems are their simplicity, fast operation, lack of moving parts, and high source efficiency. For technical reasons, EDS in general starts with elements of ordinal numbers $Z > 10$ [22]. This implies that with X-ray fluorescence (XRF), one cannot detect e.g. carbon ($Z = 6$) in carbon steels. There is a vast literature describing both practical applications as well as the solid state physics background, see for example [36, 37].

Traditionally, all hand-held XRF spectrometers used radioisotope excitation. Recent advances in X-ray tube technology, however, allow the use of these devices as an alternative means of excitation. Commercial portable systems arc available to identify atom species in about 5 s at ambient conditions, and in 60 s when operating in installations up to 500 °C providing fast and economic control. As an example, more than 20 elements between Al and U can be analyzed by a hand-held probe in less than 30 s with an equipment weight of 6.7 kg [38, 39]. Such systems are mostly operating with radio-isotopes, e.g. low energy γ-rays, and high-resolution gas-filled proportional counters or semiconductors as detectors. The typical procedure for data evaluation is pattern recognition, i.e. comparing the measured spectra with known, published or otherwise available data spectra for reference alloys which serve as standards. An internet search with the key words "Portable X-ray microanalysis" will lead to many URLs of manufacturers.

Similar to EDXRF, the elements in a sample are excited simultaneously in WDXRF spectrometers. The different energies of the characteristic radiation emitted from the sample are diffracted into different directions by an analyzing crystal or monochromator for a wavelength dispersive analysis. By placing the detector at a certain angle, the intensity of X-rays with a certain wavelength can be measured. Sequential spectrometers use a moving detector on a goniometer to move it through an angular range to measure the intensities of many different wavelengths. Simultaneous spectrometers are equipped with a set of fixed detection systems, where each system measures the radiation of a specific element. The principal advantages of WDXRF systems are high resolution (typically 5 – 20 eV) and minimal spectral overlaps (https://www.xos.com/WDXRF).

2.3.2 *Principle of X-ray Computed Tomography and Laminography*

Computed tomography is usually used for finding and evaluating defects in a component. However, it is also employed for materials characterization and increasingly for non-destructive materials characterization.

Equation (2.5) describes the attenuation of an X-ray beam as a function of propagation distance in a material having the absorption coefficient μ. The equation is valid for a homogeneous microstructure with a constant μ. If we consider a bulk material with possible inhomogeneities, we have to rewrite (2.5) in the form of the so-called Radon transformation P:

$$I(x) = I_0 e^{-\int_0^x \mu(s)ds} \tag{2.25}$$

$$P = \ln\left(\frac{I(x)}{I_0}\right) = -\int_0^x \mu(s)ds \tag{2.26}$$

It is the goal of X-ray computed tomography to invert (2.25) or (2.26), in order to obtain the spatial distribution of the X-ray absorption coefficient μ and the density ρ in a given pixel element in case of a two-dimensional reconstruction (Fig. 2.12).

The Radon transformation of (2.26) can be written in a discrete way:

$$P(j) = \sum_i w_{ij} \cdot \mu_i \tag{2.27}$$

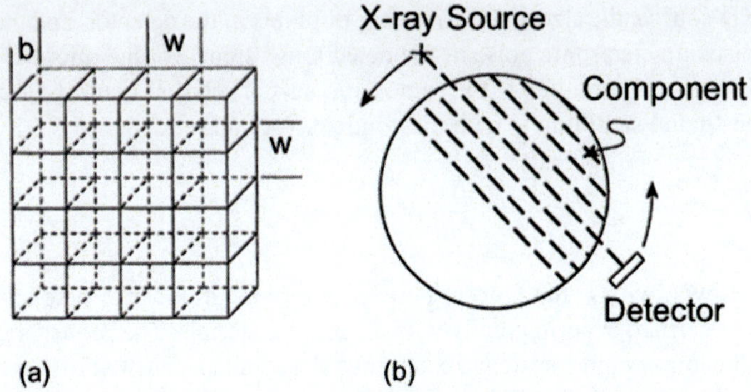

Fig. 2.12 Principle of X-ray Computed Tomography: **a** Slice of the sample of the thickness b and with square elements of width w. Each element has a weight of w_{ij} in the reconstruction process; **b** The component is scanned in a rotational fashion. However, there are many different arrangements for scanning a component

with $j = 1, \ldots, M$, $M = m \cdot n$, where m is the number of projections, and n is the number of detector elements. In a formal way (2.27) can be viewed as a multiplication of the matrix w with the vectors $\vec{\mu}$, where w_{ij} is a $N \times N$ matrix. It is determined by the X-ray system used. The vector $\boldsymbol{P}$ is the measured quantity and $\boldsymbol{\mu}$ is the object vector representing the absorption coefficient of the object.

$$\boldsymbol{P} = w \cdot \boldsymbol{\mu} \tag{2.28}$$

In the same formal manner, one has to invert (2.28) in order to obtain

$$\boldsymbol{\mu} = w^{-1} \cdot \boldsymbol{P} \tag{2.29}$$

The matrix contains $M \cdot N$ elements ($= m \cdot n^3$). With $n = 256$ and $m = 100$ projections, it contains 1,677,721,600 elements. The CPU bound for a direct inversion of the k elements can be estimated to be of order k^3 arithmetic operations. For the above example, this means 10^{27} CPU operations. Suppose that each operation last 1 ns, the total calculations last more than 10^9 years. Furthermore, one must take into account that all elements must be available in the random-access memory, so that it must at least comprise 13 Gbytes. This example is based on the capacity of computers in the early times of X-ray computed tomography. Although the computing power has progressed dramatically in the last three decades and parallel processing of data in the reconstruction process is now the norm, the example shows that despite this progress, iterative methods, for example filtered back-projection, have to be used in the reconstruction of computed tomography [40, 41]. The evaluation of neighboring "discs" of the width b, allows to reconstruct volume elements or voxels in order to obtain *3D*-reconstructions [42].

Two important aspects in X-ray imaging are contrast and spatial resolution for a given computed tomography system. They depend on many parameters like the size of the X-ray source, the size and the number of pixels in the detector, and the amount of shot noise and electronic noise of the detection system. For the micro-CT system developed at the Fraunhofer IZFP some time ago, a relation could be established linking the spatial sampling w with the contrast resolution $\Delta\mu/\mu$:

$$\frac{\Delta\mu}{\mu} w^2 = const, \tag{2.30}$$

where $w = D/N$, with D the object diameter and N the number of pixels [43]. This relation was verified experimentally with different materials. The "const" reflects the details of the tomographic system. Experience shows that a density differences of 1% can be detected. This limit still belongs to the state of the art for standard as well as for micro-computed tomography systems. Both kinds of resolution are usually intertwined and the selection of an X-ray imaging system, such as radioscopy, computed tomography (CT), X-ray microscopy, tomo-synthesis or digital laminography, for non-destructive testing of integrated circuit and ceramic parts are usually based on these two parameters. Industrial systems allow one to reach a resolution from 50 μm

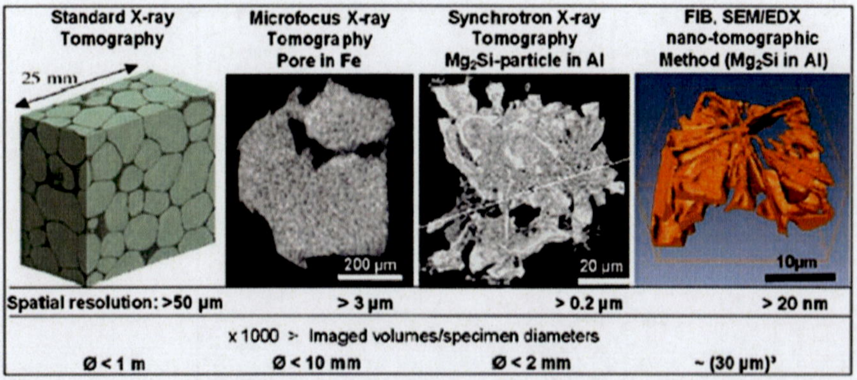

Fig. 2.13 Multi-scale X-ray tomography showing the different resolutions and object sizes accessible by cone-beam CT, by μ-CT, μ-CT using a synchrotron source, and focused ion beam (FIB) tomography (from [47], with permission from Springer Nature)

down to 1 μm and below, based on new types of X-ray sources [44], for example carbon nano-tube emitters. Materials which can be analyzed range from plastics, to ceramics to metals, with object diameters up to 300 mm. Further aspects of resolution, contrast, and scanning geometry of microtomography and its application in materials science are discussed in [45, 46].

Kastner and Heinzl [47] discussed the different size scales of the heterogeneities and the affected material volumes requiring appropriate CT methods and scanning geometries for the required resolution. In order to quantify features of interest from CT scans, a major challenge is the analysis and visualization of the generated CT data. Advanced 3D-image processing techniques are needed for the extraction and the characterization of each single feature of interest, see Fig. 2.13. The various techniques have a different spatial resolution for a given size, i.e. the thickness of the inspected volume, see Fig. 2.14.

Robot-aided scanning systems are facilitated by algebraic reconstruction methods. They allow virtually any deviation from the conventional circular path of the source around the object (Fig. 2.15). In addition, the iterative reconstruction techniques can also be combined with so-called regularization methods. In this way, it became possible to reduce the number of projections to be measured leading to a significant reduction in measuring time and rendering it possible to integrate CT in a production line [48].

Computed laminography (CL) is an image forming method of X-ray testing that yields images of object slices by simple linear translation of the object relative to the tube-detector system [49]. No relative movement of the X-ray tube to the detector is needed and in contrast to classical laminography all object layers can be obtained by one scan. Moreover, CL turns out to be equivalent to a computed tomography (CT) with a limited angular region, which allows the reconstruction of object slices using slightly modified CT algorithms, resulting in a higher resolution compared

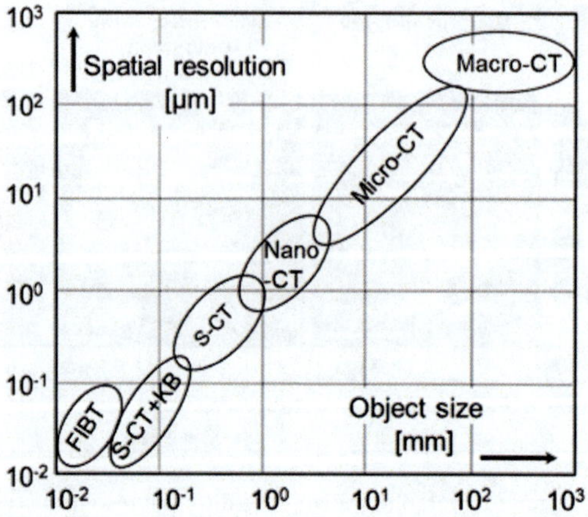

Fig. 2.14 Resolution achieved in X-ray computed tomography for a given object thickness, which are accessible by cone-beam CT systems such as standard macro-CT, various systems of μ-CT, μ-CT with a synchrotron beam as X-ray source (KB stands for mirrors allowing to focus the X-ray beam yielding a fivefold increased resolution at the corresponding ESRF beam-line), and by focused ion beam (FIB) tomography (redesigned according to [47], with permission from Springer Nature)

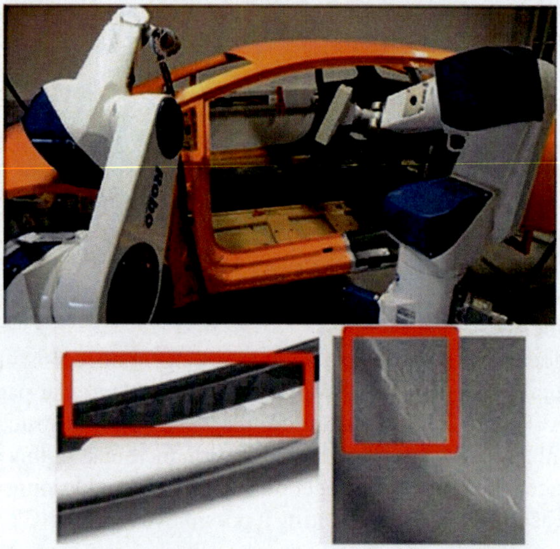

Fig. 2.15 Robot-based CT in the production process. Two robots, one of them (left) carrying the X-ray tube and the other a flat panel detector (the right one), constitute a flexible 3D X-ray system used for inspecting a body shell segment. The reconstructed volume data allow an inspection of connections as for instance adhesive bonds (lower left image) or welding seams (lower right) (from [48], with permission from Springer Nature)

with simple tomosynthesis in the case of classical laminography. CL can also be used for materials characterization, see below.

2.3.3 Applications of X-ray Computed Tomography in NDMC

Computed tomography is dominantly used to detect defects in components. Often a mix of defect detection and material questions are to be evaluated. Monitoring the soundness of the microstructure of a material requires high resolution techniques. However, CT systems have only limited space for irradiation over 360°, limiting the size of the components which can be examined. Furthermore, the detectable contrast $\Delta\mu/\mu$ is limited. Damage processes, such as fatigue, start at an atomic level or at least on a nano-scale and a contrast detectability of far less than 1% would be required to monitor these processes in the early stage. Withers and Preuss [50] discuss the progress using X-ray computed tomography to study damage accumulation in carbon-fiber reinforced plastics, creep damage in metal-matrix composites, and stress corrosion cracking within the limits of resolution just mentioned. In addition to the qualitative diagnostic studies, the authors present quantitative analysis of crack propagation in aluminum alloys, using tomography images to extract the fracture toughness values for mode I, mode II, and mode III loading. A recent survey of material science applications can be found in [51].

An example of a CT application for NDMC in a production line, using iterative reconstruction techniques combined with regularization methods, is shown in Fig. 2.15. Further applications are the monitoring of motor pistons at production speed and the rapid inspection (i.e. within a few minutes) of sea freight-containers for illegal freight, contraband trade etc. [48].

Porsch [52] describes the application of a micro-focus CT in order to determine the quality of seeds. The technique is based on the measurements of the seed's geometrical parameters, such as their volume, thickness of the perikarp, presence of twins, and other geometrical features (Fig. 2.16). The CT system with a spatial resolution of 50 μm can be operated without intervention and manual control and examines about 10^5 seeds in batches of 400 seeds within a few hours.

Fourmaux et al. [53] used the X-ray beam generated by the wake-field of laser-based synchrotron radiation to image biological objects at an energy of $\sim$ 15 keV. There were sufficient X-ray photons to realize an image with one single laser shot, with more than 50 mrad divergence, and spatial distribution homogeneity required for tomography imaging at a laser repetition rate of 2.5 Hz. The μm X-ray source size allowed a resolution of less than 20 μm and phase contrast imaging.

Similarly, Kneip et al. [54] used a plasma wake-field accelerator X-ray source to record absorption and phase contrast images of a tetra fish (Fig. 2.17), damselfly and yellow jacket, in particular highlighting the contrast enhancement achievable by phase contrast imaging, allowing one to study various aspects of biomechanics.

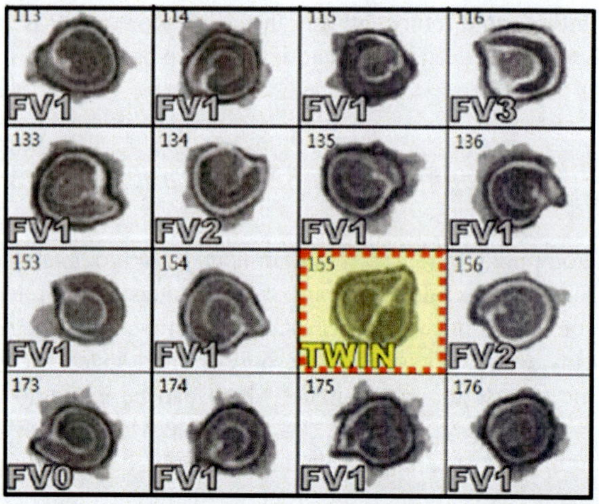

Fig. 2.16 Cross-sections of the seeds obtained by CT. The highlighted image shows a twin fruit with two seeds within the fruit. The diameter of the seeds are ≈ 2.7 mm. The images were obtained using a 50 μm microfocal X-ray tube operated at 45 kV (in modified form from [52], with permission under Creative Commons CC BY-NC licence (https://creativecommons.org/licenses/by-nc/4.0/))

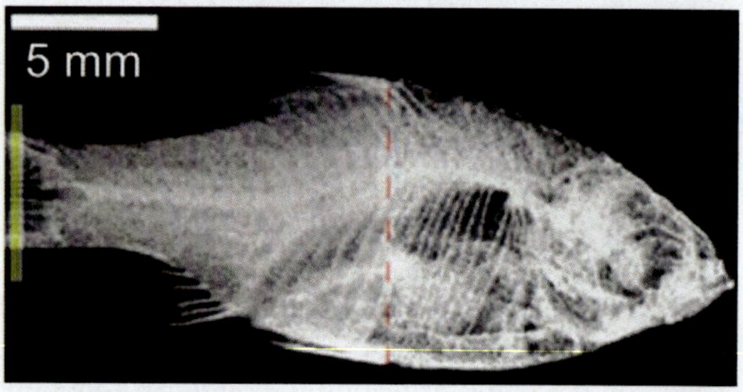

Fig. 2.17 X-ray absorption contrast image of an orange tetra. The red dashed line indicates where the image was assembled from sub-images due to the limited field of view of the detector (in modified form from [54], with the permission of AIP Publishing)

Hiller et al. [55] used an X-ray micro-computed tomography (CT) system for metrological measurements. They studied the factors in the scanning and reconstruction process, the image processing, and the 3D data evaluation, which influence the dimensional measurements based on CT.

In electronic components, like integrated circuits and power devices with plastic or ceramic encapsulation, reliability is affected by tiny defects that may be harbored within a component or assembly. The defects may grow with thermal cycling, accidental mechanical or electrical overload. When they reach a critical size, electrical performance may be affected, that is, a connection may break, or an electrical characteristic may change. At the end of the production process electrical and other tests are usually employed to ensure that only high-quality products are delivered to customers. Overviews about material issues used in microelectronics and MEMS devices are given by [56, 57]. Further information can be found in the proceedings of numerous conferences dedicated to the integrity and reliability of micro-materials, electronic devices, micro-systems and mechatronics. The field has developed and expanded rapidly since 2000, covering many different aspects which are discussed in a number of specialized conferences and journals.

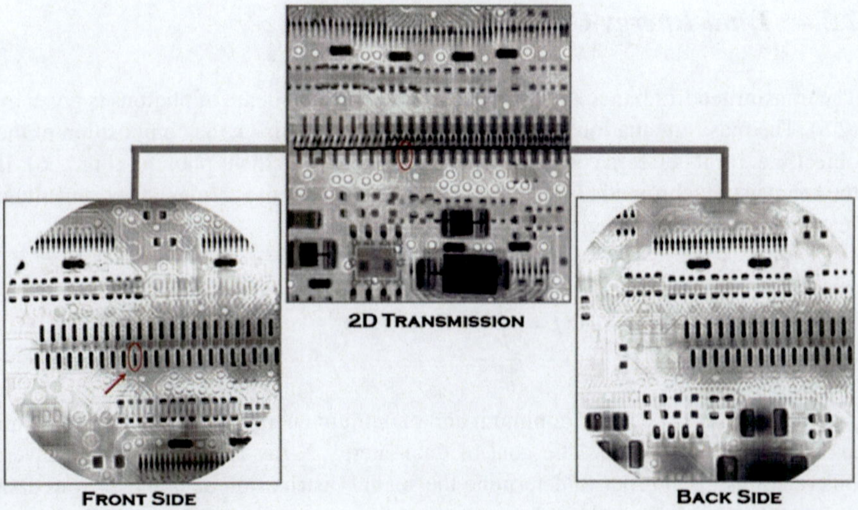

Fig. 2.18 Digital laminograms (DL) (left and right) and transmission X-ray image (center) of a Printed Circuit Board Assembly (PCBA). Left: The image shows the top side of the PCBA with a defective connection (arrow) which cannot be discerned in the 2 D transmission image (center). The back side of the PCBA shows no defect. The focus of the left DL is 200 μm *above* the upper plane of the board. For the back side, the focus of the CL is now 200 μm *below* the upper plane of the board. The board thickness itself is 2 mm. The size of both images is 75×75 mm^2. The images were taken with an X-ray microfocal system having a spot size of 8 μm. The detector employed possessed a pixel pitch of 75 μm. The laminograms contain 32 projections (image kindly provided by Dr. Gondrom-Linke, Volume Graphics, Heidelberg, Germany, see also [61])

An integrated circuit (IC) chip generates heat during its operation. To dissipate the heat and to stabilize the electrical behavior of the chip, the die is bonded to a thermally conductive substrate, such as copper, aluminum oxide, or beryllium oxide [58, 59]. An IC must be hermetically sealed from ionic contamination and moisture by plastic housing (employed in 95% of all cases) or by a sealed ceramic/metal packaging (5%). Apart from defects in the silicon chip itself, the life of the component depends on the quality of the die-attach, the integrity of the interconnection wire bonds, and the lid seal, all of which require good bonding. Printed circuit boards (PCB), in particular the solder joints, may be inspected by conventional X-ray transmission. However, digital laminography or tomosynthesis is better suited for this task because with this technique, one uses only the range of projection angles for reconstruction which contain the meaningful data. Data from close or in-plane X-rays of the PCB are excluded [49, 60] (Fig. 2.18).

2.3.4 Dual-Energy Computed Tomography

The transmitted irradiance I of a parallel monoenergetic beam of photons is given by (2.5). The mass attenuation coefficient μ_m (2.10) depends on the composition of the object, i.e. on its effective Z and the energy E of the incident photons (Fig. 2.6). If the beam is polychromatic (containing photons of different energies), the transmitted intensity of a material is given by

$$I(x) = \int_{E_{min}}^{E_{max}} I_0(E_x) e^{-\mu_m(E_x, Z)\rho L} dE_x \tag{2.31}$$

where E_{min} and E_{max} is the minimum and maximum energy of photons in the beam and L is the path length. The goal of dual-energy X-ray radiography is to invert this relationship in order to determine the spatial distribution of ρ and Z from data measured at two different beam energies. Ignoring the small contributions of coherent scattering (Fig. 2.6, (2.14)), three physical processes, as discussed in Sect. 2.2.4, are responsible for the major part of the X-ray attenuation, so that the mass attenuation coefficient can be decomposed according to (2.18). In the range of 300 keV to 4 MeV, Compton scattering dominates, whereas pair production dominates above 4 MeV. As shown in Fig. 2.19a, the mass absorption coefficient of other elements exhibits a similar overall energy dependence, but the low-energy fine details and the relative amounts of photoelectric absorption, of Compton scattering, and of pair production at a given energy depend on Z, see (2.12–2.15). Thus, for a given photon energy and atomic number, different attenuation processes will dominate the total attenuation, as summarized in Fig. 2.19b. Dual-energy radiography and CT allow one to differentiate between materials based on these properties. The performance is generally better for systems that minimize the spectral overlap between the low-and high-energy attenuation measurements, and for computed tomography in comparison to radiography.

There are a number of approaches for realizing dual-energy CT [62–64]. In order to obtain dual-energy CT images, dual X-ray sources must be available. To this end, the sources may be operated at different energies, run consecutively at different tube voltages and hence energies, and finally, the X-ray beam may be filtered at two different wavelengths. Although much faster than consecutive data acquisition, the simultaneous dual-source method requires complex hardware, has alignment issues, and cross-talk between the source and detector pairs. Instead of modifying the source output, the dual detector method acquires dual-energy data using a constant source in conjunction with detectors with different spectral responses. In order to invert (2.31), so that one obtains the spatial distribution of materials of different Z and density ρ, iterative approaches must be used.

For dual-energy CT, a calibration of the attenuation contrast of the employed X-ray system with test materials is required for different X-ray wavelengths in order to accommodate contrast differences as large as possible. For more information

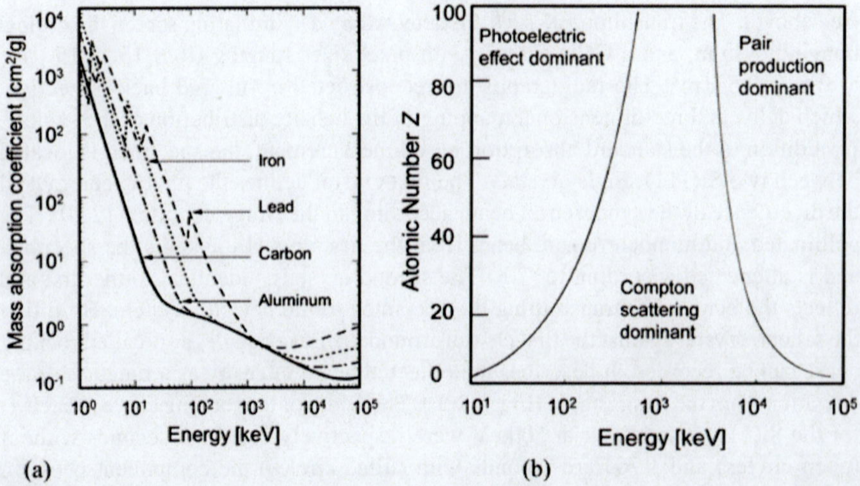

Fig. 2.19 **a** Mass absorption coefficient covering the range 20–200 keV and the range 2–9 MeV; **b** For dual-energy computed tomography using low X-ray energies, one must take into account for reconstruction only photoelectric absorption and Compton scattering, whereas for high-energy methods, one must take into account Compton scattering and pair production

on reconstruction principles, the reader is referred to [65–67], and the references contained therein. A critical appraisal of dual-energy CT for NDT applications can be found in [68]. For example applications in NDMC include the monitoring of circulating fluids and powders [69]. An application, which pushed forward the development of dual-energy CT, is the inspection of carry-on luggage in airports, in order to reliably detect explosive materials [70, 71]. There are numerous applications in medical imaging, which we will not cover in this book.

2.3.5 *X-ray Transmission Imaging and Computed Tomography using Synchrotron Radiation for NDMC*

A synchrotron beamline, specialized for microtomography and radiography, has been installed at BESSY II in Berlin, Germany) and is operated by the "Bundesamt für Materialprüfung" [72, 73]. It is used for absorption, refraction as well as phase contrast imaging. Monochromatic synchrotron light between 6 and 80 keV is attained via a fully automated double multilayer monochromator. A parallel and monochromatic beam from the double crystal monochromator (DCM), with a bandwidth of about 0.2%, is used for performing tomography. The experimental set-up is shown in Fig. 2.20a. The 50 keV beam from the DCM has a horizontal width of up to 30 mm and a vertical width of about 4 mm. Like in standard absorption-based CT, the component is irradiated by stepwise rotation through either 180° or 360°

(see above). The transmitted X-rays are detected by a scintillating screen, an optical imaging system, and a CCD camera with pixel sizes ranging from 15 × 15 μm^2 to 0.6 × 0.6 μm^2. The radiographs are reconstructed by filtered back-projection, which delivers three-dimensional mapping of the density distribution of the sample. In addition to the standard absorption-based measurement, the specimen is located between two Si(111) single crystals. The first crystal defines the photon energy and the divergence of the synchrotron beam according to the Bragg condition (2.20). The collimated and monochromatic beam from the first crystal transmits the specimen and is attenuated according to (2.8) The second crystal is identical to the first and reflects the beam after transmitting the specimen to the detector system. By tilting the second crystal against the first crystal around the Bragg angle, a so-called rocking curve can be recorded. It describes the reflected beam intensity as a function of the deviation from the Bragg angle (Fig. 2.20b). The width of the rocking curve (FWHM) for the Si(111) crystal pair at 50 keV were respectively 1.404 arc seconds without (open circles) and 1.764 arc seconds with (filled circles) the component between the two crystals. The broadening of the rocking curve is due to the refraction effect, which deflects the X-rays at all sample interfaces. Hence, all scattered X-rays will be lost at the second crystal, if the crystal pair is set to its rocking curve maximum, see Fig. 2.20b. The refracted X-rays from the inner surfaces of the specimen are blocked by the second crystal (black dotted rays in Fig. 2.20c). This leads to a contrast enhancement in the radiography of the sample (marked by black dotted stripes on the refraction image in Fig. 2.20c. If the second crystal is oriented so that it is at the half of the FWHM, only the scattered X-rays will be reflected and detected by the CCD camera system and enhanced contrast is observed, see Fig. 2.20a.

The potential of synchrotron refraction computed tomography is shown in Figs. 2.21 and 2.22. The sample examined for Fig. 2.21 was a cylindrical, 3.5 mm diameter low-cycle fatigue test sample in an aircraft project [72]. The contrast at interfaces within heterogeneous materials can be strongly amplified by X-ray refraction, which is particularly useful for materials of low absorption or mixed phases showing similar X-ray absorption properties that produce low contrast. The technique is based on small-angle X-ray scattering by microstructural elements causing phase-related effects, such as refraction and total reflection taking into account that the refractive index of materials for X-ray is very small and slightly smaller than 1 [74].

The contrast of inner surfaces is far larger than through absorption effects. Crack orientation and fiber/matrix debonding in plastics, polymers, ceramics and metal-matrix composites, after cyclic loading and hydro-thermal aging, can be visualized. The inner surface and interface structures correlate often directly to mechanical properties. The technique is thus an alternative to other attempts of raising the spatial and in particular the contrast resolution of CT machines. Due to the high photon flux, the best contrast performance is obtained when using beam-lines at synchrotrons, in particular for phase imaging. This was realized by a number of groups [72, 73, 75].

Kupsch et al. [76] used X-ray refraction in order to characterize the porosity and the pore orientation in cordierite diesel particulate filter materials. The X-ray refraction has been used in different ways: simple scattering, laboratory two-dimensional

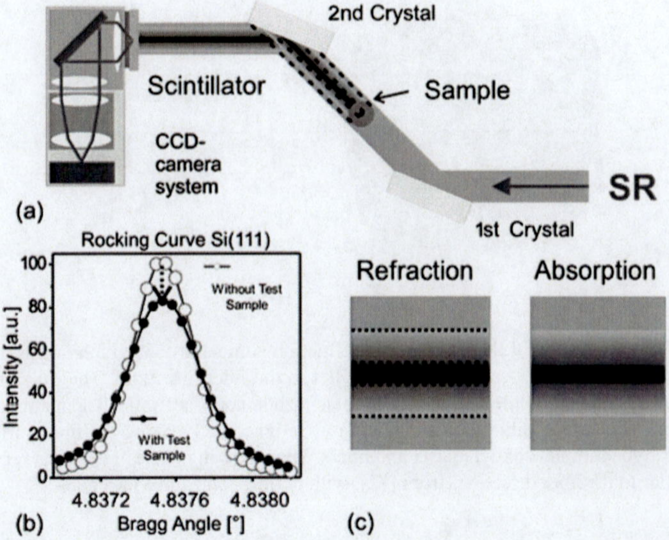

Fig. 2.20 **a**: Sketch of the synchrotron beamline specialized for microtomography and radiography at BESSY II, Berlin. The incoming beam is designated SR. The refracted rays are blocked by the second crystal (black dotted rays); **b** Rocking curve of the Si(111) single crystal pair in symmetric configuration at 50 keV with specimen (filled dots, FWHM: 1.8 arcsec) and without specimen (open circles, FWHM 1.4 arcsec) between the two crystals. The areas under the curves are set to be equal; **c** Demonstration of the refraction enhanced contrast (black-dotted lines) (in modified form from [72], with permission from Vieweg)

Fig. 2.21 Radiograph of a composite material test sample imaged at 50 keV photon energy recorded in refraction geometry and with fiber orientation parallel to the scattering plane of the analyzer crystal (2nd crystal). The pixel size was $5.6 \times 5.6\,\mu m^2$. The horizontal main crack appears dark. Around the main crack, several small cracks became visible due to the refraction geometry in the CT. Vertical stripes are caused by the reinforcing fibers in the composite material (from [72], with permission from Vieweg)

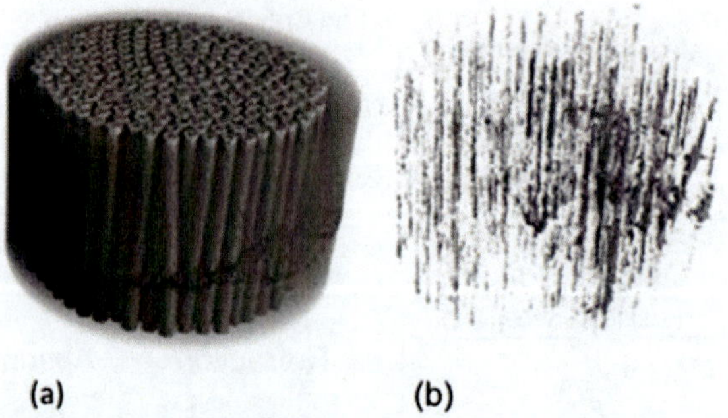

Fig. 2.22 Reconstruction of the synchrotron refraction computed-tomography measurement data (scattering plane of the analyzer crystal is parallel to the fiber direction). The titanium matrix is shown semi-transparent, while the fibers and their carbon cores are solid. The main fatigue crack in the matrix is clearly visible; **b** Visualization of fatigue cracks along the fiber surfaces of the low-cycle fatigue sample without matrix and fibers. The scattering plane of the analyzer crystal was perpendicular to the fiber direction (from [72], with permission from Vieweg)

mapping and tomography, and refraction radiography at a synchrotron source. All techniques yielded the same result: the interface between pore and solid matter has a preferred orientation along the extrusion axis. This orientation factor (~ 1.2) agrees well with reported crystal texture values, and confirms that the interface density is the quantity of interest in order to characterize the porosity in these materials. The X-ray absorption measurements yielded a porosity value in full agreement with mercury intrusion measurements. The mean pore size determined by mercury porosimetry confirmed the interface density ratio found with X-ray refraction. The refraction technique has the advantage over classic mercury intrusion that no assumption has to be made with regard to the form of the pores. X-ray refraction can yield position-sensitive information with a lateral resolution of about 100 μm, or smaller if synchrotron radiation is used.

Banhart et al. [77] used the high-intensity synchrotron X-rays for radioscopy, in order to obtain real-time images of the processes occurring during foaming of metals. Bubble generation, foam coalescence and drainage of an aluminum-based alloy foam were investigated.

Cement degradation caused by CO_2 exposure is an important environmental challenge that must be understood, for example, if underground reservoirs are to be used for CO_2 storage. When exposed to CO_2-saturated brine, cement undergoes a chemically complex carbonation process that influences physicochemical properties of the cement. It is known that under favorable conditions, fractures and voids in cement can be self-sealed by precipitation of calcium carbonate. Panduro et al. [78] reported a detailed X-ray microcomputed tomography study on the carbonation of gas pores (macropores) of approx. 1 mm diameter in cement. Cured Portland cement with sub-millimeter spherical disconnected macropores was exposed to CO_2-saturated brine at high pressure (280 bar) and high temperature (90 °C) for 1 week. High-resolution synchrotron-based μ-CT enabled visualizing the morphology of the

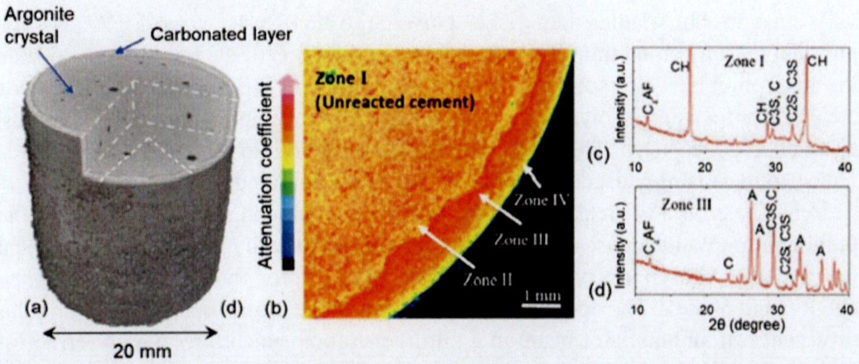

Fig. 2.23 Portland cement exposed to CO_2 saturated brine for 1 week at 90 °C and 280 bar; **a** Cut-away view of the μ-CT reconstruction (polychromatic cone-shaped beam from a tungsten target, 114 keV acceleration voltage, exposure time 1000 ms per projection, total of 3142 projections over 360°) of the CO_2-exposed core. An approximately 1.5 mm thick carbonated layer is visible as a bright rim around the core; **b** False-colored μ-CT cross-section with clearly observable regions; **c, d** X-ray diffraction patterns of material retrieved from zones I and III, respectively, with peaks assigned to crystalline phases: portlandite (CH); dicalcium silicate (C2S); tricalcium silicate C3S); tetracalcium alumino ferrite (C4AF); aragonite (A); calcite (C). The intensity gradient (lighter moving inward) in Zone I is a CT reconstruction artifact (from [78], with permission from the American Chemical Society)

precipitates inside the macropores within both unreacted and carbonated regions. Quantitative analysis of the type and amount of material deposited in the macropores during carbonation showed that the filling of the disconnected macropores involved transport of calcium ions from the cement bulk to the macropore interior, see Fig. 2.23.

2.3.6 X-ray Microscopy and its Applications in NDMC

X-ray microscopy is finding increasing applications as underlined by the large number of recent publications. It is on the verge of becoming a generally available instrument, also for industrial laboratories. X-ray microscopy reached diffraction-limited resolution of the order of several tens of nanometers. The enabling technologies were the availability of sensitive flat-panel X-ray detectors and specialized synchrotron X-ray beam lines (Sects. 2.2.3 and 2.2.2.3). As said above, they host X-ray microscopy set-ups, as well as X-ray computed tomography. In [79, 80] and more recently [81, 82], the various aspects of X-ray microscopy are discussed in great detail, also providing many references which document the scientific and technical processes. Here, we restrict ourselves to the main points.

In order to construct a microscope, one needs lenses for focusing. This is a challenging task, because the refractive index n of materials at X-ray wavelengths is very close to, but smaller than 1, i.e. 1-n = $\delta(\omega) \approx 10^{-7}$-$10^{-3}$, see Fig. 2.24. The physical reason behind this fact is the anormal dispersion of the refractive index due the absorption bands of solids in the ultraviolet region [74]. Three type of methods are used for focusing X-rays: compound lenses, reflective optics, and Fresnel lenses, also called zone-plates. The latter method has evolved into a technology, which is now also increasingly used beyond academic research in industrial applications.

Lengeler et al. [84] realized refractive X-ray compound lenses. By varying the number of individual lenses in the stack shown schematically in Fig. 2.25a, the focal length can be chosen in a typical range from 0.5 to 2 m for photon energies between about 6 and 60 keV. The aperture of the lens was about 1 mm, matching the angular divergence of an undulator beam in a third-generation synchrotron. With refractive X-ray lenses with a rotational parabolic profile (Fig. 2.25b), one can obtain focal spot sizes in the μm range with a gain factor of 1000 and more for imaging purposes in an X-ray microscope. They focus in two directions and are free of spherical aberration. As an example, Fig. 2.26 shows an X-ray micrograph of a gold mesh obtained with a refractive compound lens. Some defects are visible.

Zone plates (ZP) are based on focusing by diffraction rather than refraction [74; 85]. They have been known in optics for a long time. They consist of concentric

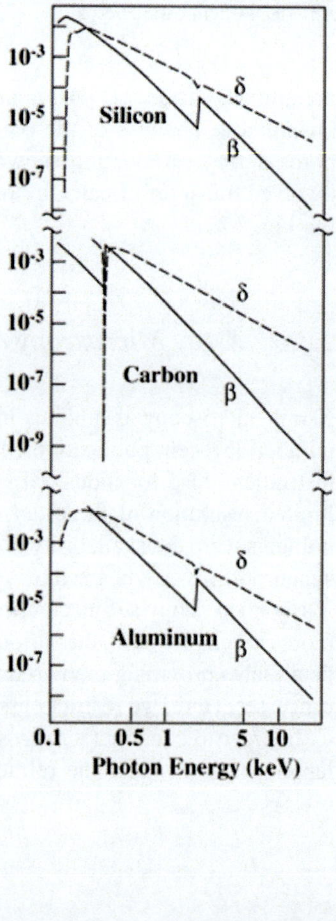

Fig. 2.24 Real (1-$\delta(\omega)$) and imaginary part $\beta(\omega)$ of the complex refractive index $n(\omega) = 1 - \delta(\omega) + i\beta(\omega)$ at X-ray energies from 0.1 to 20 keV for Al, C, and Si. An energy of 0.1 keV corresponds to a wavelength of 12 nm (2.1) (from [83], with permission from "Rivista del Nuovo Cimento" and the author)

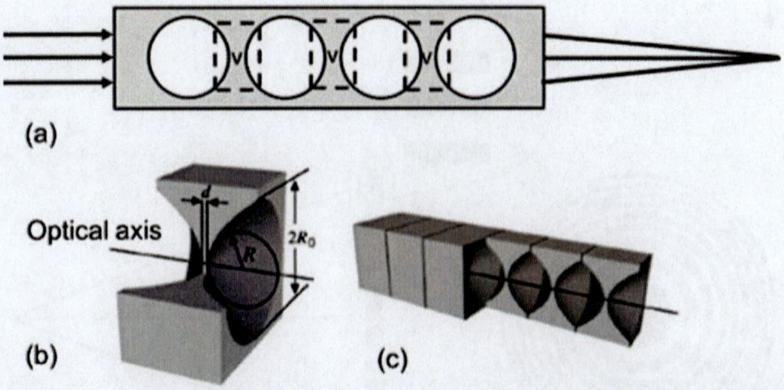

Fig. 2.25 Converging lens for X-rays; **a** Design principle: The shaded area represents a material, and the white area cavities. Because $n < 1$ for X-rays, the volume depicted by the rectangular dashed lines does not act as a diverging lens but as a converging lens in contrast to an optical lens in the visible range. Several of these focusing lenses increase the focusing power of the lens (in modified form from [14]; **b** Individual refractive X-ray lens with a parabolic profile with a radius of curvature R at the apex and with a geometrical aperture $2R_0$; **c** Stack of lenses forming a compound refractive lens (in modified form from [84], with permission from Elsevier)

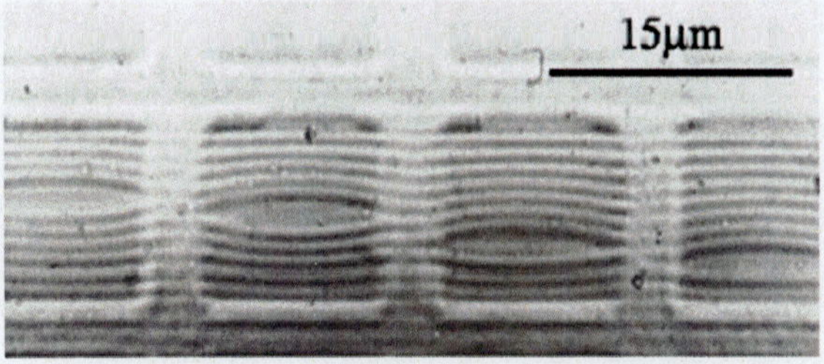

Fig. 2.26 X-ray micrograph of a gold mesh having a square period of 15 μm and a linear period of 1 μm. The image was obtained with an Al refractive lens at 19 keV. The magnification was 22 times. Some flaws in the linear array are visible (from [84], with permission from Elsevier)

annular rings which are alternatively transparent or absorbing for the radiation, here X-rays, see Fig. 2.27. The focusing effect is produced by the interference of the waves transmitted through the non-absorbing zones, which result in mutual cancellations, when m is an even number, i.e. a resultant focus exists only when m is odd. Since a zone plate is a special case of a diffraction grating, higher-order foci are to be expected. The zone radii obey the equation

$$r_n^2 = mn\lambda f_m + \frac{m^2 n^2 \lambda^2}{4} \tag{2.32}$$

Here, r_n is the radius of the nth-zone, λ is the wavelength, m is the diffraction order, and r_n is the n-th zone radius. The focal length f_m is given by:

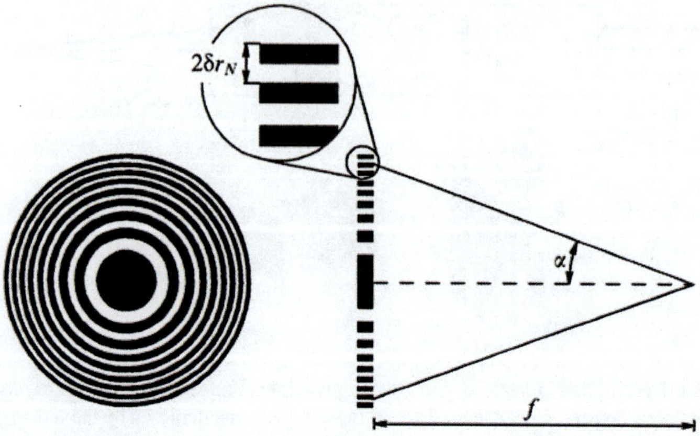

Fig. 2.27 Fresnel lens with focusing length f (from [82], with permission from Uspekhi Fizicheskikh Nauk 2017)

$$f_m = \frac{D\delta_{r_N}}{\lambda m} = \frac{D^2}{4\lambda N m} \tag{2.33}$$

where δr_n is the outer zone width, D is the diameter of the zone plate, and N is the total number of zones. The diffraction limited lateral resolution and the numerical aperture are given by

$$f_m = \frac{0.61\lambda}{NA} \tag{2.34}$$

$$NA = \sin\alpha = \frac{m\lambda}{2\delta r_N} \tag{2.35}$$

From these expressions the Rayleigh resolution of the zone plate lens can be derived:

$$d_m = \frac{1.22\delta r_n}{m} \tag{2.36}$$

In a transmission X-ray microscope (TXM), all elements of the object are illuminated, and they are simultaneously imaged by a pixel detector. Every element of the field is independently visualized when the object is irradiated by an incoherent or partially coherent beam. Therefore, laboratory sources are better suited for the TXM than an undulator source [81]. Moreover, the operation with a highly coherent beam from a synchrotron may give rise to speckles in the image, which can be reduced by placing a diffuser in front of the sample [86]. Attaining the diffraction resolution limit requires that a scanning transmission X-ray objective is illuminated by a coherent beam with a monochromaticity no worse than the reciprocal of the product of the

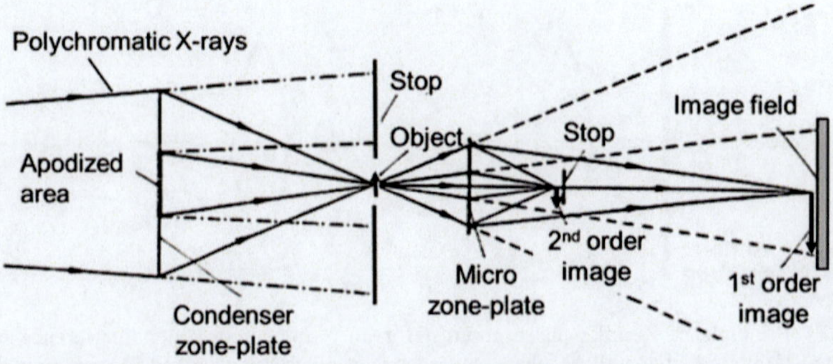

Fig. 2.28 Principle of an X-ray transmission microscope according to [87]

total number of zones and the diffraction order $\Delta\lambda/\lambda \leq 1/mN$ (diffraction order m × number of zones N) [82].

Figure 2.28 shows the beam paths for an X-ray microscope designed for use with a synchrotron [87] for amplitude imaging. The continuum radiation illuminates first a condenser plate. The purpose of this condenser consists in collecting into the object plane as many photons as possible, in order to shorten the acquisition time required for the image built-up. It also serves to focus the beam at the desired wavelength in an aperture in the object plane, filtering a small band out of the continuum radiation (2.33 and 2.34). In combination with the condenser plate, the aperture acts as a monochromator with a spectral resolution of $\Delta\lambda/\lambda \approx D/2d$, where D is the diameter of the condenser and d is the diameter of the pin-hole aperture. Typical values are $D = 9$ mm and $d = 10$ μm, hence $\Delta\lambda/\lambda \approx 450$. Furthermore, the central stop prevents the illumination of the object by the zero-order diffraction of the condenser zone plate at the central part of the apodized region. The object is magnified by the micro-zone plate and read out by a digital detector in the image plane.

For phase-contrast imaging one uses an optical arrangement as shown in Fig. 2.29 [88]. The beam from the X-ray source is limited by an annular ring aperture placed between the source and the condenser. The condenser and the objective form a lens system which produces a real image of the ring after the objective. A phase ring is then inserted in the plane of this image to match its dimensions exactly. All non-diffracted light from the sample must therefore pass through this plate. The final image is formed by interference between the non-diffracted X-rays passing through the phase ring (red lines) and the diffracted light which bypasses it (green lines). The diffracted part contains the information about the sample's properties at X-ray wavelengths, according to Abbe' theory of image formation in a microscope. The ring imposes a predetermined phase shift of 90° in order to obtain a positive contrast or 270° for obtaining a negative contrast in the non-diffracted part.

The phase-contrast image is formed by the interference of the phase-shifted non-diffracted component with the undisturbed diffracted component, translating phase modulations of the sample into intensity modulations in the image plane. For small phase shifts, these modulations are due to the differences in the real part of the

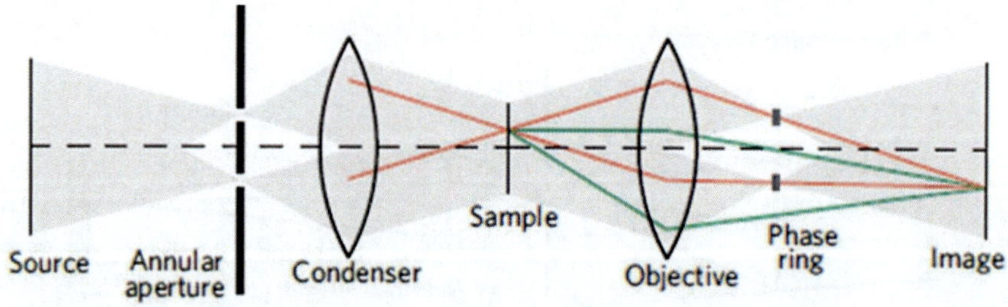

Fig. 2.29 Principle of Zernike phase contrast for an X-ray microscope. Interference of non-diffracted light (red) that will be phase shifted by the phase ring with diffracted light (green) translates phase variations of the object into intensity modulations in the image plane (from [88], with permission from Springer Nature)

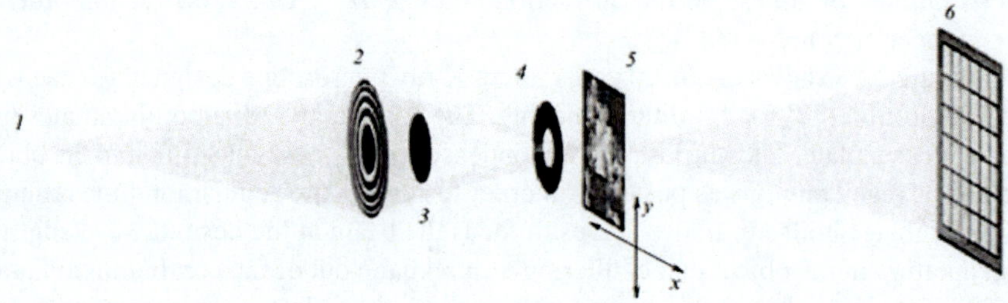

Fig. 2.30 Principle of a scanning X-ray microscope: **1** radiation source, **2** objective Fresnel zone plate, **3** stop, **4** order-sorting aperture, **5** raster-scanned sample, and **6** detector (from [82], with permission from Uspekhi Fizicheskikh Nauk 2017)

object's index of refraction, whereas the imaginary part leading to absorption contrast is usually small and can often be neglected. Phase imaging often results in improved contrast compared with absorption, especially when imaging biological samples such as small organisms, tissue sections, cells and so on are examined. For further details with regard to Zernike type X-ray imaging see [86, 88].

In a scanning X-ray microscope (SXM) (Fig. 2.30), the photon source (1) is demagnified by a zone plate (2) to form a microprobe across the specimen (5) which is raster-scanned. Since the ZP has an infinite number of diffraction orders, a so-called order-selecting aperture (4) is placed between the ZP and the specimen in order to cut all undesired diffraction orders. The direct zero-order light through the ZP and OSA is blocked by a central stop on the ZP (3). Since the lateral resolution in SXM results from the convolution with the diffraction-limited point spread function, it is maximized when the geometrical demagnification of the source is made equal to the diffraction-limited lateral resolution. This procedure allows one to keep a high photon flux. Further details of SXM and other X-ray microscope technologies can be found in [82].

Zschech et al. [89] report on the results obtained with an X-ray microscope at the ID21 beamline at the ESFR Grenoble, which operates at a photon energy of 4 keV, achieving a spatial resolution of about 50 nm. In the Zernike phase mode, the contrast of structures imaged in an X-ray microscope is increased significantly, as shown by the phase contrast X-ray micrographs of copper interconnect structures with artificial defects marked in the image (Fig. 2.31).

Using a specially designed X-ray microscope for use at the National Synchrotron Radiation Research Center in Taiwan, Yin et al. [90] report the application of X-ray tomography for imaging of so-called keyholes in tungsten plugs, i.e. of voids of 50 to 80 nm in diameter, which formed during the deposition process. A photon energy of 10.5 keV, just above a tungsten absorption edge, was chosen to ensure an appropriate image contrast. A spatial resolutions of 60 nm was attained (Fig. 2.32).

The high brightness and small source size of laser generated Betatron X-rays (see Sect. 2.2.2.4) were used for high-resolution X-ray imaging for human bones and other biological species [20]. Likewise, an irregular eutectic in the aluminum–silicon system was examined by [91]. The high spatial resolution allowed to image the lamellar spacing of the Al-Si eutectic microstructure of the order of a few micrometers. An upper bound on the resolving power of 2.7 ± 0.3 μm of the laser generated Betatron X-ray imaging system was determined, providing an alternative to conventional synchrotron sources for high resolution imaging of eutectics and complex microstructures.

Fella et al. [92] described a compact X-ray microscope in the hard X-ray regime with energies at ≈ 9 keV, using a liquid–metal-jet X-ray source. It is a hybrid instrument that works in two different modes. The first one is the μ-CT mode without optics, using a high-resolution detector to allow scans of samples in the millimeter range with a resolution of 1 μm. The second mode is a microscope, which contains a commercially available X-ray zone plate as focusing lens, in order to magnify the sample's image resolving 150 nm features (Fig. 2.33).

As an example, Fig. 2.34a shows the edge of a thin section of human tooth dentin. The exposure time was 1800 s for a pixel sampling of 147 nm and for a source of 80×20 μm^2. A tubular structure can be seen with diameters between 1 and 2 μm. In order to check the functionality of the set-up, a CT scan of a damaged injection cannula tip

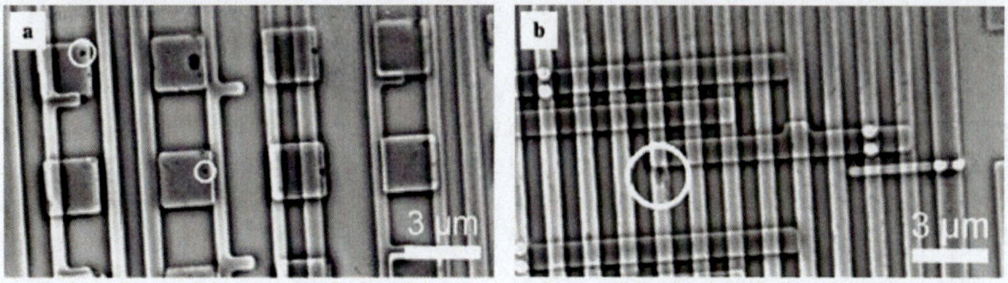

Fig. 2.31 **a** Copper interconnect structures, imaged in positive phase contrast mode in the transmission X-ray microscope at the ID21 beamline at ESRF Grenoble. Irregularities in the square structures (as indicated by the white circles) can be seen; **b** Irregularity within a conducting copper line that could be a nucleation site for electromigration processes (white circle) (in modified form from [89], with permission from Springer Nature)

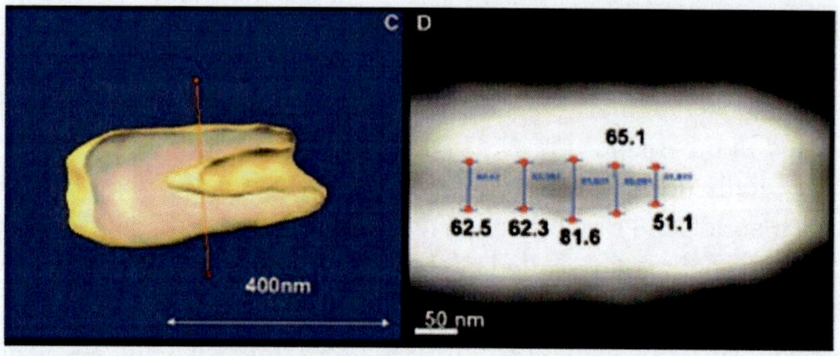

Fig. 2.32 The top view (left) and side view (right) of a tungsten. The orange line indicates the region of the cross-section plot that is shown in the right-hand image. The diameters of the keyhole range from 50 to 80 nm (from [90], with the permission of AIP Publishing)

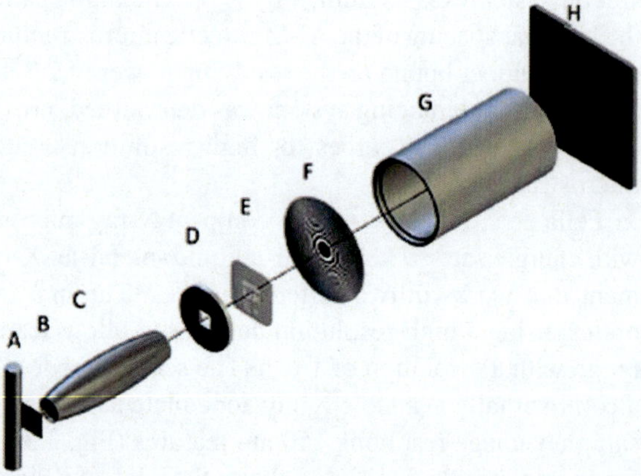

Fig. 2.33 X-ray microscope consisting of **A** a liquid–metal target source, either a Ga-In or a Ga-In-Zn alloy, **B** a Zn foil to block the Ga-K_β -line, **C** a condenser, **D** pinhole, **E** sample, **F** zone plate for magnifying the X-ray image, **G** vacuum pipe, and **H** a CMOS or CCD detector (from [92] with the permission of AIP Publishing)

was made, as depicted in Fig. 2.34b. The exposure time was set to 360 s per image with a pixel sampling of 147 nm. The pixel size on the detector was 1.33 μm (2 × 2 binning) and the X-ray magnification was ≈ 9. The measured *SNR* in the flat-field region of the projections was approx. 75 with a 46 × 16 μm^2 X-ray spot. For the CT scan, 432 projections as well as 4 flat-field and 2 dark images were recorded, leading to a total acquisition time of 44 h. It should be noted that the spatial stability was good enough for a standard reconstruction. The result demonstrates a clear twist of the cannula's tip, which occurred when the needle was dropped once on the floor. This instrument represents an important step towards establishing high-resolution laboratory-based multi-mode X-ray microscopy as a standard investigation method in NDMC.

Helfen et al. [93] used an X-ray beam from a synchrotron source for computed laminography for three-dimensional (3D) imaging of flat laterally extended objects (see Sect. 2.3.2). Spatial resolution down to the micrometer scale was attained, even for specimens having lateral dimensions of several decimeters. Operating either with a monochromatic or with a white synchrotron beam, a high throughput in small-batch

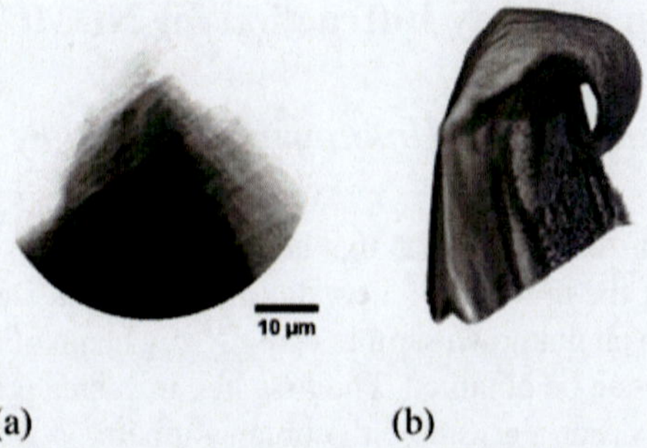

Fig. 2.34 Examples of X-ray images magnified with zone plate: **a** Edge of a thin section of human tooth dentin, imaged with a pixel sampling of 147 nm, and with an exposure time of 1800 s. The small features are the tubular structure with diameters between 1 and 2 μm; **b** Image of a bent 4 μm cannula tip (from [92] with the permission of AIP Publishing)

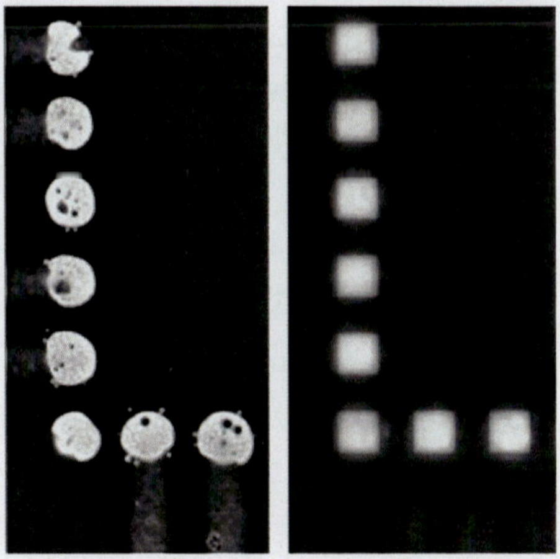

Fig. 2.35 Comparison of reconstructed cross-sectional slices by filtered back-projection of a corner of a flip-chip bonded device. Two different slices are shown, namely, the interface towards the printed circuit board (left) with its Cu conduction lines and the interface towards the integrated circuit (right). Aluminum metallization lines and connection pads can be discerned on the IC surface. In addition, the metallographic microstructure of the bumps of different phases is clearly visible. The two slices are 53 μm apart, and the solder bump pitch is 180 μm (from [93], with the permission of AIP Publishing)

industrial application could be attained. Imaging of interconnections in flip-chip and wire-bonded devices illustrates the use of laminography for the 3D inspection and quality assessment in microsystem technology. For the reconstruction, filtered back-projection was used (Fig. 2.35).

2.4 Application of X-ray Diffraction for NDMC

2.4.1 *Identification of an Unknown Specimen by XRD*

It can be understood from Sect. 2.2.5 that an XRD spectrum is unique to the crystal structure and size of the unit cell of a crystalline specimen. Hence, by obtaining an XRD spectrum from an unknown sample using X-ray beam of a known wavelength, the crystal structure can be obtained. The first step in such a procedure is to identify the crystal symmetry. For a crystal with cubic symmetry, a single peak is observed (Fig. 2.36), corresponding to a family of planes such as for (110), as the dimensions of the unit cell are exactly the same along all the three axes. However, three closely spaced (split) diffraction peaks are observed for every family of planes for crystals with orthorhombic symmetry. Further, by taking ratios of the $\sin^2(\theta)$ for the consecutive peaks, ratios of the $h^2 + k^2 + l^2$ can be obtained. This can be used to determine the crystal structure by comparing the possible peaks as per the structure factors of different planes of crystal structures. Take as an example the XRD spectrum of Cu given in Fig. 2.11. Four peaks at well separated 2θ values of 43.5°, 50.7°, 74.6° and 90.5° are observed indicating cubic symmetry. Considering that these $\theta_1, \theta_2, \theta_3$, and θ_4 are reflections from planes with Miller indices $h_1k_1l_1$, $h_2k_2l_2$, $h_3k_3l_3$, and $h_4k_4l_4$, respectively, we obtain the following equations from (2.21)

$$\frac{h_1^2 + k_1^2 + l_1^2}{h_2^2 + k_2^2 + l_2^2} = \frac{\sin^2\theta_1}{\sin^2\theta_2} = 0.749 \approx \frac{3}{4} = \frac{1^2 + 1^2 + 1^2}{2^2 + 0^2 + 0^2}, \tag{2.37}$$

$$\frac{h_1^2 + k_1^2 + l_1^2}{h_3^2 + k_3^2 + l_3^2} = \frac{\sin^2\theta_1}{\sin^2\theta_3} = 0.374 \approx \frac{3}{8} = \frac{1^2 + 1^2 + 1^2}{2^2 + 2^2 + 0^2}, \tag{2.38}$$

and

$$\frac{h_1^2 + k_1^2 + l_1^2}{h_4^2 + k_4^2 + l_4^2} = \frac{\sin^2\theta_1}{\sin^2\theta_4} = 0.272 \approx \frac{3}{11} = \frac{1^2 + 1^2 + 1^2}{3^2 + 1^2 + 1^2}. \tag{2.39}$$

From the above analysis, the crystal structure can be inferred as FCC and the size of the unit cell can be determined using (2.21) (Fig. 2.36). For more complex structures, the crystal may be identified by comparing the experimentally obtained spectrum with the Powder Diffraction Files (PDF) in a database of X-ray powder diffraction patterns maintained by the International Center for Diffraction Data (ICDD) (The International Centre for Diffraction Data -ICDD (https://www.icdd.com/)).

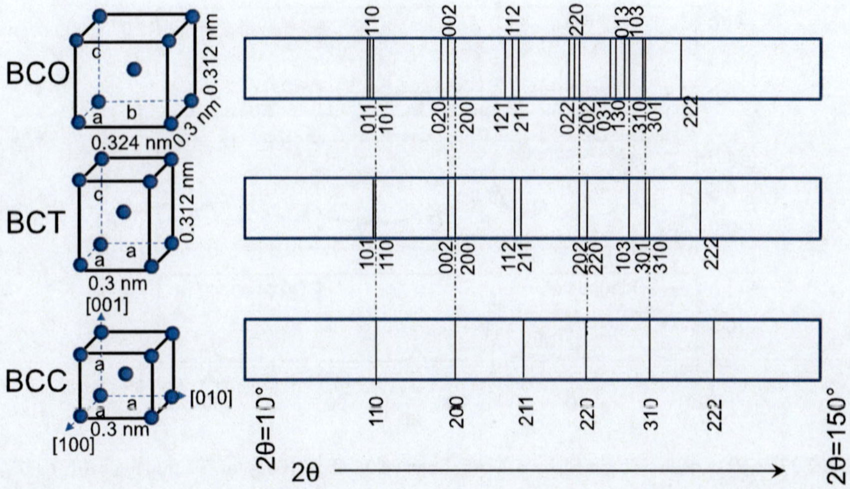

Fig. 2.36 Effect of crystal symmetry on the diffraction peaks observed in a body centered cubic ($a = b = c$), a body centered tetragonal ($a = b \neq c$), and a body centered orthorhombic crystal ($a \neq b \neq c$) with the axes perpendicular to each other

2.4.2 *Quantitative Analysis of Volume Fraction of Constituent Phases by XRD*

Quantitative analysis by XRD techniques had its origin with Hull [94], presenting the comparison of a photograph of the diffraction pattern of a mixture with the superimposed intensities of the individual components, which were exposed separately for the same length of time. Initially, quantitative estimates were made from peak intensities. Later, a more accurate relationship was found with the use of integrated peak intensities, as employed in the estimation of austenite in hardened steel [95].

The volume fraction of different phases is calculated from the integrated intensities of the diffraction peaks of the phases. However, calibrations are required to account for the different mass absorption coefficients of the different phases. Various methods for such calibrations include: (i) external standards. Here, one compares the integral intensities of phase j in the crystal mixture with the integral intensity of the pure phase under the same experimental conditions; (ii) internally prepared standards: here, a standard is added with a constant volume or mass; (iii) intensity ratio based, and (iv) employment of the Rietveld analysis method of fitting the entire spectrum. The theoretical basis for quantitative analysis was given by [96]. The intensity of the ith diffraction peak is given by the basic equation

$$I_{ij} = \frac{K_{ij} c_j}{\rho_j \mu} \tag{2.40}$$

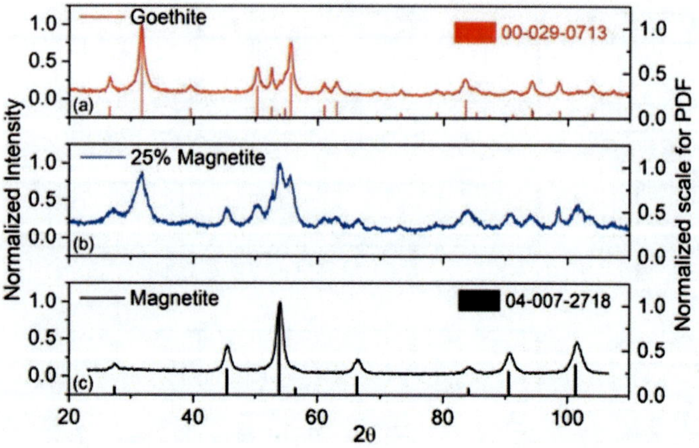

Fig. 2.37 XRD spectra obtained by a MAC Science MXP18 X-ray diffractometer with $Cr\ K_\alpha$ radiation in the angular range of 20–110° with a step size of 0.041° and a dwell time of 6 s for **a** pure geothite; **b** XRD diffraction of a mixture containing ~ 25% magnetite and 75% goethite (middle figure), and **c** pure magnetite. For the quantitative estimate of magnetite in an unknown sample, a calibration graph was obtained by mixing pure magnetite and goethite nanoparticles in different ratios (courtesy from Dr. S. Mahadevan, IGCAR, Kalpakkam; data published in [97])

where K_{ij} is a calibration constant independent of the composition of the specimen, c_j is the weight fraction of the phase j in the specimen, ρ_j is the density of phase j, and μ is the mass absorption coefficient of the specimen.

The external standard based approach was used by [97] in order to obtain the volume fraction of magnetite (Fe_3O_4) in a magnetite – goethite (α-FeOOH) mixture. Figure 2.37 shows the XRD spectra obtained with a MAC Science MXP18 X-ray diffractometer using $Cr\ K\alpha$ radiation in the angular range of 20–110° with a step size of 0.041° and a dwell time of 6 s for (a) pure goethite, (b) a mixture containing ~ 25% magnetite and 75% goethite, and (c) pure magnetite. For quantitative estimates of magnetite in an unknown sample, a calibration graph was obtained by mixing pure magnetite and goethite nanoparticles in different ratios. The (220) peak of magnetite was obtained in all the specimens by XRD measurements in the 2θ angular range of 42.4 -48.4°. The normalized integrated intensity ($I_m/I_{m\text{-}pure}$) of this peak was used to correlate with the weight % of magnetite in the mixture by the following equation:

$$\frac{I_m}{I_{m-pure}} = \frac{c_m \mu_m}{c_m(\mu_m - \mu_g) + \mu_g} \tag{2.41}$$

where $I_{m\text{-}pure}$ denotes an intensity of 100% magnetite and I_m corresponds to an intensity of magnetite with weight fraction of c_m. The mass absorption coefficients are $\mu_m = 86.6$ cm^2/g for magnetite and $\mu_g = 79.5$ cm^2/g for goethite. As the mass absorption coefficients of the two phases are very close, a linear relation between the normalized integrated intensity and magnetite weight % could be obtained, as shown in Fig. 2.38.

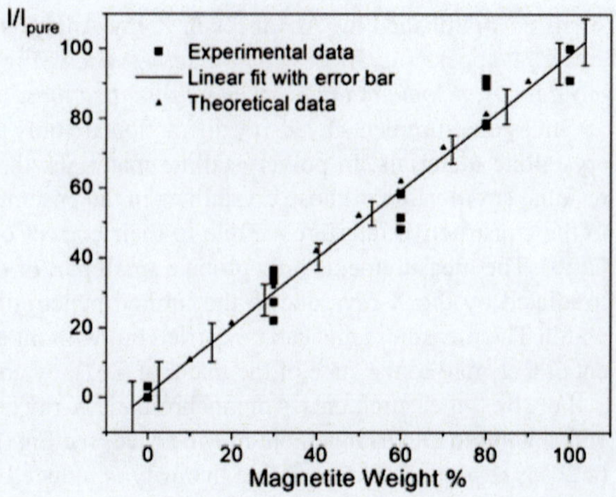

Fig. 2.38 Calibration graph of relative intensities of (220) peak vs magnetite weight % (courtesy by Dr. Mahadevan, IGCAR. Kalpakkam; data published in [97], with permission from Elsevier)

X-ray diffraction is used frequently to estimate the volume fraction of austenite in a ferritic steel as per the equation (ASTM-E975-13):

$$V_\gamma = \frac{I_\gamma / R_\gamma}{I_\gamma / R_\gamma + I_\alpha / R_\alpha} \tag{2.42}$$

where, I_γ and I_α are the integrated peak intensities corresponding to austenite and ferrite, respectively. The parameter R is proportional to the theoretical integrated intensity that depends upon interplanar spacing (*hkl*), the Bragg angle θ, the crystal structure, and the composition of the phase being measured (ASTM-E975-13). Values of R for ferrite and austenite, reported in (ASTM-E975-13) for different planes and different input X-ray radiation, can be used for obtaining volume fraction of austenite in a ferritic steel specimen with near random crystallographic orientation.

2.4.3 Stress Measurements by XRD

X-ray diffraction measurements are used to determine residual macro-stresses as well as micro-stresses. These are specialized techniques whose physical and mathematical foundations are described in a number of textbooks and edited books [98–105] as well as in reviews, peer-reviewed articles, numerous conference papers, and practical guide reports [106]. A number of them are cited in the following pages, however, it is far beyond the scope of this book to give a comprehensive overview. Many aspects of progress made in X-ray diffraction for stress determination are discussed in the conference series "Proceedings of the Annual Conferences on Applications of X-ray

Analysis", published in "Advances in X-ray Analysis", see www.dxcicdd.com/advances/advances.htm. *We restrict ourselves discussing the material physics and the basic ideas of some of the experimental realizations.*

Stress measurements by X-ray diffraction are only possible in crystalline or polycrystalline materials. In polycrystalline materials, they yield the mean external or residual strain value of those crystallites in the ensemble of grains which contribute to the constructive interference due to their correct orientation as per Bragg's law (2.20). The measurements only probe a small part of the surface volume which was irradiated by the X-rays due to the limited penetration depth and the finite beam width. The measurements can be carried out without external markers. They do not entail a change of the state of the material, i.e. they are non-destructive.

For stress measurements, a monochromatic X-ray beam is preferably used. To this end, a goniometer arrangement may be used, see Fig. 2.10 where a Bragg reflection (2.20) is exploited. The available intensity is reduced by a factor of 10^3 and above, which increases the measurement time and reduces the available depth range, because a minimum intensity is required for a given signal-to-noise ratio [107]. Besides this reference, a number of other arrangements for stress measurements are discussed in [108] and many other references.

If there is a stress acting on the lattice planes, the interatomic distances change according to Hooke's law from d_0 to d. This in turn changes the incident angle in order to maintain constructive interference. Measuring the angle change relative to the undisturbed lattice is the principle of stress measurement by X-ray diffraction. The equilibrium lattice distance serves as a marker. Suppose that the lattice experiences a strain $\varepsilon = (d - d_0)/d_0$ due to external force or internal stress (Fig. 2.7b). Then, from (2.21 and 2.22) it follows that

$$\varepsilon = \frac{\Delta a}{a} = \frac{\Delta d}{d_0} = -cot(\theta)\Delta\theta \tag{2.43}$$

As can be seen from 2.43, changes of the lattice constants due to stresses can be better resolved for higher diffraction angles. Due to the always present measurements inaccuracies, this means in practice that Bragg angles $2\theta > 100°$ are preferable.

If measurements are carried out at a fixed wavelength given by the characteristic lines of the X-ray tube target or by the frequency-selective element inserted in the synchrotron radiation beam, the Bragg angle must be varied for different measurements. This is called the angle-dispersive technique. If however, the wavelength is varied by changing the X-ray energy according to (2.1), one speaks of the energy-dispersive technique in order to determine strains by X-ray diffraction. If their stress-free spacings d_0 are known, the strain $\varepsilon = (d - d_0)/d_0$ can be calculated from the diffraction data, which in turn can be converted to stresses using Hooke's law. The inversion of Hooke's law in Voigt's notation:

$$\sigma_i = c_{ij}\varepsilon_j (i, j = 1, 2, 3, 4, 5, 6) \tag{2.44}$$

i.e. $\boldsymbol{s} = \boldsymbol{c}^{-1}$, yields

$$\varepsilon_k = s_{kl}\sigma_l \tag{2.45}$$

where $\boldsymbol{s}$ is the compliance tensor and $\boldsymbol{c}$ is the tensor of elastic constants (c_{ij}). The compliance tensor s_{kl} [mm^2/N] connects the stress tensor with the elastic constants c_{ij} [N/mm^2] yielding the dimensionless strain tensor. In X-ray diffraction measurements, the elements of s_{kl} are called the X-ray elastic constants (XEC), despite the fact that they possess the inverse dimension of an elastic constant. Experimentally, the compliance values or XEC's must be determined by X-ray diffraction measurements for each material as a function of stress σ. This is by no means a simple task and is discussed in detail in [109]. Theoretically they can be calculated from the elastic constants of single crystals, respectively from the averaged values for polycrystals [99]. Some XECs are listed in Table 2.3 for α-iron and for ferritic-pearlitic steels and for other metals in Table 2.4. Further data can be found in [102, 104, 110–114], and in the references therein.

If the elastic properties of the materials are homogenous within the penetration depth, the strains measured with X-rays diffraction are the same as the macroscopic values. In this case one obtains for the mean compliances $s_1{}^m$ and $s_2{}^m$:

$$s_1^m = -\frac{\nu}{E} \tag{2.46}$$

$$\frac{1}{2}s_2^m = \frac{1+\nu}{E} \tag{2.47}$$

Here, E is Young's modulus. The weight of the grains contributing to constructive interference as a function of depth $\boldsymbol{x}$ decreases because of X-ray attenuation (2.8):

$$d = \frac{\int\limits_V d(x)e^{-x\mu}dx}{\int\limits_V e^{-x\mu}dx} \tag{2.48}$$

Due to the strong attenuation of X-rays in particular in metals, stress measurements are limited to a depth of some 10 μm, see Tables 2.3 and 2.4. In reality, stresses are inhomogeneous on a microscopic scale. It is convenient to differentiate between three types of residual stresses according to [116], see Fig. 2.39 (and Sect. 1.11 in Chap. 1):

(1) Type I stresses, also abbreviated as σ_I, are macroscopic stresses which extend over distances covering many grains;
(2) Type II stresses, σ_{II}, are microscopic stresses within single grains.
(3) Type III stresses, σ_{III}, are localized stresses caused by grain boundaries, small inclusions, and dislocations.

The local total stress σ_{RS} is the sum of the above defined stresses:

$$\sigma_{RS} = \sigma_I + \sigma_{II} + \sigma_{III} \tag{2.49}$$

Table 2.3 Penetration depths and X-ray elastic constant S_2^{xr} for α-iron and for ferritic-pearlitic steels. In the Table, Γ is the so-called orientation factor, see Sect. 2.4.7 (from [6]). The theoretical values are based on the theory of [115] for the elastic constants of polycrystals

Crystal plane	Γ	Target Material	λ K_α [Å]	Energy [keV]	2θ	$(\mu/\rho)_{Fe}$ [cm^2/g]	δ_{Fe} [μm]	$S_{2,exp}^{xr}/2$ [10^{-6} mm^2/N]	$S_{2,th}^{xr}/2$ [10^{-6} mm^2/N]
(211)	0.25	Cr	2.291	54.2	156.4	115	5.5	5.68 ± 0.08	5.76 ± 0.12
(220)	0.25	Fe	1.937	64.1	145.8	72.8	8.7	5.36 ± 0.23	5.76 ± 0.12
(310)	0.09	Co	1.790	69.3	161.88	59.5	10.7	7.02 ± 0.12	6.98 ± 0.16

Table 2.4 X-ray elastic constants and other data required for the stress measurement using X-ray diffraction. The data are compiled from tables in [99]. The penetration depths refer to certain goniometer refraction geometries. Note that S_1 is always negative (2.46)

Material	Radiation	Plane (*hkl*)	Penetration depth for $\sin^2\psi$ = 0.3 Ω-Diffr. [μm]	Penetration depth for $\sin^2\psi = 0.3$ Ψ-Diffr. [μm]	$S_2{}^{rö}/2$ [10^{-6} MPa^{-1}]	$-S_1{}^{rö}$ [10^{-6} MPa^{-1}]
BCC iron Ferrites Martensites	Cr-K$_\alpha$	(211)	4.4	4.5	5.76	1.25
	Co-K$_\alpha$	(310)	8.7	8.8	6.98	1.66
	Mo-K$_\alpha$	(732 + 651)	13.2	13.5	6.05	1.34
Austenitic steels	Mn-K$_\alpha$	(311)	5.5	5.7	6.98	1.87
	Cu-K$_\alpha$	(220)	3.8	4.2	6.05	1.56
Aluminum and Al-based materials	Cu-K$_\alpha$	(511 + 333)	31.1	31.4	19.77	5.22
	Cu-K$_\alpha$	(422)	27.7	27.9	19.07	4.97
Copper and Cu-based materials	Cu-K$_\alpha$	(420)	8.1	8.5	11.22	2.92
Nickel and Ni-based materials	Cr-K$_\alpha$	(220)	2.7	3.0	5.88	1.24
Titanium Ti-based materials	Cu-K$_\alpha$	(006) (302)	4.3	4.3	9.96 11.90	2.25 2.91

In order to measure stress, strain, and texture by X-ray diffraction, the X-ray wavelengths must be comparable to the lattice constants entailing energies of some 10 keV, see (2.1, 2.21 and 2.22). For the corresponding energies, penetration depths δ of some 10 μm are obtained, see Table 2.3. In contrast, X-ray imaging of components and weldments require much higher X-ray energies for penetration in the centimeter to decimeter or even meter range.

2.4.4 *The $\sin^2\psi$ Law for Stress Measurements by XRD - Angle Dispersive Measurements*

Let us assume that in the isotropic material to be measured, the principal strains are ε_1, ε_2, ε_3 and the affiliated principal stresses are σ_1, σ_2, σ_3 (note that the Voigt notation (2.44)) is used. Let the principal stresses σ_1, σ_2, be parallel to the surface

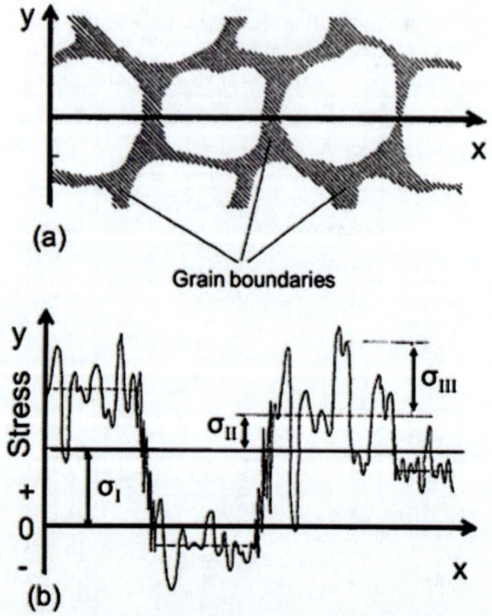

Fig. 2.39 Left: **a** Schematic view of grain boundaries; **b** Definition of stresses of the first (σ_I), second (σ_{II}), and third kind (σ_{III}) (in modified form [116], with permission from Springer Nature)

plane of the material. Using well-known trigonometric relations, the corresponding strain amplitude $\varepsilon_{\varphi,\psi}$ in the direction given by the angles (φ, ψ) of the coordinate system of Fig. 2.40a is given by [101, 117]:

$$\varepsilon_{\varphi,\psi} = \varepsilon_1 cos^2\varphi \; sin^2\psi + \varepsilon_2 sin^2\varphi \; sin^2\psi + \varepsilon_3(1 - sin^2\psi) \tag{2.50}$$

Hooke's law provides the relation between the principal strains and the principal stresses:

$$\begin{aligned} &\varepsilon_1 = \sigma_1/E - (\upsilon/E)(\sigma_2 + \sigma_3), \\ &\varepsilon_2 = \sigma_2/E - (\upsilon/E)(\sigma_3 + \sigma_1), \\ \text{and} \quad &\varepsilon_3 = \sigma_3/E - (\upsilon/E)(\sigma_1 + \sigma_2). \end{aligned} \tag{2.51}$$

Inserting these relations into (2.50) yields

$$\varepsilon_{\varphi,\psi} = \frac{\upsilon + 1}{E}\left(\sigma_1 cos^2\varphi + \sigma_2 sin^2\varphi - \sigma_3\right) sin^2\psi - \frac{\nu}{E}\left(\sigma_1 + \sigma_2 - \frac{\sigma_3}{\upsilon}\right) \tag{2.52}$$

As said above, due to the small penetration depth of the X-ray in metals, about 10 μm (see Tables 2.3 and 2.4), only a near-surface region can be examined. Because the stress at the surface in the 3-direction must disappear, $\sigma_3 = 0$ is usually set for the

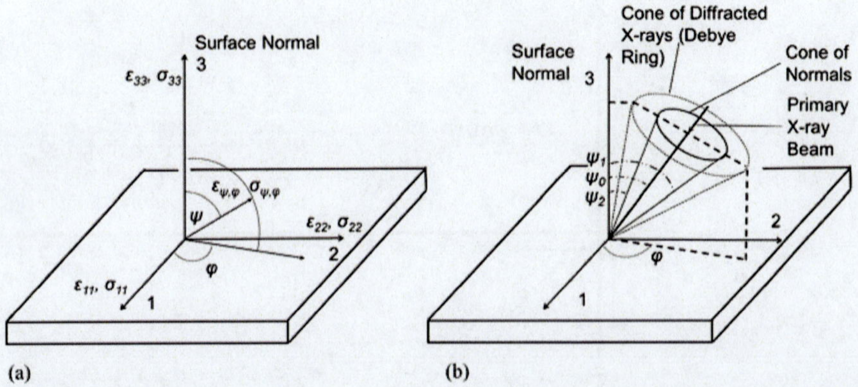

Fig. 2.40 **a** Sample defined coordinate system to describe the strain and the stress state. The principal stress components σ_I (σ_{II}) and σ_2 (σ_{22}) are in the surface of the component under test. Note that the following tensor-notation is used: Tensor 11 → Index 1; 22 → 2; 33 → 3; 23,32 → 4; 13, 31 → 6; and 12, 21 → 6; **b** Individual crystal diffracting the incoming oblique X-ray. The crystallites diffract into a direction which lies on the diffraction cone symmetric to the incoming X-ray beam. Those crystallites with the Miller indices (*hkl*) contribute to the interference whose normal vector lies on the normal cone symmetric to the primary beam. In case there are mechanical stresses, the interference and normal cones are no longer rotationally symmetric to the primary X-ray beam. The diffracted spots may be detected by single spot-detectors, counters, films, or pixel array detectors. The dashed line defines the cutting plane of the biaxial stress state

whole penetration depth. This means that the data evaluation based on this assumption can yield only plane surface stresses (see however, Sect. 2.4.7). Defining

$$\sigma_\varphi = \sigma_1 cos^2\varphi + \sigma_2 sin^2\varphi \tag{2.53}$$

and with (2.46 and 2.47), (2.52) yields for the strain in the direction (φ, ψ) of a plane stress-state:

$$\varepsilon_{\varphi,\psi} = \frac{1}{2} s_2 \sigma_\varphi sin^2\psi + s_1(\sigma_1 + \sigma_2) \tag{2.54}$$

Setting the measured lattice strain $(\Delta d/d_0)_{\varphi,\psi}$ equal to the theoretical strain $\varepsilon_{\varphi,\psi}$, yields

$$\varepsilon_{\varphi,\psi} = \left(\frac{\Delta d}{d_0}\right)_{\varphi,\psi} = -cot(\theta)\Delta\theta_{\varphi,\psi} \tag{2.55}$$

Combining (2.54) and (2.55), one obtains the basic equation for stress measurements based on X-ray diffraction:

$$\varepsilon_{\varphi,\psi} = -cot(\theta_0)\Delta\theta_{\varphi,\psi} = \frac{1}{2} s_2 \sigma_\varphi sin^2\psi + s_1(\sigma_1 + \sigma_2) \tag{2.56}$$

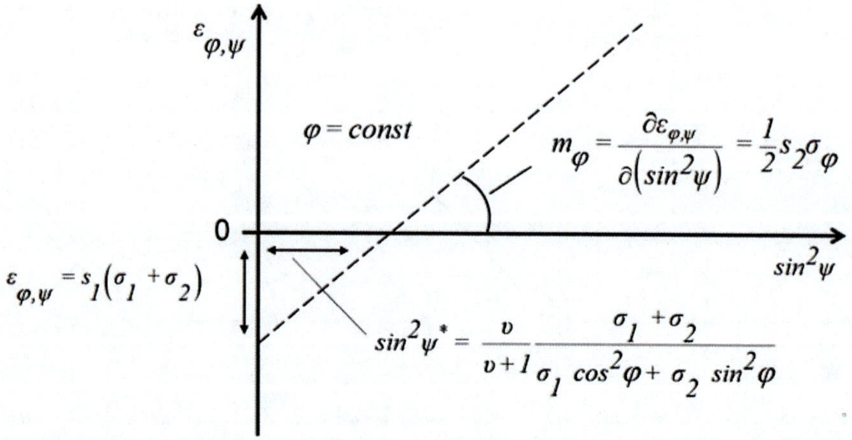

Fig. 2.41 Geometrical representation of the $sin^2\psi$ law for stress measurements; in modified form from [117], with permission from Springer Nature)

The strain amplitude given by (2.56) is a cut through the strain ellipsoid at $\varphi =$ const. of a biaxial stress-state. From (2.56) and from Fig. 2.41 one can derive that

1. The lattice distortions $\varepsilon_{\varphi,\psi}$ for a given azimuth angle φ depend linearly on $sin^2\psi$.
2. The slope of the $\varepsilon_{\varphi,\psi} - sin^2\psi$ curve

$$m_\varphi = \frac{\partial \varepsilon_{\varphi,\psi}}{\partial\left(sin^2\psi\right)} = \frac{1}{2}s_2\sigma_\varphi \tag{2.57}$$

is proportional to the surface stress component σ_φ which acts in the azimuth angle φ. The slope m_φ is determined by taking data for different ψ with the angle-dispersive technique. The measurements are always taken perpendicularly to the reflecting lattice planes, see Fig. 2.40b.

3. The ordinate section of the $\varepsilon_{\varphi,\psi}$ versus $sin^2\psi$ curve is independent of the cutting plane and proportional to the sum of the principal stresses $(\sigma_1 + \sigma_2)$:

$$\varepsilon_{\varphi,\psi} = s_1(\sigma_1 + \sigma_2) \tag{2.58}$$

4. For a given azimuth angle φ, the lattice distortions $\varepsilon_{\varphi,\psi}$ are zero for the direction $\psi*$ given by

$$sin^2\psi^* = \frac{-s_1}{s_2/2}\frac{\sigma_1 + \sigma_2}{\sigma_\varphi} = \frac{\nu}{\nu + 1}\frac{\sigma_1 + \sigma_2}{\sigma_1 cos^2\varphi + \sigma_2 sin^2\varphi} \tag{2.59}$$

In recent years technologies were developed to obtain X-ray diffraction data by area detectors allowing to measure residual elastic strains in a material along a set of scattering vectors in a single measurement as discussed in the following section.

2.4.5 Stress Measurements by Two-Dimensional X-ray Diffraction Data

It was recognized early on that area detectors like X-ray sensitive photographic films can be used to measure the positions of the diffracted spots on the Debye rings in laboratory as well as in portable equipment [112]. Nowadays, films are replaced by area detectors which allow to record and to read out incident X-ray intensities electronically from its pixels [107, 118, 119]. These techniques are sometimes called "Two-dimensional X-ray diffraction" [100]. In contrast to the $sin^2\ \psi$-technique used with single detectors, which requires the measurement of strain in at least two distinct sample orientations for plane stress states [99, 103], area detectors allow one in principle to determine the residual stress in a single exposure. This has resulted in commercial portable apparatus designed for in-line and field measurement applications, especially for ferrous metals [120]. An advantage of an area detector is the possibility to record the entire Debye Ring and thus determine texture and grain size as well [121]. Different approaches for the determination of the stress state from the information contained in a single Debye exposure exist, for example the $cos\ \alpha$ method [122–124] or the direct least-squares fitting of the measured strain [119]. Despite the attractiveness of these portable devices in both industrial and scientific applications, there is a debate among experts on the most efficient, precise, and accurate stress determination formalism.

In order to relate the positions of the diffraction spots (ψ, φ) in the coordinate frame of the area detector with the stress state in the sample, one must know the relations between the scattering vectors, the Miller indices (*hkl*) of the diffracting polycrystalline grains causing the diffraction spot, the sample coordinates, and the laboratory coordinates. Furthermore, let the orientation of the area vector of the detector be parallel to the primary X-ray beam. Then, those crystallites which subtend the angle η with the X-ray beam will interfere constructively under the angle $2\eta = \pi$ -2θ, and the spots lie on the Debye ring indicated in Fig. 2.42. The scattering vector will reside in the cone of normals, also shown in Fig. 2.42 with the opening angle η. Since each point in the ring corresponds to a different orientation of the scattering vector, there exists a transformation that relates its coordinates in the laboratory system to those in the sample system. We refrain from developing the relations just mentioned, because a complete analysis is given by [125], where it is shown that to the vector (ψ, φ) belongs the scattering vector (η, α) (see Figs. 2.40b and 2.42) and that for this scattering geometry the $sin^2\psi$ law of (2.59) is recovered.

One of the methods that allows the calculation of stress from a single measurement with an area detector is the so-called $cos\alpha$ method described by [122] and [126] and

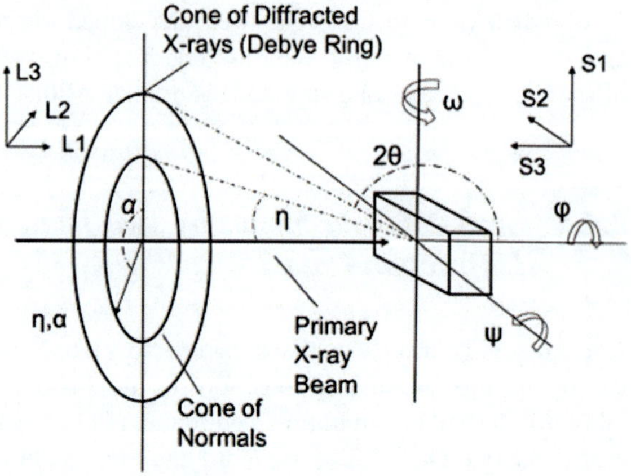

Fig. 2.42 Scattering geometry using an area detector with the sample coordinate system, the angles (ω,φ,ψ) of the diffracting crystallites with the Miller indices (*hkl*), and the laboratory coordinate system according to (in modified form from [125], with permission from Springer Nature)

implemented in commercial instrumentation. The basic idea is to start from the expression of the scattering vector in sample coordinates and in terms of Debye ring coordinates with an arbitrary angle ψ (Fig. 2.42). Restricting to the sample-detector geometry of Fig. 2.42, only a rotation along S_2 in sample coordinates is then allowed. The strain projection along (η,α) coordinates can be written in terms of the scattering vector and strain components. Eventually one obtains two functions [125]:

$$a_1(\varphi) = \frac{1}{2}((\varepsilon(\alpha) - \varepsilon(\pi + \alpha)) + (\varepsilon(-\alpha) - \varepsilon(\pi - \alpha))) \tag{2.60}$$

$$a_2(\varphi) = \frac{1}{2}((\varepsilon(\alpha) - \varepsilon(\pi + \alpha)) - (\varepsilon(-\alpha) - \varepsilon(\pi - \alpha))) \tag{2.61}$$

Applying plane stress conditions relating ε_{ij} to the stresses, (2.60 and 2.61) can be written as

$$a_1(\varphi) = \frac{1+\upsilon}{E} sin2\psi\ sin2\eta\ cos\alpha(\sigma_{11}(1 + cos2\varphi) + \sigma_{22}(1 - cos2\varphi) + 2\sigma_{12} sin2\varphi) \tag{2.62}$$

$$a_2(\varphi) = \frac{1+\upsilon}{E} sin\psi\ sin2\eta\ sin\alpha(\sigma_{22}\ sin2\varphi - \sigma_{11}\ sin2\varphi + 2\sigma_{12}\ cos2\varphi) \tag{2.63}$$

For the special case of $\varphi = 0$, one obtains:

$$a_1(\varphi = 0) = \frac{1+\upsilon}{E}\sigma_{11} sin2\psi\ sin2\eta\ cos\alpha \tag{2.64}$$

$$a_2(\varphi = 0) = \frac{1+\upsilon}{E}\sigma_{12} sin\psi\ sin2\eta\ sin\alpha \tag{2.65}$$

Thus, by plotting the parameters a_1 and a_2 as a function of $cos\alpha$ and $sin\alpha$, one obtains two linear relationships whose slopes are proportional to σ_{11} and σ_{12}. Generalizing these equations, it can be shown that there is a linear relationship between the experimentally determined strain projection ε_α and the strain components ε_{ij} in the sample coordinates where the coefficients are the components of the scattering vector:

$$\varepsilon_\alpha = q_i q_j \alpha_{ij} \tag{2.66}$$

The q_i depend only on the angle positions of the sample and the coordinates on the Debye ring, i.e. the angles ψ, φ, η, and α. Equation (2.66) represents a set of linear equations for each polar angle α at which the strain is determined. The four unknowns are the α_{ij} because ε_α is determined experimentally. Since the number values probed is more than four (typically several hundred), the system is overdetermined and can be solved numerically by least-squares fits. Experimentally, the Debye ring is fitted to obtain $2\theta(\alpha)$ and then the strain is calculated using

$$\varepsilon_\alpha = ln\left(\frac{sin\theta}{sin(\theta(\alpha))}\right) \tag{2.67}$$

Concerning the further development of the area detector methods, details can be found in [123, 125, 127–130] including calibration issues, standard materials, measurement accuracies, analysis procedures, comparison to the $sin^2\psi$ method, and software issues. Further information can be found in the references mentioned at the beginning of this section.

2.4.6 *Deviations from the Linear Behavior of the $sin^2\psi$ Law*

The linear $\varepsilon_{\varphi,\psi}$ versus $sin^2\psi$ law (2.59) (Figs. 2.41, 2.43a) is not always observed, as already mentioned above. In the case of biaxial stress with shear (i.e. only $\sigma_{33} = 0$), the shear strain results in a deviation from the linear relation of the $\varepsilon_{\varphi,\psi}$ -$sin^2\psi$ plot, as shown in Fig. 2.43b. The deviations are both positive and negative from the linear relation, depending on the sign of the ψ tilt angle. Therefore, it is also referred to as ψ-split. The shear stress can be measured from the amount of ψ split [99]. Figure 2.43c displays the situation when stress gradients perpendicularly to the surface are present. At low values of $\varepsilon_{\varphi,\psi}$ versus $sin^2\psi$, a linear slope is observed, whereas at higher values of $sin^2\psi$ the curve starts to bend. Figure 2.43d shows the situation when a texture parallel to the residual stress is present. In this case, the measured strain ε as a function of $sin^2\psi$ resembles a sinusoidal curve. In summary if stresses overlap with inhomogeneities, deviations from the linear $\varepsilon_{\varphi,\psi}$ versus $sin^2\psi$ law will occur. A detailed discussion of these effects can be found in [99, 100].

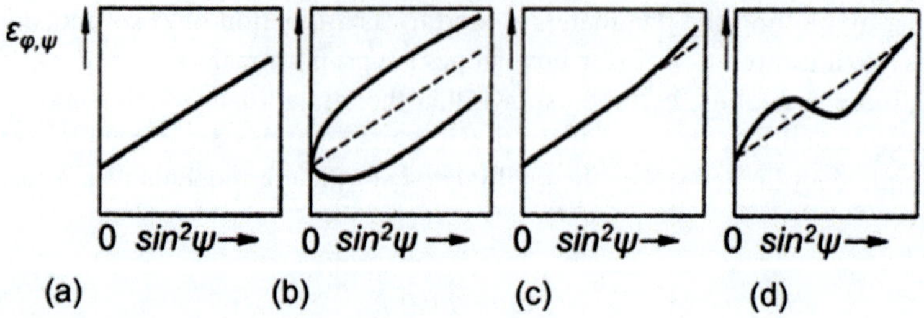

Fig. 2.43 Schematic shapes of $\varepsilon_{\varphi,\psi}$-$\sin^2\psi$ curves deviating from linearity. Measured data showing this behavior can be found in [99, 100, 117, 131]

2.4.7 *Stress Measurement by XRD - Energy Dispersive Measurements*

Stress gradients may be evaluated quantitatively by etching surface material away step by step with electrolytic solutions, which is a destructive method. However, by employing the energy dispersive method in X-ray diffraction measurements, stress gradients can be recognized and evaluated. Combining (2.1, 2.2 and 2.20) leads to the equation underlying principle of the energy dispersive X-ray diffraction method:

$$E_x d sin(\theta_0) = \frac{hc}{2} \approx 6.199\ [keV\text{Å}] \tag{2.68}$$

Equation (2.68) defines hyperbolic surfaces which are determined by the energy E, lattice plane distance d, and diffraction angle θ. Using (2.21) allows to draw these surfaces in two-dimensional diagrams, one of which is shown in Fig. 2.44. In contrast to the angle-dispersive technique, one uses for the energy dispersive technique polychromatic X-ray radiation at constant θ. In analogy to (2.55) one can write:

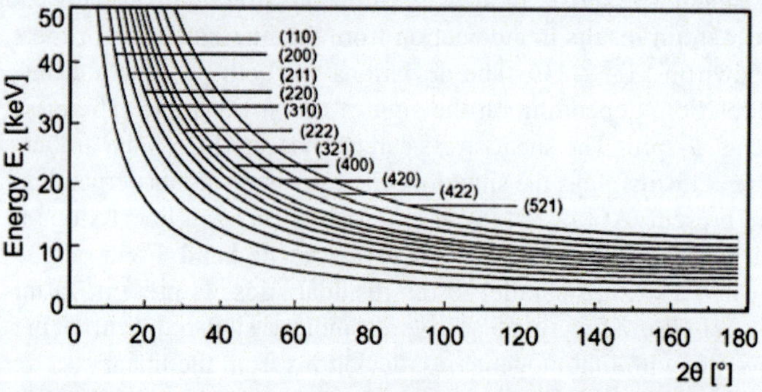

Fig. 2.44 X-ray energy as a function of diffraction angle for reflexes for ferrites

$$\varepsilon = \frac{\Delta a}{a} = \frac{\Delta d}{d_0} = \frac{\Delta E_x}{E_{x,0}} = -cot(\theta)\Delta\theta \quad (2.69)$$

Here, ΔE is the energy resolution of the detector and $\Delta\theta$ is the angle divergence of the radiation. The lattice strains are determined by the shift of the peaks of given reflexes in the energy spectra of the diffracted beam. In energy-dispersive stress-measurement determined by the angles (φ, ψ) and θ, all reflexes are measured which are possible by the energy spectrum of the incident radiation according to the Bragg condition. Because there is an individual penetration depth for each reflex, the gradient of the stress in 3-direction becomes measurable using conventional X-ray radiation from X-ray tubes [132]. This was exploited in a number of experiments in order to further develop the energy dispersive methods for applications.

In a series of publications, Ruppersberg et al. [132] first examined single-phase polycrystalline materials using energy dispersive X-ray diffraction measurement. They discussed the relation between the elastic strain $\varepsilon(\psi,\varphi)$ obtained and the polynomial expansion of the stress components $\sigma(z)$ and $\sigma(z)_{\varphi+90°}$ as a function of depth z from the surface of the specimen leading to a universal plot for the averaged strain curves $\overline{\varepsilon}(\psi, \varphi, \delta(hkl)) = 1/2(\varepsilon(\varphi) + \varepsilon(\varphi + 90°))$ measured for various X-ray penetration depths δ and for different reflecting lattice planes. The relation also allowed absolute scaling of the ε values. Furthermore, the authors applied their formalism in order to discuss $\varepsilon(\psi, \varphi, \delta(hkl))$ curves obtained with synchrotron radiation for a cold-rolled nickel plate. The data indicate strong non-linear variations of stress in the surface of 2 μm thick layer. Furthermore, Ruppersberg and Detemple [133] also showed that energy-dispersive X-ray diffraction measurements are a tool in order to investigate the complex stress-fields of the surface region of polycrystalline materials after plastic deformation, here for a ground steel plate, covering a penetration depth of the X-rays from 0.04 to 60 μm, yielding the variations in the residual stress components $\sigma_{11}(z)$ and $\sigma_{22}(z)$.

White cast iron consists of the two phases cubic ferrite (with traces of C and other impurities) and orthorhombic cementite (Fe_3C, which may also contain impurities). On cooling strong thermoelastic stresses $\Delta\sigma_{jj}$ develop which have different signs in the two phases [134]. According to the $sin^2\psi$ -law of conventional X-ray stress analysis, the corresponding thermoelastic strains of the lattice plane spacings, if plotted versus $sin^2\psi$, should yield straight lines with slopes proportional to the $\Delta\sigma_{jj}$ components (2.57). The slopes observed for the two phases should, therefore, have different signs. Experimentally, the opposite was observed. It was shown that this unexpected behavior is related to the anisotropic microscopic thermal expansion coefficients of the cementite combined with anisotropic macroscopic thermal expansion coefficients of the specimen. The latter is a result of the texture of the cementite phase.

In a further work, Ruppersberg et al. [135] outlined the use of synchrotron radiation for the evaluation of stress fields in polycrystalline materials with the individual crystals being elastically anisotropic. The residual stress-fields $\sigma_{11}(z)$ and $\sigma_{22}(z)$ were examined which varied significantly within the X-ray penetration depth δ. However, on a macroscopic scale the specimens were supposed to be isotropic and homogeneous within the irradiated volume when neglecting effects related to plastic deformations. Normalized strain curves $\varepsilon, \varphi, \psi$ were obtained from the diffraction experiments. From the ε-data a "universal curve" was calculated which was independent of the reflecting lattice planes and the wavelength. Furthermore, [135] demonstrated that the non-linear ε-$\sin^2\psi$ curves observed for the (531) reflections of a cold-rolled nickel plate are related to strong stress gradients (see also Fig. 2.43c). Finally, Ruppersberg [136] analyzed the triaxial thermoelastic stresses in white cast iron by X-ray diffraction. The transition to the biaxial stress field in the surface region of the polyphase specimen was studied using synchrotron radiation. The study was based on the continuum theory of elasticity applied to the case of inclusions in a half space. An important result was the observation of stress gradients which drop to zero in the surface, thus inducing gradients of the remaining biaxial stresses. The conclusion was that it is not justified to assume that X-ray diffraction experiments in the surface layer of polyphase materials alone allow one to detect the biaxial stresses in the bulk material.

2.4.8 Stress Measurement by XRD in Anisotropic Polycrystalline Materials

For elastically isotropic materials, the $sin^2\psi$ -law is well obeyed and the relations between strain and stress are determined by the elastic compliances, see (2.54, 2.56, and 2.58) For elastically anisotropic materials the situation is different. If it is assumed that there is a uniform strain in the polycrystalline ensemble, i.e. the Voigt boundary condition holds [137], the X-ray elastic constants do not change. However, if the Reuss condition [138], i.e. if uniform stress is assumed, the orientation factor Γ

$$\Gamma = \frac{(hk)^2 + (kl)^2 + (hl)^2}{\left(h^2 + k^2 + l^2\right)}. \tag{2.70}$$

must be taken into account in order to obtain the X-ray elastic constant replacing those of (2.46 and 2.47) [6, 105]:

$$S_1^{Reuss} = S_{12} + \Gamma S_0 \tag{2.71}$$

$$\frac{1}{2} S_2^{Reuss} = S_{11} - S_{12} - 3\Gamma S_0 \tag{2.72}$$

Here, $S_0 = S_{11} - S_{12} - 0.5S_{44}$ and S_{11}, S_{12}, and S_{44} are the single crystal elastic constants. A detailed discussion of the orientation factor for different crystal lattices can be found in [99]. The X-ray elastic constants (XECs) depend upon the *(hkl)* reflexes, see Table 2.4. They are found to agree well with the average of the Voigt and Reuss result, known as the Kröner-solution [115], similar to the situation in

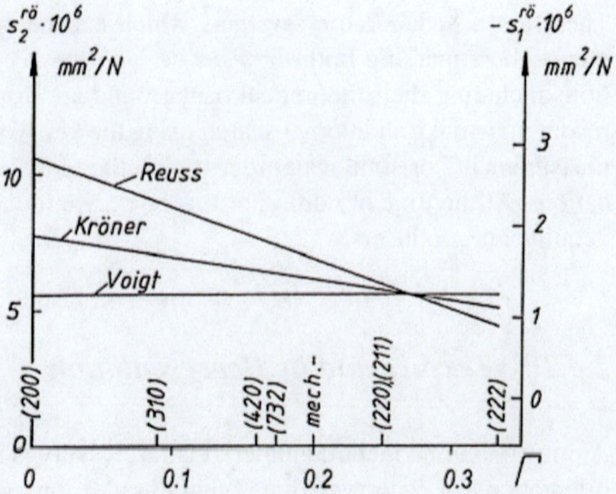

Fig. 2.45 X-ray elastic constant for α-iron as a function of Γ (from [6], with permission from Springer Nature). Data for ferrite, austenite, aluminum, copper, nickel, and other materials can be found in [99] and in [140]

ultrasonics, see Chap. 1, Sect. 1.9. This is shown in Fig. 2.45 for α-iron, where the dependence of the X-ray elastic constants is shown as a function of the orientation Γ according to (2.70). This was further discussed by [113, 139] by considering grain to grain interactions for different crystal symmetries.

2.4.9 Applications of X-ray Stress Measurements for NDMC

Navas et al. [141] applied X-ray stress measurements using the $\varepsilon_{\varphi,\psi}$ versus $sin^2\,\psi$ law to study surface stresses caused by machine turning of steels. The parameters examined were the cutting speed, feed, tool-nose radius, geometry of the tool chip breaker, and the coating of the cutting tool in AISI 4340. The steel bars were subjected to turning tests using cutting speeds between 200 and 300 m/min and two cutting feeds 0.075 and 0.2 mm/rev. Also the type and the geometry of the chip breakers were varied. In all cases surface residual stresses were more or less tensile, which consequently meant to be more or less detrimental to the service life of the machined component, depending on the cutting conditions and the characteristics of the cutting tool such as tool-nose radius. Navas et al. [141] found that as feed increases, the likelihood of tensile stresses increased due to an increase in cutting temperature, which was accompanied by an increase in surface roughness. For lower cutting speeds the trend was opposite. Using the Debye–Scherrer ring method (see Sect. 2.2.5), Gupta [142] studied the triaxial stress-state of a helical coil-spring of a passenger car, which led to fracture, both with the conventional X-ray diffraction analysis as well as with two-dimensional X-ray diffraction "$cos\alpha$" method.

There are a number of portable X-ray diffraction systems on the market which are specially designed for field measurements, in particular based on the "$cos\alpha$" method.

There are also laboratory systems which can be used for large sample measurements encompassing both the "$sin^2\psi$" and the "$cos\alpha$" method. Practical applications including their theoretical background are discussed by [120, 125, 128]. The reader may make an internet search using the key words "X-ray diffraction for stress measurement" or similar in order to find the most suitable equipment for his applications. At the time of editing of this book, we found about 6 URLs of commercial #equipment producers.

2.4.10 Texture and its Representation

A polycrystalline metallic material is rarely truly isotropic with completely random orientations of the crystallites or grains. Various stages of processing in making a component, such as casting, forging, rolling, extrusion and heat treatment may introduce specific preferred orientations of grains due to crystallographic direction dependent physical and mechanical properties. Preferred orientations of grains in a component lead to anisotropic properties of the final product. Sometimes, components are desired to have anisotropic properties, such as directionally oriented nickel based superalloys, which are used in turbine blades as their creep rupture resistance is enhanced by orienting the grain boundaries parallel to the applied stress direction [143]. Similarly, a special pilgering process is employed to have more than 80% of basal poles oriented along the radial direction of zircaloy clad tubes used in nuclear reactors in order to avoid the formation of deleterious hydrides in the radial direction [144]. Furthermore, the randomization of a crystallographic texture formed during the rolling of uranium fuel rods is essential to avoid their irradiation growth during operation in a nuclear reactor [145].

The texture in a rolled sheet material is commonly represented by {*hkl*} <*uvw*>, which means that the {*hkl*} planes of these grains lie parallel to the sheet plane, whereas their < *uvw* > directions point parallel to the rolling direction [34, 146]. For materials formed into shapes like extruded rods, wires or thin films, any one of these two sets ({*hkl*} or < *uvw* >) are used to describe texture [147]. Important texture components and fibers in FCC and BCC metals are listed in Table 2.5. In the "cube texture" in FCC or BCC materials, the grains are oriented so that the {001} planes lie nearly parallel to the plane of the sheet and the < 100 > directions point approximately in the rolling direction. This texture, as well as another texture known as Goss texture {110} < 001 > , are usually desired in magnetic materials, in which it is easy to magnetize in the cube edge < 100 > direction.

The orientation of a crystal in a polycrystalline material with respect to the specimen's orientation is described quantitatively by a set of three Euler angles [150]. The Euler angles in ϕ_1, Φ and ϕ_2 refer to the three rotations when performed sequentially, transform the specimen coordinate system X (rolling direction), Y (transverse direction), and Z (normal directions) onto the crystal coordinate system [100], [010] and [001] (Fig. 2.46).

Table 2.5 Texture components and fibers in FCC and BCC metals ([148, 149]. ND: normal direction: RD: rolling direction

Fiber	Miller index description	
α (FCC)	< 110 > //ND	
α (BCC)	< 110 > //RD	
γ	< 111 > //ND	
θ	< 100 > //ND	
τ	< 110 > //TD	
Component	Miller indices {hkl} <UVW>	Euler angles Φ_1, Φ, Φ_2
Copper	{112} < 111 >	90, 35, 45
S	{123} < 634 >	59, 37, 63
Goss	{011} < 100 >	0, 45, 90
Brass	{011} < 211 >	35, 45, 90
Dillamore	{4 4 11} < 11 11 8 >	90, 27, 45
Cube	{001} < 100 >	0, 0, 0

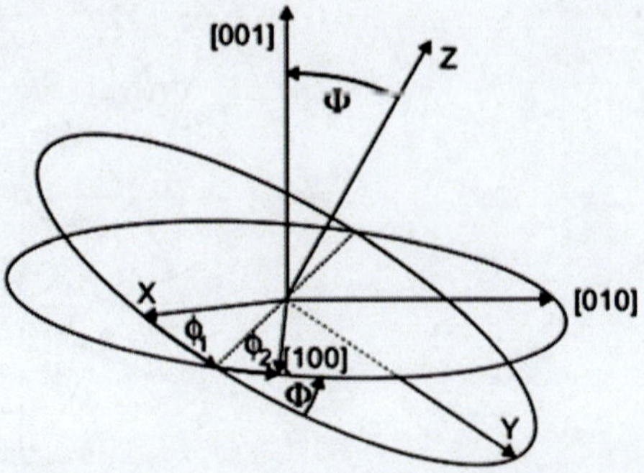

Fig. 2.46 Schematic of Euler angles according to [150]) showing three sequential rotations ϕ_1, Φ and ϕ_2 about Z, X and Z axes, respectively in order to transform the specimen coordinate system (X, Y, Z) onto the crystal coordinate system ([001], [010] [100])

A texture in a specimen is presented pictorially by mapping the spatial distribution of the crystallographic orientation of individual grains in a so-called pole figure map. It displays the two-dimensional stereographic projections that show the variation of pole density with pole orientation for a selected set of crystal planes $\{hkl\}$ relative to the specimen's geometrical orientation. The generation of a pole figure map involves the following two steps: (i) finding the intersection of the plane normals (poles) with a sphere. Here, it is thought that all crystallites of the specimen are located in the center of this sphere (Fig. 2.47a); (ii) connecting the south pole to the intersection on the sphere in the northern hemisphere to find its projection (intersection) on the equatorial plane of projection (Fig. 2.47b). The stereographic projections of the plane poles can be visualized on the great circle of equatorial plane as shown in Fig. 2.47c. Pole figures of a specimen can be generated individually for specific Bragg reflection planes.

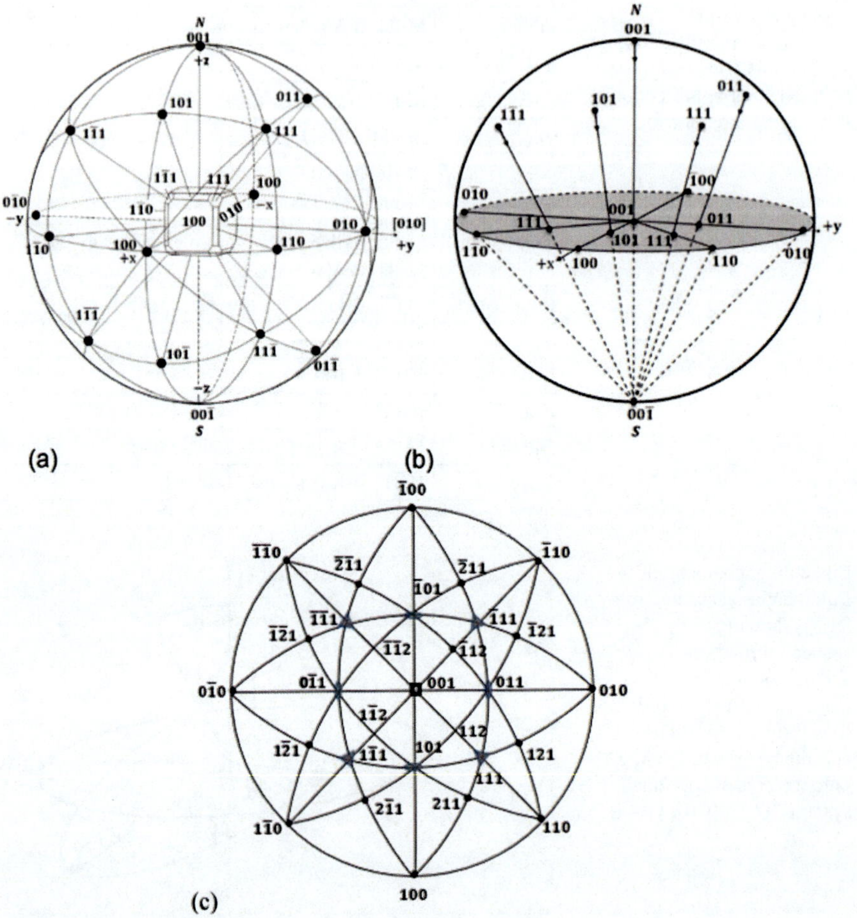

Fig. 2.47 **a** A cubic crystal showing faces of the form {100}, {110} and {111} at the center of the stereographic sphere with face normals intersecting the surface of the sphere; **b** Connecting the south pole to the intersection points on the surface of the sphere provides the projection lines. The intersections of the projection lines with the equitorial plane provide the poles of planes in the northern hemisphere; **c** A plan view of the stereographic plane of projection showing the poles of various planes in **a**. Great circles in **a** project as the arcs of circles in **c**

Figure 2.48a represents the < 100 > pole figure for the single crystal shown in Fig. 2.47a with the crystallographic directions aligned with the specimen's coordinate system (RD-rolling direction, TD-transverse direction and ND-normal direction (center of the pole figure circle)). The poles corresponding to < 100 > are seen only on the circumference at the top [-100], bottom [100], right [010], left [0–10] and center [001] of the stereogram. In case of a polycrystalline specimen with grains having completely random directions, the < 100 > poles can be observed at all the locations in the pole figure randomly distributed (Fig. 2.48b). Figure 2.48c shows

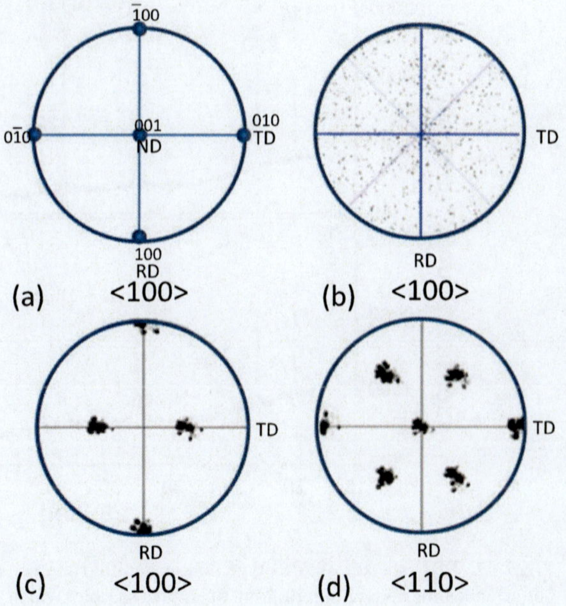

Fig. 2.48 **a** < 100 > pole figures for **a** a single crystal with crystallographic < 100 > directions aligned with the specimen's coordinate system (RD-rolling direction, TD-transverse direction and ND-Normal direction (center of the pole figure circle)); **b** < 100 > pole figure of a polycrystalline specimen with randomly oriented grains; **c** < 100 > pole figure of a Goss texture present in a grain oriented electrical steel; **d** < 110 > pole figure map for the same specimen with Goss texture (**c** and **d** according to [151])

the < 100 > pole figure map obtained for a grain oriented electrical steel exhibiting a strong Goss texture {011} < 100 >, as shown in Table 2.5. The < 100 > poles along the RD direction and along the horizontal diameter can be clearly observed. If a < 110 > pole figure is generated for the same sample, i.e. the center of the pole figure map represents [110] reflection and all other < 110 > reflections are utilized, then one set of [1–10] can be observed in the transverse direction and the [101], [011], [01–1], and [10–1] can be observed in the 1st-4th quadrants, respectively.

2.4.11 *Principle of Texture Evaluation by XRD*

Texture measurements by XRD are based on the fact that the intensity of any diffraction peak depends on the volume fraction of crystals/crystallites which are oriented in diffraction position in the considered sample direction [33]. This can be seen in Fig. 2.49 that shows the XRD spectra for AISI type 304 austenitic stainless steel (FCC) specimens, which were mill-annealed and warm-rolled at 473 K in order to obtain a 12% thickness reduction. The increase of the {220} peak intensity, as compared to the intensity of the {311} peak in the normal direction, is clearly evident after rolling. The increase in {220} peak amplitude is in line with the development of the {110} < 112 > rolling texture reported in FCC materials with low stacking fault energy [148].

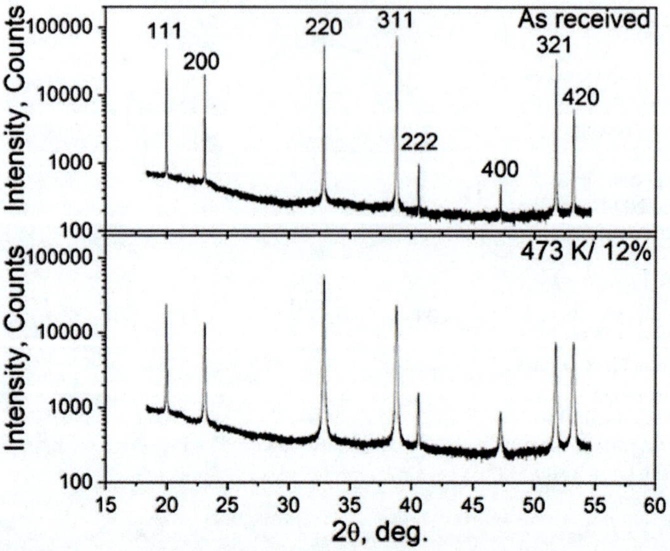

Fig. 2.49 XRD spectra of AISI type 304 austenitic stainless steel specimens. Top: XRD spectra before annealing and rolling; Bottom: Spectra of samples which were mill-annealed and warm rolled at 473 K in order to obtain a 12% thickness reduction. The XRD measurements were conducted using angle dispersive XRD at the Indian synchrotron source Indus-2. The spectra were recorded with a "Mythen" detector in the angular range of 18°-55° with a resolution of 0.003° for an X-ray beam of 0.719 Å wavelength having a spot size of 500 × 500 μm^2. The increase in the (220) peak amplitude after rolling as compared to the {311} peak amplitude in the normal direction is clearly evident (in modified form from [152], with permission from Elsevier

Stresses change the diffraction angle θ_{hkl} due to the change in the interplanar distance d_{hkl}, see (2.43)*, whereas a texture does not change θ_{hkl}, but changes the relative amplitudes of the reflection peaks. It causes a skewing of the grain orientation distribution function (ODF),* (2.73) *and Fig.* 2.50b.

Changes in the relative amplitudes of the peaks for different planes indicate the presence of a texture qualitatively. With the introduction of X-ray detectors, semi-quantitative measurements of texture through X-ray pole figures became possible. A pole figure is a two-dimensional distribution function representing the orientations of a specific crystallite direction ((*hkl*) pole). Detailed descriptions of the X-ray methods for determining pole figures can be found in [154–156].

In current X-ray texture goniometer arrangements, the geometry of Schulz [153] is used. They are equipped with detectors which allow the measurement of the position of the diffracted X-ray beam. The schematic of the texture measurement is shown in Fig. 2.50. In order to generate a pole figure of a given specimen for a specific

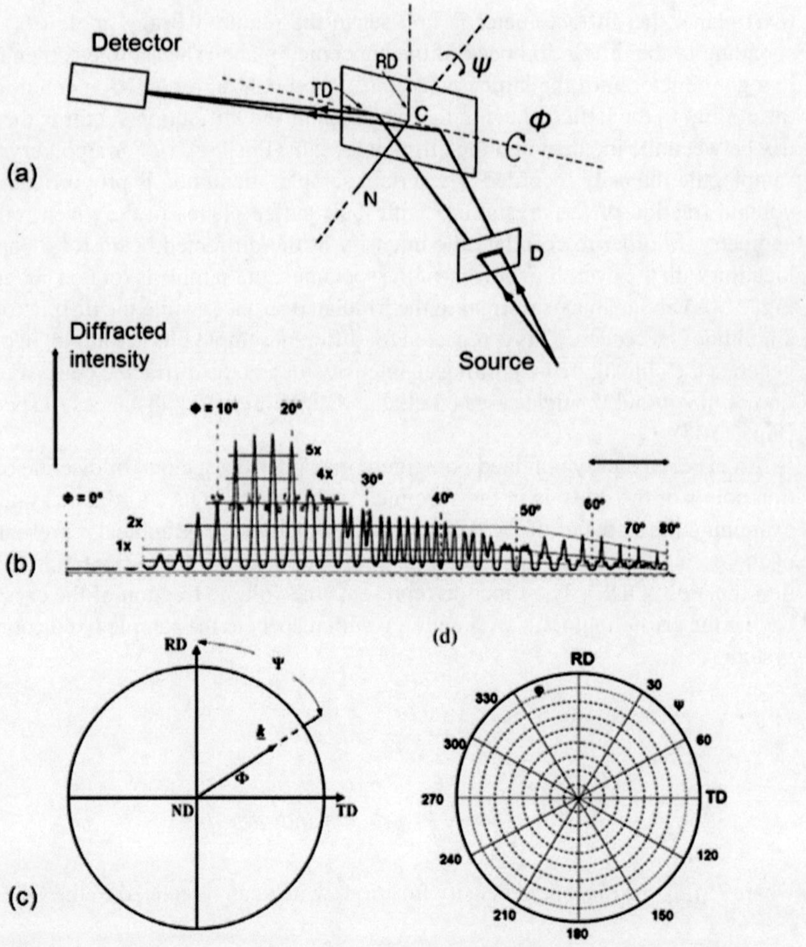

Fig. 2.50 Principle of texture measurements in reflection geometry [153]: **a** Depiction of diffraction geometry. The refraction angle θ is kept constant selecting a plane (*hkl*) for diffraction. When acquiring the data in order to obtain a pole figure, the sample is scanned in the plane of the rolling (RD) and transverse direction (TD). In addition, it is rotated in small increments around the axis normal to the surface (angle ψ) and tilted around the transverse direction TD (angle Φ). In this way the diffraction intensities according to the orientations of the crystallites can be measured for a given peak (*hkl* as shown in **b**; **c** Position of a pole for rotation angle Ψ and tilt angle Φ from ND (diffraction vector ***k*** described with reference to the specimen coordinates); **d** Trace of a pole on a concentric chart. Several slits (D) in the path of the beam serve for collimation. (Figure **a** in modified form from [33] with permission of Springer Nature and **b** and **c** from [147] under open source GPL v2 license)

(*hkl*) plane, the diffractometer is first set to the required Bragg angle (θ_B) corresponding to the diffraction peak of the concerned plane. The sample is then rotated in a goniometer until the lattice plane (*hkl*) is brought into reflection condition, i.e. the normal to the lattice plane in the specimen or the diffraction vector is the bisectrix between the incident and the diffracted beam (Fig. 2.50a). For a polycrystalline sample, the intensity recorded at a certain sample orientation is proportional to the volume fraction of the crystallites with their lattice planes in the given reflection geometry. In order to correlate the intensity of the diffracted beam for a particular location with the overall geometry of the specimen, the sample is rotated (ψ-angle in Fig. 2.50a) about an axis normal to the irradiated surface, while the diffraction peak amplitude is recorded. This is repeated for different sample tilts (Φ-angle) in order to generate the pole figure by plotting an intensity map of the diffracted peaks as a function of the Φ-and ψ-angles as the radial and circumferential distances, respectively (Fig. 2.50d).

An experimentally obtained pole figure map provides a cloud of discrete orientation points of the crystals in the specimen with respect to its coordinate system. By examining the statistics of the collected data, functional or graphical representations of the orientation distributions can be obtained in form of an orientation distribution function (ODF). The function represents the volume fraction of the crystallites having the crystallographic orientation g with respect to the sample fixed coordinate system [150]:

$$\frac{dV(g)}{V} = f(g)dg \tag{2.73}$$

$$\frac{dV(g)}{V} = P_{hkl}(\Phi, \Psi) sin\Phi \; d\Phi \; d\Psi \tag{2.74}$$

where P_{hkl} is the (*hkl*) pole density function, which can be related to the ODF by

$$P_{hkl}(\Phi, \Psi) = \frac{1}{2\pi} \int f(\varphi_1^{hkl}, \Phi^{hkl}, \varphi_2^{hkl}) d\varphi_2^{hkl} \tag{2.75}$$

where Φ^{hkl} corresponds to Φ and ${\varphi_1}^{hkl}$ corresponds to Ψ.

As said above, a pole figure is a two-dimensional distribution function depending only on the orientation of one specific crystal direction ((*hkl*) pole). It is independent of a rotation of the crystal about this direction. The three-dimensional orientation distribution function $f(g)$ can be obtained from pole figure inversion of several of its two-dimensional projections $P_{(hkl)}$ by using a suitable mathematical method, such as the series expansion method (harmonic method) [150]. The three-dimensional ODF is given by

$$\frac{dV}{V} = f(\varphi_1, \Phi, \varphi_2) \frac{\sin \Phi}{8\pi^2} d\varphi_1, d\Phi, d\varphi_2 \tag{2.76}$$

where, the factor sinΦ represents the size of the volume element in the orientation space and the factor $8\pi^2$ takes care of the usual normalization of the texture function in multiples of the random orientation density.

The ODFs $f(\varphi_1,\Phi,\varphi_2)$ $f(g)$ are expanded in a series of generalized spherical harmonic functions T_{lmn} (g) with the series expansion coefficients C_{lmn} ($l_{max} = 22$), which can be determined by the X-ray diffraction measurements [154, 157]:

$$f(g) = \sum_{l=0}^{\infty} \sum_{m=-l}^{l} \sum_{n=-l}^{l} C_{lmn} T_{lmn}(g) \tag{2.77}$$

A detailed description of the procedure to obtain the ODF from pole figures can be found in [150, 158].

For a polycrystalline fine-grained material with an average grain size of 5 μm, a volume V of about 100 μm side length and a depth of 20 μm contains about 500 grains of different, unknown orientations. Thus, an X-ray beam of a diameter of 100 μm interacts with many grains and many different lattice planes.

2.4.12 Applications of XRD for Texture Evaluation

In Chap. 1, Sect. 1.15, we discussed the measurement of the parameters characterizing deep-drawn metal sheets. These are the Lankford-parameters R, R_m, and ΔR [159, 160] and are usually measured by destructive tests. Tensile specimens are cut out of the metal sheet (Fig. 2.51) and subjected to strains ε from 15 to 20%. If the ductility of the material is small, the strain is limited so that uniform straining is maintained. Then, the dimensions of the specimens are measured and compared to their dimensions before loading. The R parameter is defined as:

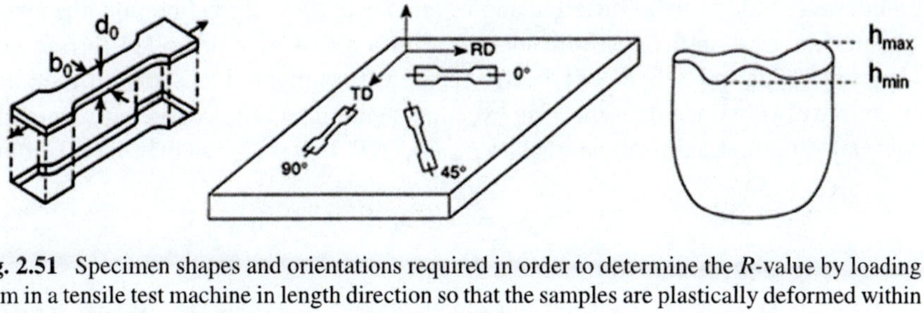

Fig. 2.51 Specimen shapes and orientations required in order to determine the R-value by loading them in a tensile test machine in length direction so that the samples are plastically deformed within the regime of uniform loading

$$R = \ln(b_0 / b_\varepsilon) / \ln(d_0 / d_\varepsilon) \tag{2.78}$$

Here, b_0 is the width of the test specimen before the test, b_ε is the width after plastic deformation at the applied length strains, and d_0 and d_ε are the corresponding thicknesses. If $R > 1$ the material flows more in width direction and less in thickness direction, and for $R < 0$ vice versa.

If the metal sheet is planar anisotropic, a mean value R_m is used for characterization which is composed of three values measured in three directions (Fig. 2.51):

$$R_m = \left(\frac{R(0°) + 2R(45°) + R(90°)}{4} \right) \tag{2.79}$$

Finally, the planar anisotropy value of the plate, ΔR, is defined as

$$\Delta R = \left(\frac{R(0°) - 2R(45°) + R(90°)}{2} \right) \tag{2.80}$$

Large anisotropy of metal sheets leads to earing when deep-drawn, see Fig. 2.51. In directions of large R-values humps are generated, for directions of small R-values valleys. Cup-tests yield the ear heigths $\Delta h = (h_{\max}\text{-}h_{\min})$, the earing parameter Z_e (Eq. 2.81), and the concurrent determination of R_m and ΔR in tensile tests. They allow to obtain the parameters which result in minimum earing [161]. From practice, it turns out that $R_m > 1$ is favorable for deep-drawing but this entails in general that $\Delta R > 0$, which leads to the occurrence of ears, see Figs. 2.51 and 2.53. The number of ears formed (4, 6, or 8) depends on the polycrystalline crystallographic structure of the metal sheet. In order to minimize the occurrence of ears in cup forming, the planar anisotropy value should be zero, i.e. $\Delta R = 0$. Within certain limits this can be controlled by alloying, drawing temperature, heat cycles, and alternate drawing in length and thickness directions [159].

$$Earing\ Z_e(\%) = 100\left(\frac{h_{max} - h_{min}}{h_{max}} \right) \tag{2.81}$$

As discussed in Chap. 1, Sect. 1.15, by measuring the velocities of various ultrasonic waves modes under different angles relative to the rolling direction, the texture coefficients C_{411} and C_{413} were measured in a series of cold-rolled ferritic steel sheets by Spies and Schneider [162, 163] and were compared to X-ray diffraction or destructively obtained data showing excellent agreement. The coefficients were then correlated to the Lankford parameters R_m and ΔR. It was shown that the relations

$$R_m = a + bc_{411} \tag{2.82}$$

$$\Delta R = c + dc_{413} \tag{2.83}$$

hold. The parameters *a*, *b*, *s*, and *d* were determined by calibration tests. The parameter Z_e (Eq. 2.81) was obtained from time-of-flight data $\Delta t/t$ of Lamb mode a_l in the metal sheets [164]. It was measured in the direction of maximum planar anisotropy and correlated with the earing parameter Z_e (Eq. 2.84) in an almost 1:1 relation, i.e.

$$Z_e \cong \Delta t/t \tag{2.84}$$

Similarly, Serebryany [165] obtained the texture coefficients C_{400} and C_{402} in sheets of low carbon steel and aluminum alloys using ultrasonic velocity data (see Chap. 1, Sect. 1.15). The ultrasonic data were then used to calculate from the angle-dependent Taylor factor $M(q,\theta)$:

$$M(q,\theta) = \frac{\sum_{l=0}^{6}\sum_{n=0}^{l}\sum_{p=0}^{N} m_{0,nlp} C_{0ln} q^p cos(n\theta)}{2l+1} \tag{2.85}$$

the anisotropy factor *R*. Here $m_{0,nlp}$ are the coefficients of the expansion into a series of spherical functions, C_{0ln} are the coefficients obtained by the ultrasonic measurements, θ is the angle between the rolling and the transverse direction, and q is the orientation-dependent Taylor factor for single crystals. The *R*-value is a function of the expansion coefficients $m_{0,\mathrm{nlp}}$ up to the sixth order ($l = 6$). This leads to *R* by

$$R = \frac{q_{min}}{1 - q_{min}} \tag{2.86}$$

where q_{min} is the so-called contraction ratio, the value of q in (2.86) when $M(q,\theta)$ achieves its minimum value. Based on this formalism, Serebryany [165] determined the *R*-parameter and the coefficients C_{400} and C_{402} for steel and aluminum test samples and obtained good agreement with X-ray data.

For the analysis of the ultrasonic and the X-ray data, there are the Hill and Taylor models relating the texture anisotropy to the R_m and ΔR values, see Fig. 2.52 [166]. Such measurements allow one to develop techniques to obtain the parameters R_m and ΔR in situ in a production line [167, 168]. Eventually such measurements can be enhanced by modeling efforts [169].

Gao et al. [170] studied annealing textures for different thickness layers and the *R*-value in 2507 duplex stainless steel by tensile testing, X-ray diffraction, and electron backscatter diffraction. The results showed that 2507 duplex stainless steel has poor formability, especially poor deep drawability and noticeable planar anisotropy after cold-rolling with an average *R*-value of 0.86, a planar anisotropy ΔR of -0.38, and an earing parameter of $Z_e = 7.43\%$ (2.81). The orientation distribution functions (ODFs) from the different thickness layers show that the texture mainly exists in the ferrite α-phase, and there is a significant texture gradient in the thickness direction. The results of the grain orientations after tensile deformation in different directions show that the maximum volume fraction ratios for the (001) and (111) orientation are the main cause of the largest *R*-values in the 45° direction (Fig. 2.53).

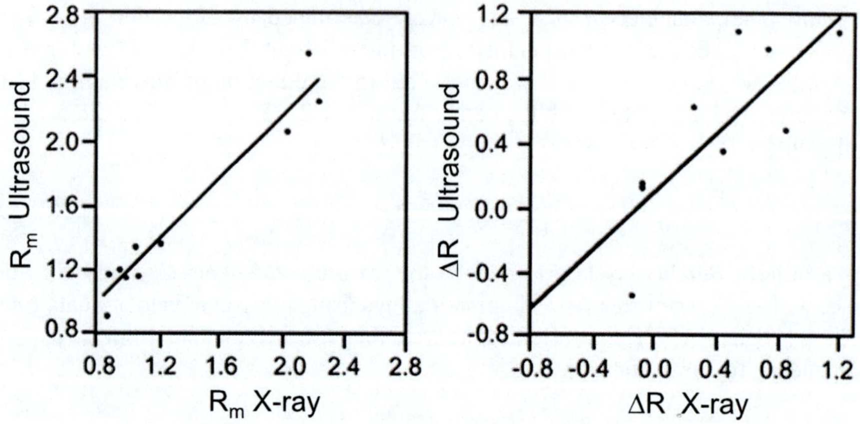

Fig. 2.52 Correlation between the standard value of R_m and ΔR determined by ultrasonic and X-ray diffraction based on the Taylor model. The linear relation between ΔR and R_m for both measurements indicate that their microscopic origin is the same (in modified form from [166], with permission from Springer Nature)

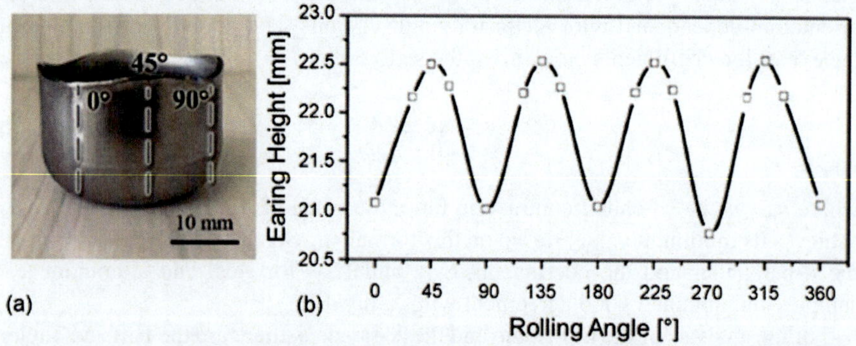

Fig. 2.53 Drawn cup and the earring height profile in a duplex stainless steel. **a** The profile depicts a characteristic four–ear profile with ears under 45°, 135°, 225°, and 315°. Because of the orthotropic symmetry of the sheet, the drawn cup has four 45°ears, as shown in **a, b**, which is consistent with the result for a planar anisotropy $\Delta r < 0$ (in modified form from [170], with permission from Wiley)

2.4.13 Line-Widths of X-ray Diffraction Spots

To obtain accurate stress data, it is necessary to understand which parameters contribute to the line-widths in X-ray diffraction measurements. The widths of the diffraction spots or diffraction lines are firstly determined by the instrumental width of the equipment, such as the wavelength dispersion in the beam, typically about 0.1 nm. A typical intensity profile, which is obtained from a reflecting grain under the Bragg-angle θ_{hkl}, is shown in Fig. 2.54. Its width $HW\theta_{hkl}$ is defined at half-height between the intensity maximum and the noise level [131]. The instrumental

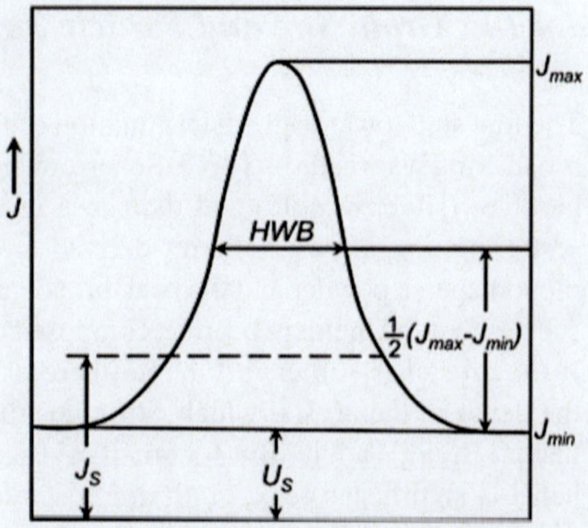

Fig. 2.54 Line shape of an X-ray diffraction spot with the definitions: **1** Full width *HWB* at half-height; **2** J_{min} minimum intensity; **3** J_{max}, maximum intensity; **4** J_s threshold; and **5** U_s background noise

contribution to the measured width can be determined using a strain free powder standard.

Secondly, various effects on the nanoscale contribute to the broadening such as the presence of dislocations, stacking faults, inclusions, strains fluctuations, and inhomogeneous residual stresses. These distortions may be viewed as being caused by an effective local stress σ:

$$HW\theta_{distortion} = \sigma \frac{4tan\theta}{E} \tag{2.87}$$

Thirdly, coherent scattering from the volume elements of grains of size d contribute to the width as:

$$HW\theta_{grain\ size} = \frac{\lambda}{d} \frac{1}{cos\theta} \tag{2.88}$$

The assumption that the different contributions to broadening add up linearly may not always be correct, but it is an accepted procedure based on experience.

$$HW\theta_{measured} = HW\theta_{grain\ size} + HW\theta_{equipment} + HW\theta_{distortion} \tag{2.89}$$

Because of the different angle dependencies in (2.87 and 2.88), their separation becomes feasible by measurements at different angles θ. Further details are discussed in [131, 171]. The integrated width of the diffraction peak (area under the peak divided by the height) is often used in place of the half-width to describe the peak broadening.

2.4.14 Grain Size and Particle Size Measurements by XRD

The first step towards the determination of the crystallite size from XRD line-profile broadening was made by [172]. Scherrer worked on the structure analysis of colloidal particles (silver & gold) and the cause of broadening was attributed to coherently scattering crystallites. Scherrer derived a relation between the crystallite size of a monodisperse powder and the peak broadening by assuming certain ideal conditions, such as that the incident beam is composed of parallel and monochromatic radiation. If the crystal is sufficiently large, there exists a set of planes from the surface to the depth in the crystal which cause destructive interference in all directions other than at Bragg's angle. But for small crystals, there is no complete cancellation, and hence a significant peak width due to finite crystallite size is observed. This peak width increases with decrease in crystallite size. From the condition of destructive interference, a relation between crystallite size and peak broadening was derived which is known as the Scherrer equation:

$$D = \frac{k\lambda}{\beta_S cos\theta} \tag{2.90}$$

where D is crystallite size, λ is the wavelength, $\theta/2$ is the diffraction angle, β_S is the peak width, and k is the Scherrer constant, frequently referred to as the crystallite-shape factor [96]. Langford and Wilson [173] discussed in detail the shape factor values. The shape factor k can be between 0.62 -2.08, depending on the definitions of the average crystallite size and the peak width.

The effect of the grain size on the X-ray diffraction profiles was exploited in order to determine not only the grain size in the nm range, but also the grain size distribution for nano-crystalline palladium. The agreement with data obtained by Transmission Electron Microscopy was excellent [174]. Comparable results were found for iron with a grain size of 6 nm [175]. The data analysis of these studies can be applied for other X-ray diffraction measurements as well.

Using a polychromatic tungsten X-ray source, diffraction contrast tomography (DCT) has been used by [176] for mapping the grain orientations in three-dimensions. They describe a novel laboratory-based X-ray DCT instrument for routine use and in-depth studies of temporal changes in crystallographic grain structure in a titanium alloy (*Ti-*β*21S*) with the smallest grains of $\approx 40\,\mu$m. The technique provides insights to the properties of recrystallization and grain growth of materials supporting the 3D modeling of theses processes. In this instrument a micrometer spot size X-ray source is used with cone-beam geometry and a high-resolution optically-coupled detector in order to monitor the transmitted and the diffracted X-rays, see Fig. 2.55. Diffraction reflections manifest themselves at specific scattering angles (2θ) corresponding to the lattice spacing (d), the orientation of the crystal planes, and the specific X-ray energy according to the Bragg condition, (2.20). With a polychromatic X-ray spectrum diffraction events are not observed at discrete angles. Instead, different energies are diffracted at varying diffraction angles and each grain may contribute

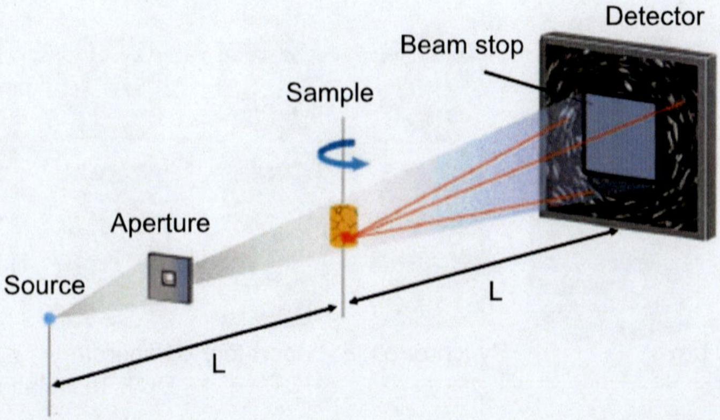

Fig. 2.55 Schematic showing the experimental set-up of a diffraction laboratory X-ray microscope (from [176], with permission from Springer Nature under Creative Commons CC BY license (https://creativecommons.org/licenses/)

multiple reflections to a single diffraction pattern relaxing the requirement on the number of projections needed for the reconstruction of the grain orientation, thus reducing the overall acquisition times needed, which are comparable to absorption tomography. Furthermore, the approach of McDonald[176] takes advantage of the fact that for a point X-ray source with a divergent beam, a crystal grain diffracts X-rays such that they are focused in the plane of diffraction at a distance equal to the source-sample distance L. Two consecutive scans are performed. One scan collects a diffraction pattern using a detector with a beam stop. These data were used to calculate the orientations of grains and their positions in relation to the sample reconstruction obtained from a tomographic absorption scan. The result of the reconstruction of the grain size distribution was compared to EBSD measurement and to X-ray phase contrast data acquired at the beamline ID 19 of the ESRF in Grenoble, see Fig. 2.56.

2.4.15 *Estimation of Dislocation Density*

Dislocations in a crystalline structure can be directly visualized under transmission electron microscopy (TEM) [177]. However, TEM studies can only provide localized information about the dislocation structure and its characteristics. The presence of dislocations causes distortion in the matrix, which in-turn can be used to calculate the average dislocation density. Zhao et al. showed [178] that the dislocation density Λ can be determined from

$$\Lambda = \frac{2\sqrt{3\langle\varepsilon^2\rangle}}{Db} \tag{2.91}$$

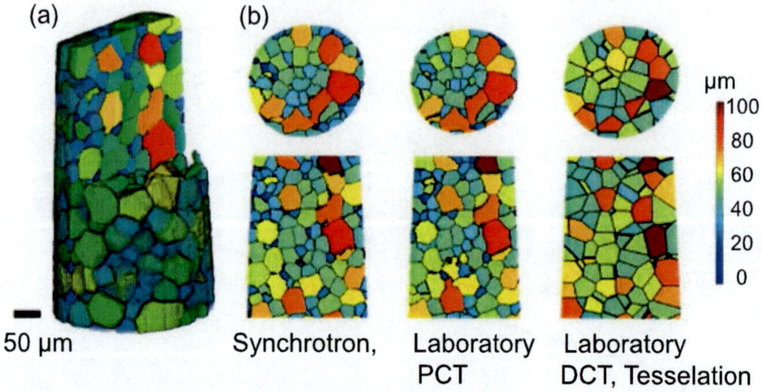

Fig. 2.56 a Grain morphology in a slice inside the sample volume; **b** Grain shapes extracted from reconstructed volumes acquired using synchrotron phase contrast, laboratory phase contrast, and laboratory diffraction contrast tomography. The color scale represents the grain size (from [175], with permission from Springer Nature under Creative Commons CC BY license (https://creativecommons.org/licenses/)

where b is the Burgers vector, $\langle\varepsilon^2\rangle^{1/2}$ is the micro-sttrain, and D is the crystallite size, which are determined from XRD profile analysis based on the Warren–Averbach method [179] after correcting for instrumental broadening.

The XRD method is of particular interest as a complement to TEM, since it provides reliable information especially for high dislocation densities, which are not easily investigated by the TEM technique. Further, it allows the investigation of macroscopic volumes in bulk material. For instance, XRD peak broadening has been observed upon mechanical processing such as rolling of various materials including aluminum [180], ultrafine grained 7075 Al alloy [178] and austenitic stainless steel. The evolution of dislocation density has been accurately correlated with TEM observations [181].

2.4.16 Precipitation of Secondary Phases

In a wide range of alloy systems, mechanical properties can be tailored by controlled precipitation of secondary phases. A few important alloys based on precipitation strengthening include aluminum-copper alloys, maraging steels, and precipitation hardened nickel base alloys. These alloys attain their strength essentially from the coherent strains produced by fine precipitates in the matrix. The strains are obtained by measuring the broadening of the XRD line profile. Furthermore, a reduction of the unit cell size is normally observed with precipitation, which can also be monitored by XRD analysis.

Rai et al. [182] demonstrated that the parameters micro-strain, lattice parameter, and crystallite size, obtained from the XRD line profile analysis, can be used in a complementary way to study the precipitation/dissolution of various intermetallics and carbides in nickel base superalloy Inconel 625, without extracting the precipitates from the matrix. While the change in the lattice parameter can provide the information about the change in composition of the matrix due to formation of precipitate of either coherent or incoherent types, the micro-strain is influenced only by the coherent precipitates. Further, in the specimens with grain size in the range of a few tens or hundreds of microns, the crystallite size does not reflect the true grain size but a coherently scattering domain size.

Mahadevan et al. [183] used X-ray diffraction (XRD) studies in order to characterize the aging behavior of M250 grade maraging steel samples subjected to isothermal aging at 755 K for varying durations of 0.25, 1, 3, 10, 40, 70, and 100 h. Earlier studies had shown typical features of precipitation hardening, wherein the hardness increased to a peak value due to precipitation of intermetallics and decreased upon further aging due to reversion of martensite to austenite. Intermetallic precipitates, while coherent, are expected to increase the micro-strain in the matrix. Hence, an attempt has been made by [183] to understand the microstructural changes in these samples using XRD line profile analysis. The anisotropic broadening with diffraction angle observed in the Williamson–Hall (WH) plot has been addressed using the modified WH approach, which takes into account the contrast caused by dislocations on line profiles, leading to new scaling factors in the WH plot. The normalized mean square strain and crystallite size estimated from the modified WH approach have been used to infer early precipitation and to characterize aging behavior. The normalized mean square strain has been used to determine the Avrami exponent in the Johnson–Mehl–Avrami (JMA) equation, which deals with the kinetics of precipitation. The Avrami exponent thus determined has matched well with values found by other methods, as reported in literature.

2.5 NDMC using Neutrons

2.5.1 *Properties of Neutrons Relevant for NDMC*

Neutrons are neutral particles and have a mass of $m_n = 1.673 \times 10^{-27}$ kg. They possess a spin $\hbar/2$ and their magnetic moment is $\mu_n = -1.91\mu_N$, where $\mu_N = e\hbar/2M_p$ is the nuclear magneton with the proton mass M_p. This property makes them sensitive to the location and orientation of magnetic moments in materials and they became therefore the probe of choice for many studies of magnetism. The mass of the neutron has the consequence that they transfer in a scattering process momentum and energy to the atoms of a sample, making it very useful for the study of fundamental vibrations within materials such as phonons and magnons yielding information on the strengths of chemical bonds and the magnetic interactions.

Neutrons are not stable as a free particle. Their radioactive decay time is about 17 min and they disintegrate into a proton, an electron, and an antineutrino. In stable nuclei, this decay process is absent due to energetic reasons. Neutrons of different flux density can be created by decay processes in isotopes, nuclear reactions in reactors, and accelerators:

(1) Isotopes: $10 - 10^4$ n/(cm^2 s).
(2) Accelerators: $10^3 - 10^6$ n/(cm^{2}s) where e.g. electrons of energies > 10 meV hitting target materials are setting free neutrons.
(3) Subcritical assemblies are mixtures of radioactive materials. They contain a material which serves as source for α- or γ-radiation (e.g. $Sb^{124)}$ and another material (e.g. Be^9) which sets free neutrons under this radiation: $10^4 - 10^6$ n/(cm^2 s).
(4) Nuclear research reactors: $10^5 - 10^{15}$ n/(cm^2 s). Due to their design und purpose, nuclear power plant reactors are usually not used as neutron source for research experiments and applications in non-destructive testing and materials characterization.
(5) Spallation sources: $10^{10} - 10^{18}$ n/(cm^2 s). These are the most intense accelerator-based neutron sources. They consist of a high-powered accelerator that injects protons with energies of greater than 0.5 GeV into a heavy metal target, such as mercury or tungsten. These metals "spall off" free neutrons in response to the impact. As accelerators can be readily pulsed, spallation sources are generally pulsed neutron sources, unlike most reactors that generate neutrons constantly.

Neutrons are detected by nuclear reactions, where particles with electrical charges are set free, e.g. α-particles or electrons. Their visualization is made by ionization or scintillation. There are gas proportional detectors, scintillation neutron detectors with liquid organic scintillators, crystals, plastics, glass and scintillation fibers, and finally semiconductor neutron detectors. Details of their operational principles can be found in [184].

Following the equivalence of mass and energy and the concept of de Broglie waves, the wavelength of a neutron is given by [7]:

$$\lambda_n = \frac{h}{p} = \frac{h}{m_n \mathrm{v}} = \frac{h}{\sqrt{2 m_n E_n}} \tag{2.92}$$

with λ_n the de Broglie wavelength, h the Planck-constant (6.626×10^{-34} Js), and E_n the neutron energy. This is an analogous equation to photons relating their frequency and energy:

$$\upsilon = \frac{E_n}{h} \tag{2.93}$$

2.5.2 *Interaction of Neutrons with Materials Relevant for NDMC*

Neutrons can be used for the non-destructive materials characterization, similar to X-rays [185]. However, the mass absorption coefficients μ_m for neutrons and X-rays in different materials behave quite differently. Whereas μ_m for X-rays continuously increases with the atomic number, see Fig. 2.57, the attenuation for neutrons varies vastly with the atomic number, which is caused by the scattering and absorption processes of neutrons with the nuclei [29, 186]. Neutrons interact in general weakly with a material and hence they penetrate deeply into the bulk. The penetration depth (2.97) can be an order of magnitude higher than that for X-rays, see Fig. 2.58, allowing to inspect large volumes in contrast when using X-rays. For example, the variations in contrast between elements make it possible to image hydrogen in neutron radiography.

Whereas the interactions leading to the attenuation and scattering of X-rays takes place with the electrons of an atom, neutrons are attenuated by their interaction with the nuclei and the magnetic moments of the electrons. Many isotopes in nuclear materials exhibit strong peaks in neutron absorption cross-sections in the thermal energy range and beyond (1–1000 eV). These peaks, often referred to as resonances, occur at energies specific to particular isotopes, providing a means of isotope identification and concentration measurements, see (Fig. 2.59).

Neutrons may be absorbed by a nucleus forming a compound nucleus, which may decay by emitting an elastic or inelastic neutron, a γ-quant or two neutrons or may lead to processes like fission. Hence, the cross-section for the attenuation of neutrons contains several contributions. These are coherent, incoherent, elastic, and inelastic scattering, and absorption [188, 189]:

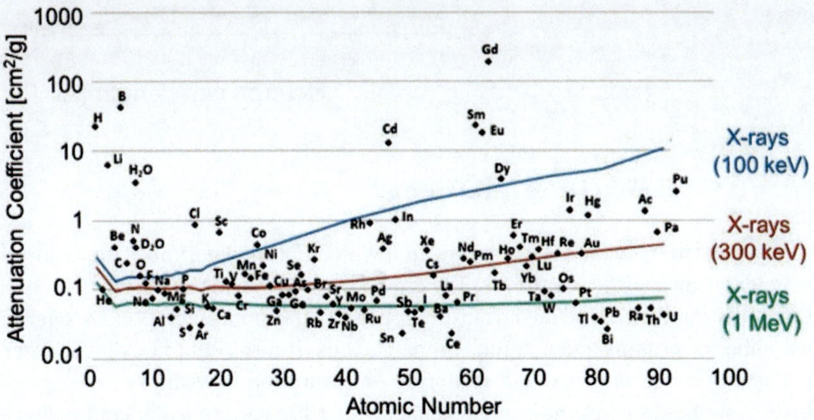

Fig. 2.57 Mass absorption coefficient μ_m for neutrons in comparison to X-rays of different energies (in modified form from [187], with permission from OpenTech under Creative Commons Attribution 3.0 License (https://creativecommons.org/licenses/by/3.0/)

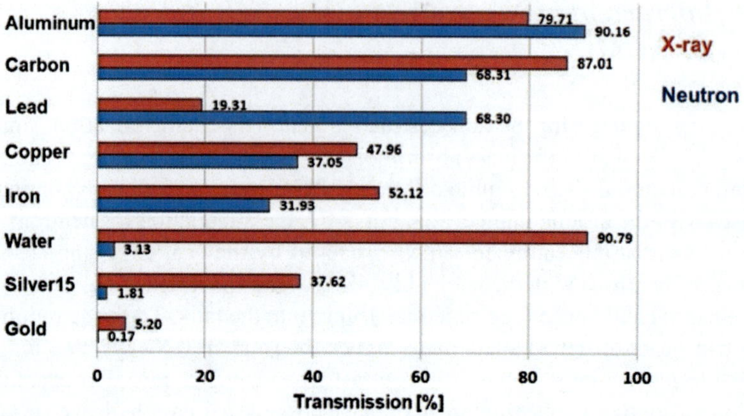

Fig. 2.58 Comparison of 25.3 meV neutron and 0.5 MeV X-ray transmission through materials having a thickness of 1 cm (from [187], with permission from OpenTech under Creative Commons Attribution 3.0 License (https://creativecommons.org/licenses/by/3.0/))

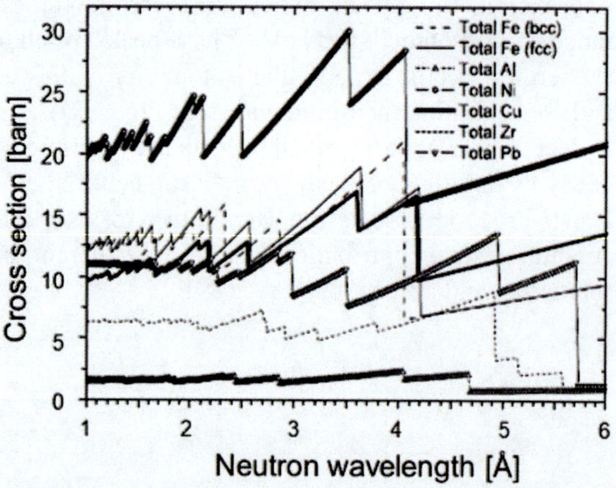

Fig. 2.59 Total neutron cross-sections for seven polycrystalline materials. The dominant elastic coherent scattering with characteristic Bragg edges enables energy selective neutron imaging with a high contrast due to the Bragg edges (in modified form from [188]), with permission from Springer Nature)

$$\sigma(\lambda) = \overline{\sigma}_0\left(S_{coh}^{el}(\lambda) + S_{coh}^{inel}(\lambda)\right) + \overline{\sigma}_{inc}\left(S_{inc}^{el}(\lambda) + S_{inc}^{inel}(\lambda)\right) + \sigma_{abs} \tag{2.94}$$

They contribute to the total cross-section which in turn determines the absorption coefficient in an analog way to (2.9) The cross-section depends on the neutron energy, on the order number Z, and on the mass number A of the atom since Z and A determine the number of protons and neutrons in the nucleus. It turns out to be on the order of (femtometer)^2 = 1 barn, i.e. the scattering or absorption is weak. For radiography, mostly slow neutrons in the energy range up to 1000 eV are used, comprising the range of thermal neutrons ($\hat{=}$ 25 meV). These interactions are discussed in detail in a number of textbooks on nuclear physics. Related to NDMC, the reader is referred to [186; 190].

For NDMC measurements, one must take into account the spatial distribution of the elements in the component under test as well as the detector efficiency $\varepsilon(E_n)$ in order to arrive at the signal amplitude after the detector. Let us assume that we have a parallel beam of neutrons, the signal at the output of the detector depends on the intensity $I_n(x,y)$ behind the component under test and on the efficiency $\varepsilon(E_n)$, and is given by [188]:

$$I_n(x, y) = D \int I_0(x, y, E_n)\varepsilon(E_n)\left(e^{\int -\sum(x,y,z,E_n)dz}\right)dE_n. \tag{2.95}$$

Here, $I_0(x,y)$ is the initial intensity, E_n is the neutron energy, and D is a conversion constant. Furthermore, the attenuation coefficient is given by

$$\sum(x, y, z, E_n) = n(x, y, z)\sigma(E_n) \tag{2.96}$$

where $n(x,y,z)$ is the number of atoms per volume, and $\sigma(E_n)$ is the neutron cross-section as a function of energy. The mass attenuation coefficient $\sum(E_n)/\rho$ is a function of position (x,y) and is based on the atomic mass attenuation coefficient μ_m/ρ, which is plotted in Fig. 2.57 for the elements of the periodic table.

In neutron radiography and scattering experiments, the penetration length $t_{50\%}$

$$t_{50\%} = \frac{\ln 2}{\mu} = \frac{0.69}{\mu}. \tag{2.97}$$

is often used. For a given neutron source with a given wavelength, it defines the distance t after which the intensity of the neutron beam is diminished to 50% of the initial value. For multiphase materials the corresponding total attenuation is given by

$$\frac{1}{t_{50\%}^{total}} = \sum_i \frac{c_i}{t_{50,i}} \tag{2.98}$$

where c_i is the percentage content of the phase i.

The transmission of neutrons through a polycrystalline material displays sudden increases in intensity as a function of the neutron wavelength (Fig. 2.59). These so-called Bragg edges are caused by the reflections at a given (*hkl*) lattice plane. They depend on the neutron energy because the Bragg angle increases as a function of increasing neutron wavelengths (2.99) until 2θ becomes equal to 180°. At a wavelength larger than this value, no scattering at the specific (*hkl*) lattice plane can occur and there is a sharp increase in the transmitted intensity [191]. This occurs at $\lambda = 2d_{\mathrm{hkl}}$, yielding a measure of the d_{hkl} spacing in the direction of the beam. The edges can be accurately determined by the time-of-flight technique ($\Delta d/d \approx 10^{-5}$), making it possible to obtain information about the stress state of the material, texture and phases present.

2.5.3 Neutron Radiography for NDMC

As pointed out by Tremsin et al. [192], the characterization of the cladding of nuclear fuel elements can be achieved by exploiting neutron resonance absorption imaging using spatially resolved neutron time-of-flight detectors. In this technique, the neutron transmission of the sample is measured as a function of the beam's location and neutron energy. In the region that borders the resonance energy for a particular isotope, the change in transmission is used to acquire an image revealing the 2-dimensional distribution of that isotope within the sample. An isotope-specific location map can be acquired simultaneously for several isotopes by correspondingly changing the neutron energy. Such radiographs can be utilized to measure isotope concentration, and with computed tomography three-dimensional isotope distributions can be obtained.

Using dynamic neutron radiography El-Ghany El Abd and Milczarek [193] studied the kinetics of the wetting process in fired-clay and siliceous bricks as porous construction materials. The technique provided accurate experimental data concerning the beginning as well as the advanced stages of the water infiltration process for both materials. It was shown that the time dependence of the wetting front deviated from the classical $\sqrt{t}$ versus t behavior. A theoretical analysis of the anomalous diffusion during water infiltration in porous building materials was based on a non-linear generalization of the Fick's law. Likewise, Tumlinson et al. [194] studied by neutron computed tomography soil–water content values. The sensitivity to small changes in soil volumetric water content allowed the estimation of the spatial distribution of soil water, roots, and root-water uptake.

The Kamini reactor at IGCAR, Kalpakkam is a U^{233} fueled, demineralized light water moderated and low power (30 kW) nuclear research reactor. This reactor functions as a neutron source with a flux of 10^{12} n/(cm^{2}s) at its core center, with facilities for carrying out neutron radiography, neutron activation analysis, and neutron shielding experiments. The availability of high neutron flux, coupled with a good collimated beam, provides high quality radiographs with short exposure time. The reactor being a national facility of India for neutron radiography, has been utilized in the examination of irradiated components, aero-engine turbine blades, riveted plates, automobile chain links and for various types of pyro devices used in the space program. Figure 2.60 shows neutron radiographs of gas turbine blades cast in a ceramic investment shell mold. The turbine blades have internal cooling passages extending through the aero-foil and root portions through which compressor bleed air is conducted for cooling while the engine is in operation. Neutron radiography helped to reveal the presence of core materials in a cast blade [195].

Energy selective neutron radiography experiments were performed at two cold neutron imaging beam lines, one at the Paul-Scherrer Institute (PSI, Switzerland) and the other at the Hahn-Meitner Institute (HMI, Germany) [188]. Both institutes provided facilities for energy selective radiography measurements. The velocity selector device employed at the PSI enabled wavelength band widths of $\Delta\lambda/\lambda \sim 0.15$, and a double monochromator device at the HMI, wavelength bands of $0.1 > \Delta\lambda/\lambda >$

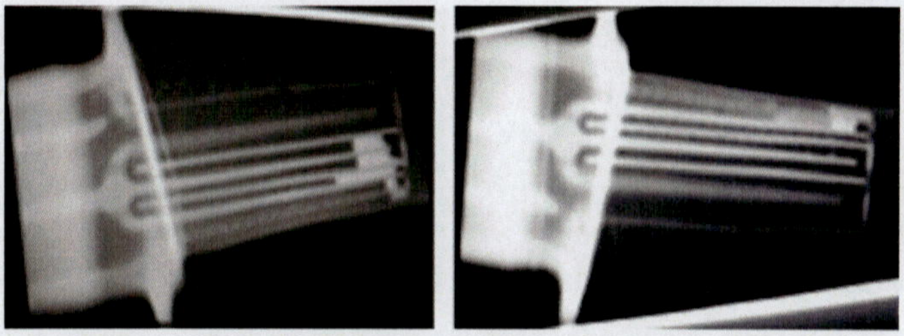

Fig. 2.60 Radiographs of HP turbine blades revealing entrapment of core particles within the cooling passage (from [195], with permission from American Institute of Physics)

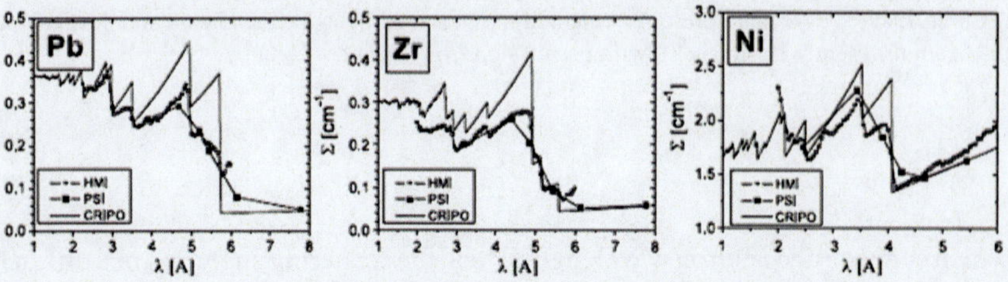

Fig. 2.61 Attenuation coefficient for Pb, Zr, and Ni polycrystalline materials obtained at the Paul-Scherrer Institute (full symbols) and at the HMI (open symbols) together with the theoretical values, solid lines (in modified form from ([188], with permission from Springer Nature)

0.01 widths. The detection system used at the PSI and the HMI consisted of a cooled CCD camera with 13.5 μm pixel size and neutron absorbing Gd and Li scintillators that in combination with the collimation, size, homogeneity and divergence of the neutron beams assured a spatial resolution of ~50 μm. The aim was to derive the total cross-sections for neutron interaction with different polycrystalline materials, here Al, Zr, Pb, Cu, Fe and Ni (Fig. 2.61), and to visualize spatially resolved structural parameters of complex material structures, such as welded steel, by neutron imaging (Fig. 2.62) exploiting Bragg edges (Fig. 2.61).

Finally, a short, yet very informative review on the neutron imaging in material science can be found in [196].

2.5.4 Stress Measurements by Neutron Diffraction

Thermal neutrons possess an energy of ≈ 25 meV (equivalent to $k_B T = 300$ K) having a velocity $v \approx 2200$ m/s. They have wavelengths of the order of atomic distances in crystalline lattices. Hence, diffraction of thermal neutrons at atomic lattices can be carried out in an analog way to X-rays [197, 198]. The Bragg equation holds correspondingly to (2.20):

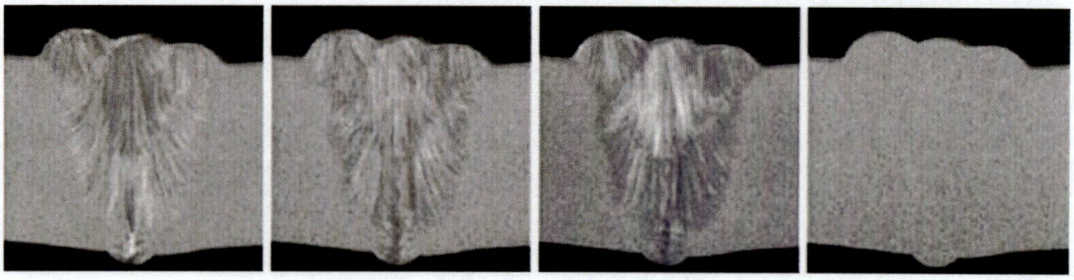

Fig. 2.62 Maps of attenuation coefficients of a welded steel sample at neutron wavelengths of respectively 0.34, 0.38, 0.4, and 0.44 nm. The images depict the weld and the adjacent bulk material. Long thin vertical stripes are visualized for each wavelength and hence energy, except for 0.44 nm where apparently no Bragg edge exists for the materials in the weld. The contrast in the welds in the first three images from the left was interpreted by the presence of different crystallographic textures. Each texture has a preferred orientation in respect to the incident beam and thus attenuates the beam differently (from [188], with permission from Springer Nature)

$$d_{hkl} = \frac{\lambda_n}{2\sin(\theta_{hkl})}. \tag{2.99}$$

which connects the neutron wavelength λ_n and the scattering angle θ_{hkl} of a diffraction peak with the interplanar spacing d_{hkl}, with the Miller indices (hkl). Neutron stress analysis is analogous to X-ray stress analysis, i.e. it is based on precise measurements of interplanar lattice spacings in different directions of a component or a sample. For X-rays the atomic scattering power is proportional to the square of the atomic charge, whereas the neutron cross-section varies in an irregular manner with Z, as discussed above, see Fig. 2.57. As a consequence, different reflection lines may give maximum intensities for X-rays or neutrons.

An overview on the possibilities of stress measurements by neutron diffraction is given by [98, 197, –200]. With neutron diffraction one registers only elastic strains. Strong plastic flow may show up by side effects such as reflection line broadening and/or texture similar to X-ray diffraction. Neutron diffraction is sensitive to residual as well as to applied stresses. As neutron spectrometers can host bulky experimental set-ups, it is not difficult to load a component under investigation. Stresses can be determined for each crystalline phase of a multiphase material provided this phase is sufficiently abundant. Neutrons are also sensitive to macro-and micro-stresses. The large penetration depth of neutrons into materials (typically a factor of 1000 larger than X-rays) allow residual stress measurements by neutron diffraction under scattering angles of 90° in the interior of components. For example, measuring internal strains throughout the interior of a 2 cm thick steel plate is possible with neutrons within reasonable data acquisition times. In addition to absorption, also scattering and incoherent background noise determine the data acquisition. Concerning the diffraction geometry, the 90° scattering angle not only leads to a near-cubic test volume, but also makes strain measurements possible even in large parts with a complicated geometry.

Stress measurements using neutron diffraction can be performed by angle-dispersive [201] or by energy-dispersive techniques [202]. In angle dispersive measurements, the strain is obtained by

$$\varepsilon = \frac{d - d_0}{d_0} = -\frac{(\theta - \theta_0)}{\tan(\theta_0)} \tag{2.100}$$

where d_0 is the stress-free lattice parameter. In case of an isotropic polycrystal where Young's modulus E and Poisson's ratio ν are known, the strains measured in three perpendicular directions yield for the stress [198, 201]:

$$\sigma_i = \frac{E(1-\nu)}{(1+\upsilon)(1-2\upsilon)}\varepsilon_i + \frac{\nu E}{(1+\upsilon)(1-2\upsilon)}\sum_j \varepsilon_j \tag{2.101}$$

Using the neutron elastic constants (NEC) S_{1NEC}, S_{2NEC} [mm^2/N] (which are the same as the X-ray elastic constants) [197], one obtains for the stress [198]:

$$\sigma_i = \frac{2(4S_1 + S_2)}{S_2(6S_1 + S_2)}\varepsilon_i - \frac{4S_1}{S_2(6S_1 + S_2)}\sum_j \varepsilon_j \tag{2.102}$$

Like in X-ray diffraction measurement, the NECs are determined by appropriate measurements under load. Alternatively, the unstrained lattice parameter d_o of (2.100) is measured at a sample or coupon taken from the initial material.

Angle-dispersive neutron diffraction experiments are conceptually simple [202]. A polychromatic or monochromatic neutron beam is directed on a sample, and the neutrons scattered at defined scattering angles are discriminated in energy by a detector (Fig. 2.63a). For a polycrystalline sample, the spectrum recorded in the detector consists of several diffraction peaks at the incident wavelengths that fulfill the Bragg condition. Pulsed sources are in particular suited for energy-dispersive diffraction experiments, because they emit neutrons having a wide energy range. The neutron speed v defines the wavelength λ_n by the de Broglie relation, 2.92. Because all neutrons leave the source at the same time, the wavelength of a neutron traveling the distances L_1 from source to sample and L_2 from sample to detector, is defined by its time-of-flight t (Fig. 2.63b):

$$\lambda_n = \frac{h}{mv} = \frac{h}{m}\frac{t}{(L_1 + L_2)} \tag{2.103}$$

and hence the Bragg relation can be written as

$$d_{hkl} = \frac{h}{m}\frac{t_{hkl}}{2\sin\theta_{hkl}(L_1 + L_2)} \tag{2.104}$$

Using neutron diffraction Taran et al. [203] studied the elastoplastic properties of a stainless steel with an austenitic matrix and martensitic inclusions induced during cyclic tensile–compressive fatigue loading. Specimens of annealed and quenched

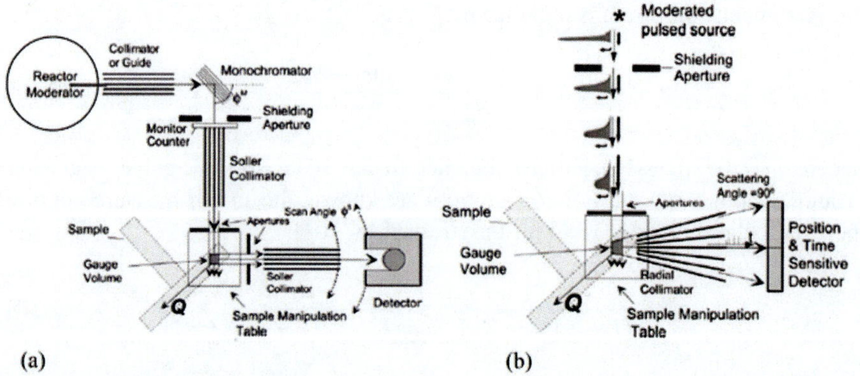

Fig. 2.63 Principle of neutron diffraction experiments; **a** For the continuous and monochromatic beam generated by a nuclear reactor, the wavelength λ is fixed by the monochromator resulting in the angle-dispersive measurement set-up similar to the X-ray diffraction method. The scattering angle of $\approx \pi/2$ is generally used for neutron diffraction because it gives a cubical gauge volume; **b** For the pulsed neutron beam from a spallation source with a white wavelength spectrum, the diffraction angle $2\theta_{hkl}$ is fixed and the arrival time t of neutrons at detectors gives λ, see (2.103 and 2.104). In both cases the strain is measured in the direction of the Q -vector (from [199], with permission from Elsevier Masson SAS.)

AISI type 321 steel were subjected to low cycle fatigue (strain amplitude of 1% at 0.5 Hz). Subsequent in-situ loading tests provided the elastoplastic responses of both austenitic and martensitic phases via Rietveld refinement of neutron diffraction spectra. A clear trend of increasing elastic modulus with increasing fatigue level was noted in the austenite matrix. The results of modified refinements accounted for the elastic anisotropy in polycrystalline materials under load. The residual strains in the austenitic matrix and the deviatoric components of both phases' residual micro-stresses were determined as a function of fatigue cycling.

In a further work, Taran et al. [204] measured the triaxial residual strains in a composite tube from an austenitic stainless steel as a parent material and a shape welded ferritic steel by the time-of-flight neutron diffraction method on the POLDI instrument at the Paul-Scherrer-Institute SINQ neutron pulsed facility in Villigen/Switzerland. The shape weld is used to build compressive stresses and, as a result, to suppress stress corrosion. Investigations of the residual stresses in such composite tubes are important for developing optimal welding techniques. A calculation of the residual stresses was performed using measurement results with a comb-sample, machined from the tube by the electro-discharge method, as the stress-free reference sample. The results were compared to the results of the destructive turning-out method and to finite elements calculations.

Withers [199] reviewed neutron diffraction measurements of residual stresses. Steuwer et al. [205] from the same group examined the residual stresses in dissimilar friction stir welds by neutron and synchrotron X-ray diffraction in two dissimilar aluminum alloys (Fig. 2.64). The effect of tool traverse and rotation speeds on the residual stresses was quantified for welds between non-age-hardening AA5083 and age hardening AA6082. The region around the weld lines showed significant tensile residual stress-fields which were balanced by compressive stresses in the parent material. The rotation speed of the tool was found to have a substantially greater influence than the transverse speed on the properties and residual stresses in the welds. The changes in residual stress are related to microstructural and hardness changes.

Furthermore Thiebault et al. [206] studied residual stresses by neutron diffraction on 13%Cr–4%Ni welds made by 410NiMo weld filler metal. The transverse, longitudinal and normal stress components were determined by neutron diffraction employing 2.104. Finally, Gnaupel-Herold [207] demonstrated that with special beam arrangements, small volumes of side-lengths of 0.1 mm can be achieved for stress measurements.

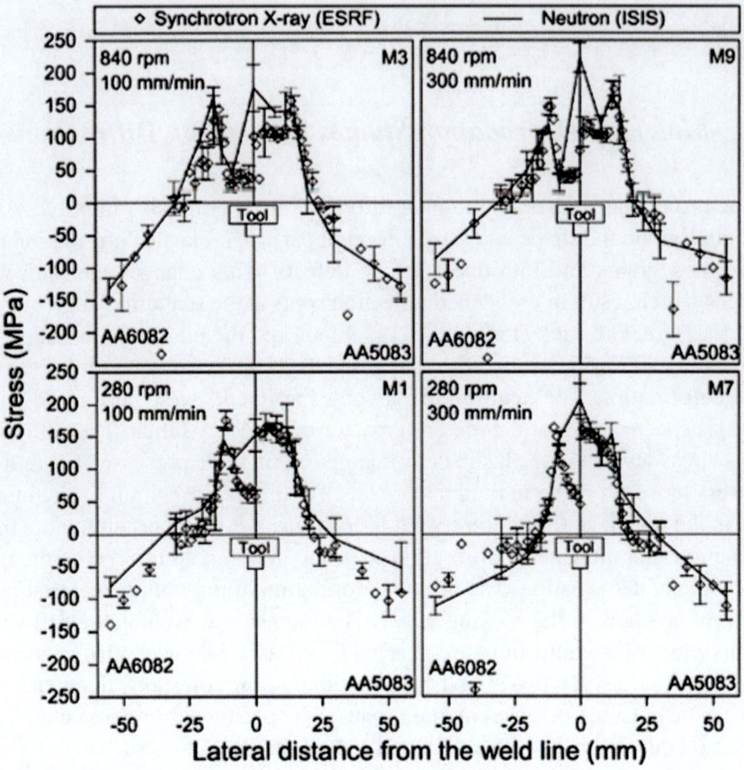

Fig. 2.64 Longitudinal residual stresses measured by neutron scattering in 3 mm thick dissimilar friction stir weld (AA 5083 and AA 6082). The weld was prepared at a rotating speed of 840 rev/min and a welding speed of 5 mm/s. The figures display the variation in the longitudinal residual stress as a function of distance from the weld line in four dissimilar welds with AA5083 on the advancing side. The residual stress values obtained at the ISIS (UK) and NFL (Sweden) facilities (lines) are in excellent agreement with those obtained by X-ray scattering at the ESRF in Grenoble (France) (diamond symbols) (from [205]), with permission from Elsevier

2.5.5 Texture Measurements by Neutron Diffraction

Neutron diffraction is especially useful for the determination of stress and texture and materials degradation [208]. They allow to determine the stresses to be differentiated between the load levels in different phases. This refers to mono-crystalline materials as well as to multi-phase and composite materials. Usually, texture is visualized by neutron measurements in pole figures like with X-rays [209].

STRESS-SPEC is a neutron diffractometer instrument installed at the "Forschungsreaktor" Munich II in Garching, Germany [210]. It is available for routine operation and open for both national and international users. It is a set-up suitable for global and local texture measurements. Crystallographic texture analysis can be combined with micro-and macro-strain measurements. A robotic system was installed in order to handle large industrial components for texture and strain mapping.

2.5.6 Materials Degradation Studies by Neutron Diffraction

With neutrons one is also able to gain information on grain size in polycrystalline materials, and on the shape and size of foreign particles. Plastic deformation results in deformed grains and introduces lattice defects. This changes the lattice plane distances which result in broadened reflection spots at the scattering angles θ_{hkl}., see Fig. 2.65. Here, Crostack et al. [208, 211] monitored the plastic deformation of the Ni-base alloy 939 by measuring the changes in the (200) reflection profiles caused by plastic deformation. The neutron diffraction experiments were performed with the triple axis spectrometer at the research reactor of the KFA Jülich. The data taken at various stress levels in the elastic deformation region were processed to yield strain and stress tensors for the individual crystals. As the interpretation of the measurements in the plastic deformation region is concerned, it is important to note that the deformation inhomogeneity from grain to grain amounts up to 30% of the macroscopical mean deformation. The plastic deformation of the sample material leads to significant changes in the rocking curve. The intensity as well as the full-width at half maximum of the reflections are affected (Fig. 2.65). Likewise, also by measuring the fullwidth values $HW\theta_{hkl}$ at half-height of the diffraction spots, fluctuating strain levels can be detected in different grains caused by plastic deformation due to creep, fatigue, and micro-cracks with collimated neutron beams [211]. Such a measurement allows one to characterize critical surface areas and near-surface volumes in early stages of degradation.

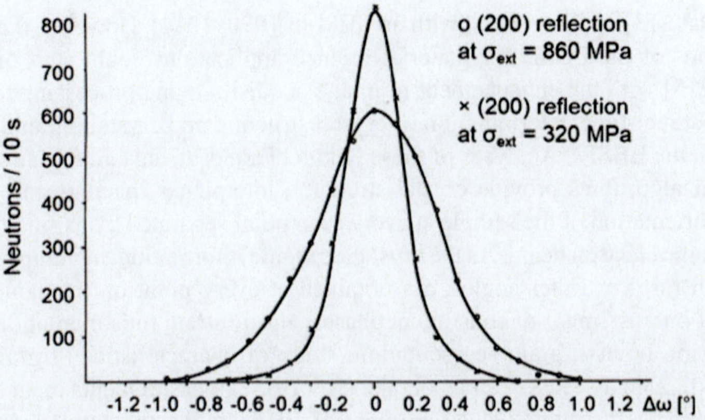

Fig. 2.65 Comparison of the ω line-profiles of the (200) reflection obtained by neutron diffraction in the elastic and plastic deformation region of the Ni-based alloy 939 (from [208], with permission from Springer Nature)

2.6 NDMC using Electrons and Positrons

2.6.1 Electron Backscatter Diffraction

There are several widely used material characterization techniques/tools, such as scanning electron microscope (SEM), transmission electron microscope (TEM), electron probe micro-analyzer (EPMA), electron spectroscopy for chemical analysis (ESCA) and electron backscatter diffraction (EBSD), utilizing different interactions of electrons with materials. Due to the requirement of very high vacuum, these techniques are mostly not amenable for NDMC and hence are out of scope of the present book. As the principle and application of Electron Backscatter Diffraction (EBSD) technique is very similar to that of X-ray diffraction or Neutron diffraction, this is briefly discussed here.

EBSD is an SEM based crystallographic characterization technique which can provide three dimensional crystallographic orientation of the material in the measurement point with a spatial resolution of ~ 75 nm and a probing depth of ~ 15 nm [212]. The electron beam in an SEM is directed to the material under test. The electrons are scattered simultaneously at all lattice planes. The back-scattered electrons leaving the material generate an image, which is displayed by a detector comprising either a scintillator screen, or a fiber optic bundle, or by a photon sensitive imaging detector like a charge coupled device (CCD) camera, to form an electron backscatter diffraction pattern. The detector is generally 40 mm^2 in size and is positioned close to the sample so as to subtend a relatively large angle ($\approx 60°$) for the electrons emanating from the sample. The sample is also tilted ($\approx 70°$) towards the detector. Each pattern consists of many bands, termed Kikuchi bands. These patterns named as electron back-scattering patterns (EBSPs) were examined in detail by Alam et al. as early as the 1950s [213] and integrated with an SEM in 1970s [214]. However, the effective utilizations of the EBSD for material science applications could start only in the 1990s [215] with the enhancement in image acquisition and processing techniques for almost real-time determination of crystal structure and crystallographic orientation from the EBSPs. Analyses of these Kikutchi bands using suitable autoindexing computer algorithms provide crystal structure, interplanar spacings and crystallographic orientation of the sample at every scan point separated by as small as 50 nm based on the electron beam. As the crystallographic information, including 3D orientations in form of Euler angles, are obtained at every point on the sample in the region of interest, maps of constituent phases, elastic strain (misorientation between neighboring points), grain segmentation, study of characteristics of grain boundaries, texture analysis etc. can be obtained by EBSD measurements in an SEM. For appreciable back-scattered electron intensity, EBSD is mostly coupled with a Field Emission Gun (FEG) SEM.

The evaluation of the EBSD images delivers the grain-orientation allowing to determine the angle between neighboring grains, and in the ensemble of grains the presence of texture or specific mis-orientations. In recent years, EBSD has become the most popular tool for texture analysis, as it provides not only the quantitative information about the bulk texture but also the orientation map of individual grains. Pole figure maps of individual points or a collection of points on the specimen surface, inverse pole figure maps, ODFs etc. can be obtained from the crystallographic orientation information obtained by EBSD. Wright et al. [216] reported that EBSD measurements in about 10,000 grains provide a bulk texture characterization equivalent to that obtained by X-ray diffraction in moderately textured and fine-grained materials, and less for strong textures. The EBSD technique was successfully applied to characterize the deformation of the microstructures and textures during cold rolling of pure Mg and magnesium alloys Mg–0.2Ce and Mg–3Al–1Zn (also known as AZ31) [217]. The Mg–0.2Ce alloy turned out to be considerably more rollable than pure Mg and the Mg–3Al–1Zn alloy. This was attributed to the texture sharpness and the severity of the shear banding observed through EBSD studies. The dominant features of the microstructure were twins and shear bands. The frequency of the former decreased while that of the latter increased with rolling reduction. Each alloy displayed a fiber texture in which the c-axis was closely aligned with the sheet normal direction. Much of the deformation appeared to be concentrated in the shear bands. It is speculated that the striking effect of alloying addition on cold rollability can be understood in terms of differences in severity, frequency and lifetime of shear bands.

Kumar and Pollock [218] used EBSD based strain mapping in order to determine the dislocation density in femto-second laser induced damage in a single crystal nickel-based superalloy. It was observed that while the so-called Kernal average misorientation, providing crystallographic misorientation of a point with respect to the surrounding points, is used for calculating the dislocation density, a grain reference orientation deviation map provides a better visualization of the strain distribution (Fig. 2.66). Furthermore, aspects for the application of EBSD at minerals were given in [219], while an extended review of EBSD covers the complete theme [220].

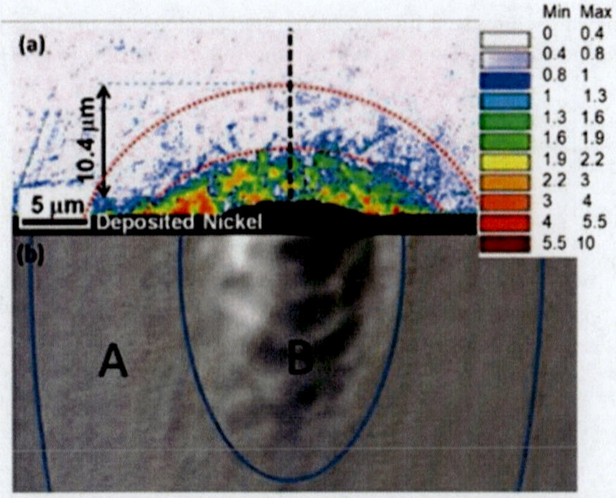

Fig. 2.66 **a** EBSD grain reference orientation deviation map (in degrees) obtained beneath the crater produced by a single fs laser pulse providing an energy density of 5.1 J/cm^2; **b** Corresponding secondary electron image of a crater obtained from the ablated surface. Regions A and B in **b** denote the low and the high damage regions, respectively, on the laser ablated surface. The two red dotted lines in **a** indicate the boundaries of the intense and the low damage regions underneath the crater (in modified form from [218], with permission from AIP Publishing)

2.6.2 NDMC by Positron Annihilation Spectroscopy

Positron annihilation spectroscopy (PAS) has emerged as a NDMC tool for the characterization of open volume lattice defects such as vacancies, dislocations and grain boundaries. A positron is the antiparticle of the electron with an electric charge of + 1e, a spin of ½, and the mass is the same as of an electron. The sources of positrons include cosmic rays, radioactive decays, nuclear reactions and pair-production by interaction of high energy electromagnetic waves with a heavy atom. For example, the beta (β$^+$) decay of ^{22}Na (half lifetime of 2.6 years), as given below, is an efficient source of positron emission and hence used in many laboratory based PAS studies:

$$^{22}\mathrm{Na} \rightarrow ^{22}\mathrm{Ne} + \beta^+ + \nu + \gamma\left(\beta^+ \mathrm{decay},\, \mathrm{p}^+ \rightarrow \mathrm{n} + \mathrm{e}^+ + \nu\right)$$

^{22}Na decays to an excited state of ^{22}Ne by the emission of an energetic proton (β$^+$) and an electron neutrino. The daughter nucleus of ^{22}Ne de-excites by emission of a γ-ray with an energy of 1.27 MeV. Since the lifetime of the excited state of ^{22}Ne is only a few picoseconds, the γ-ray is emitted practically simultaneously with the positron and provides, thereby, important information at what time the positron was born. Positrons emitted by a ^{22}Na radioisotope have a continuous spectrum of energies with the end-point-energy of $E_{max} = 0.545$ MeV.

When positrons enter a metallic material, they thermalize in a few ps to the thermal energy (3/2 $kT \cong 40$ meV at room temperature) by losing their kinetic energy by ionization, excitation of electrons, and scattering at phonons [221]. In a perfect metal

Table 2.6 Calculated positron bulk lifetimes τ for elemental and group-IV semiconductors. The arrangement of the elements is adapted to the periodic Table (according to [222]

Li 305	Be 137												
Na 337	Mg 237											Al 166	Si 221
K 387	Ca 297	Sc 199	Ti 146	V 116	Cr 101	Mn 103	Fe 101	Co 97	Ni 96	Cu 106	Zn 134		Ge 228
Rb 387	Sr 319	Y 219	Zr 159	Nb 122	Mo 111	Tc 95	Ru 90	Rh 93	Pd 103	Ag 120	Cd 153		Sn 275
Cs 407	Ba 315	Lu 199	Hf 149	Ta 117	W 100	Re 91	Os 86	Ir 87	Pt 94	Au 107			Pb 187

lattice free of open volume defects, the positron annihilate in the bulk in about 100–400 ps, depending on the electronic configuration of the metal [222, 223] as shown in Table 2.6.

As the positron is repelled by the positively charged ion cores, structural defects where material is missing or the density is reduced, provide an attractive potential for positrons. Hence, if the metal contains defects such as vacancies, vacancy clusters (including voids and bubbles) and dislocations i.e. regions of less than the average atomic density, positrons may get trapped at these defects, increasing their lifetime before annihilation by electrons. The life-time is highly sensitive to open-volume defects, with sensitivity starting from atomistic size to large vacancy clusters or voids up to a few hundreds of microns [224]. For example, an increase in positron lifetime from 197 to 386 ps was reported with increasing number of vacancies in a cluster from 2 to 15 [225].

Annihilation of a thermalized positron with an electron is marked by the conversion of the electron–positron pair into two annihilation γ-rays of > 0.511 MeV (= m_0c^2), where m_0 is the mass of an electron/positron and c is the speed of light) energy as shown schematically in Fig. 2.67.

The life-time of a positron is determined by measuring the time difference between the 1.27 MeV γ-ray emitted during the creation of positrons and the 0.511 MeV γ-ray emitted during the annihilation of positrons. Various experimental techniques in positron spectroscopy are detailed by Coleman [226].

Many applications of positron life-time in non-destructive characterization of defect structures in metals and semiconductors can be found in the book of Krause-Rehberg and Leipner [227] and the review articles [221, 223, 225, 228]. The measurement of the positron life-time has been utilized for the characterization of defect structures evolved due to radiation [229], plastic deformation [230], precipitation [231, 232], fatigue damage [233], and creep damage as well as for the annealing of defect structures by heat treatment [231, 234]. Figure 2.68 shows a typical positron lifetime spectrum exhibiting the influence of plastic deformation [227]. The positron life-time is obtained from the inverse of the annihilation rate (slope of the curve). The bulk life-time of 218 ps was observed in a defect free zone of the silicon crystal

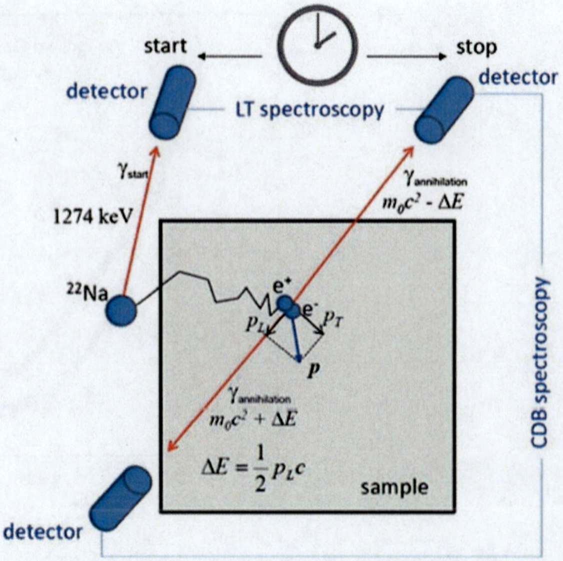

Fig. 2.67 Schematic of positron annihilation spectroscopy showing life-time (LT) spectroscopy and co-incidence Doppler broadening (CBD) (from [221], with permission from Elsevier)

grown by the Czochralski method. The value is similar to the 221 ps reported by [223].

In the plastically deformed Si sample, three annihilation rates corresponding to the bulk (τ_I -not shown), di-vacancies ($\tau_I = 320$ ps) and vacancy clusters ($\tau_3 = 520$ ps) are observed. A positron trapping model [235], as shown in Fig. 2.69, can be used to explain the trapping and annihilation scheme in a plastically deformed material with dislocations and vacancy clusters. Multiple annihilation rates (λ_i) and associated life-time ($\tau_I = 1/\lambda_i$) corresponding to bulk and different trapping defects sites can be theorized. However, such a decomposition of an experimental spectrum is often not possible and the average positron lifetime can only be reliably determined. Various algorithms have been reported for the analysis of experimental positron annihilation spectra [236, 237] as a sum of decaying exponentials for the extraction of multiple annihilation rates.

Positron life-time measurements have been used for studying precipitation behavior in various alloy systems including maraging steel [232], austenitic stainless steel [231] and aluminum alloys [238]. In general, an increase in positron life-time is observed with the precipitation of secondary phases due to the increasing defect density at the precipitate-matrix interface. Figure 2.70 shows the variations in positron lifetime as a function of annealing temperature in a 20% cold worked Ti-bearing stainless steel and a Ti-free model FeCrNi alloy [231]. The solution annealed state shows the positron life-time of 110 ps which is similar to that reported by [223] for iron, as shown in Table 2.6. The positron life-time increases to about 165 ps for both steels after imparting 20% cold work. The Ti-free alloy shows a typical defect annealing behavior in the temperature range of 300 to 800 K in form of a continuous reduction in the positron life-time. This defect recovery stage is also seen

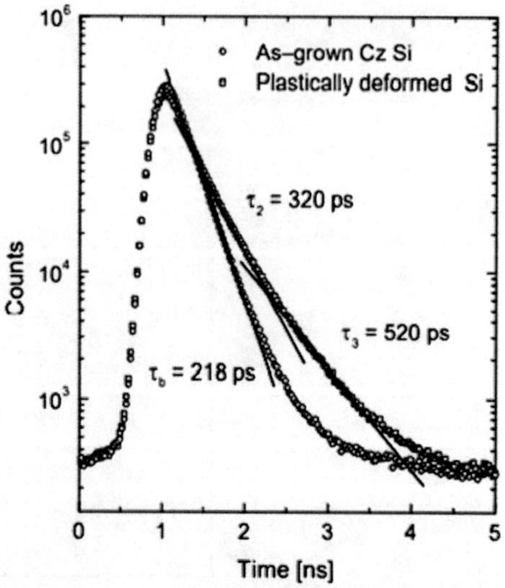

Fig. 2.68 Positron lifetime spectra for defect-free Czochralski grown silicon and for plastically deformed silicon. For times longer than 1.5 ns, the slope of the spectrum of the plastically deformed sample (3% strain, deformation temperature 775 °C) is smaller and the curve is above that of the defect-free silicon indicating the longer life components of the positron spectrum. The components obtained in the decomposition of the spectra are shown as solid lines. In the as grown silicon, only the lifetime for defect-free silicon of $\tau_{bulk} = 218$ ps is shown. Three components can be resolved in the spectrum of the deformed silicon. The Gaussian function-like shape of the curve for times less than 1 ns reflects the resolution function of the spectrometer arrangement ([227], with permission from Springer Nature)

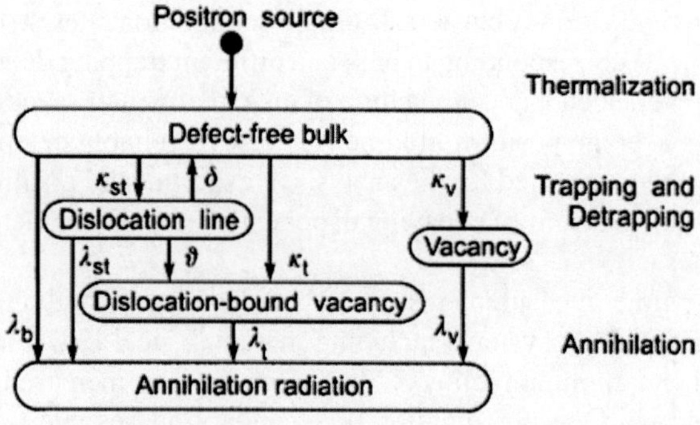

Fig. 2.69 Schematic showing trapping and annihilation of positrons in a plastically deformed semiconductor with dislocations, vacancy clusters and vacancies. The dislocation lines are regarded in the trapping pre-state as a shallow positron trap with a trapping rate of κ_{st}, a de-trapping rate of δ and an annihilating rate of λ_{st}. A transition rate from the shallow state to the deep ground state is ϑ, which is correlated between the dislocations and the vacancy defects. κ_v and κ_t are, respectively, the trapping rates of the vacancy clusters and the dislocation-bound vacancy, λ_v, λ_t and λ_b are taken to be the annihilating rates of the vacancy clusters, the dislocation-bound vacancy and the bulk, respectively (from [235], with permission from TransTech Publisher)

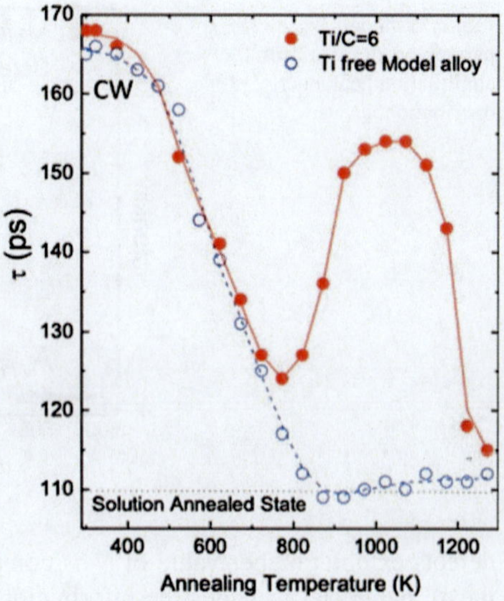

Fig. 2.70 Variation of positron lifetime with annealing temperature in 20% cold worked D9 and 20% cold worked Ti-free model alloy (in modified form from [231], with permission from Wiley)

for Ti-bearing steel. However, the Ti-bearing steel shows significant defect trapping beyond 800 K, which ultimately recovers back beyond 1200 K. The increase in lifetime between 800 to 1000 K is attributed to the formation of an increasing number of TiC precipitates and the final fall arises from TiC precipitate coarsening during grain growth above 1000 K, which reduces the number density of TiC precipitates [231].

The positron annihilation radiation also carries information concerning the momentum of the electron–positron pair. Consequence of this momentum is the Doppler broadening of the profile of the positron annihilation radiation, as indicated in Fig. 2.71. The annihilation profile can be measured using a high energy-resolution germanium detector. At the moment of annihilation, a positron is normally in the thermalized state. However, the electron may have different momentum, depending upon the annihilation site, i.e. on whether the positron is annihilated in the bulk by a core electron or by a valence electron at an ion-free defect site. For a non-zero total energy of the annihilating electron–positron pair, the energy of the annihilation γ-rays may deviate from 0.511 MeV by $\pm \Delta E$ ($= \frac{1}{2} cp_L$), where p_L is the momentum of the annihilating pair into the direction of the annihilation γ-ray) as necessitated by energy and momentum conservation during annihilation. Due to the different momentum of the electrons in the defect sites and in the bulk, the Doppler broadening can be effectively used for the characterization of defect density in a material. Figure 2.71 defines the two parameters, S and W, obtained from the annihilation peak energy distribution. The ratio of the intensity in the central region of the peak area (A_S, from ~ 0.5102 to 0.5118 meV) to the total intensity A_{tot} of the peak function, is defined as the S parameter. The ratio of the total intensity in the tails of the peak function to the total intensity of the peak function is defined as the W parameter [225]. It is obvious from these definitions and the above discussions that a material with open

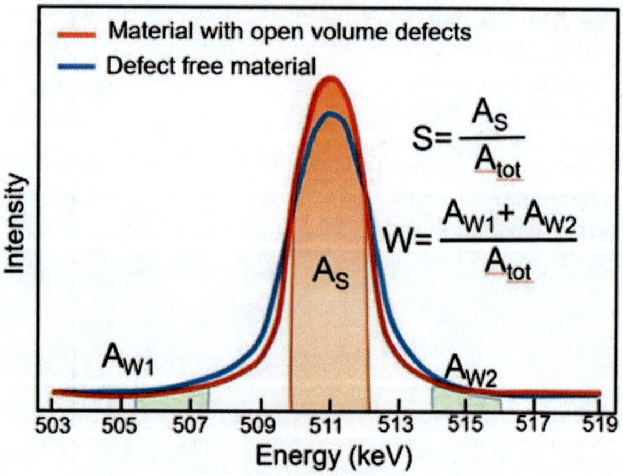

Fig. 2.71 *S* and *W* parameters derived from the annihilation peak energy distribution

defects exhibits higher value of *S* as compared to a defect-free material. Further, the *W* parameter provides the information about the chemical environment at the annihilation site and has been a powerful tool to study the segregation of impurities in irradiated materials [229].

References

1. K.V. Anisovich et al., X-ray Testing, in *Nondestructive Testing*. ed. by V.V. Kluev (Publishing House Spektr, Moscow, 2010), pp.633–991
2. D.E. Bray, R.K. Stanley, *Nondestructive Evaluation: A Tool for Design, Manufacturing, and Service* (McGraw-Hill Inc., New York, NY, 1989)
3. R. Halmshaw, *Industrial Radiology* (Springer, Dordrecht, 1995)
4. C.T. Ulrey, An experimental investigation of the energy in the continuous X-ray spectra of certain elements. Phys. Rev. **11**, 401–410 (1918)
5. C.G. Barkla, Polarised röntgen radiation. Philos. Trans. R. Soc. Lond. A **204**, 467–480 (1905)
6. H. Ruppersberg, Röntgenspannungsmessung, in Handbuch Spannungsmesspraxis, ed. by A. Peiter (Vieweg & Teubner, Braunschweig, 1992), pp. 138–182
7. P.A. Tipler, G. Mosca, in *Physics for Scientist and Engineers* (W.H. Freeman and Company, New York, 2007)
8. H.G.J. Moseley, The high-frequency spectra of the elements. Philos. Mag. **26**(6), 1024–1034 (1913)
9. D. Laguitton, W. Parrish, Experimental spectral distribution versus Kramers law for quantitative X-ray fluorescence by fundamental parameters method. X-ray Spectrom. **6**, 201–203 (1977)
10. D.H. Larsson, P.A.C. Takman, U. Lundstrom, A. Burvall, H.M. Hertz, A 24 keV liquid-metal-jet X-ray source for biomedical applications. Rev. Sci. Instrum. **82** (2011)
11. M. Feser et al., Sub-micron resolution CT for failure analysis and process development. Meas. Sci. Technol. **19** (2008)
12. A. Tkachuk, F. Duewer, H. Cui, M. Feser, S. Wang, W. Yun, X-ray computed tomography in Zernike phase contrast mode at 8 keV with 50-nm resolution using Cu rotating anode X-ray source. Z. Kristallogr. **222**, 650–655 (2007)
13. J.D. Jackson, *Classical Electrodynamics* (Wiley, New York, 1998)

14. J. Als-Nielsen, D. McMorrow, *Elements of Modern X-ray Physics* (John Wiley & Sons Ltd., New York, 2011)
15. R.D. Owen, Portable linear accelerators for X-ray and electron-beam applications in civil engineering. NDTE Int. **31**, 401–409 (1998)
16. R.G. Schonberg et al., Portable, X-band, linear accelerator system. IEEE Trans. Nucl. Sci. **32**, 3234–3236 (1985)
17. V.B. Sorokin, A.S. Lutsenko, V.G. Gentselman, Use of a Slit collimator for determining the focal-spot size of betatron bremsstrahlung. Instrum. Exp. Tech. **61**, 192–197 (2018)
18. S.V. Chakhlov et al., Betatron application in mobile and relocatable inspection systems for freight transport control. J. Phys: Conf. Ser. **671**(012024), 1–5 (2016)
19. S. Cipiccia et al., Gamma rays from harmonically resonant betatron oscillations in a plasma wake. Nat. Phys. **7**, 867–871 (2011)
20. J.M. Cole et al., Laser-wakefield accelerators as hard X-ray sources for 3D medical imaging of human bone. Sci. Rep. **5** (2015)
21. K.G. Dong et al., Micro-spot gamma ray generation based on laser wakefield acceleration. J. Appl. Phys. **123** (2018)
22. R. Jenkins, *Quantitative X-ray Spectrometry* (CRC Press, Boca Rota, 2014)
23. G.F. Knoll, *Radiation Detection and Measurement* (Wiley, New York, 2010)
24. M. Körner, C.H. Weber, S. Wirth, K.J. Pfeifer, M. Reiser, M. Treitl, Advances in digital radiography: physical principles and system overview. Radio Graph. **27**, 6675–6687 (2007)
25. P. Allé, E. Wenger, S. Dahaoui, D. Schaniel, C. Lecomte, Comparison of CCD, CMOS and Hybrid Pixel X-ray detectors: detection principle and data quality. Phys. Scr. **91** (2016)
26. J.P. Moy, Recent developments in X-ray imaging detectors. Nucl. Instrum. Methods Phys. Res. Sect. A **442**, 26–37 (2000)
27. J.G. Rocha, S. Lanceros-Mendez, Review on X-ray detectors based on scintillators and CMOS technology. Recent Adv. Electr. Electron. Eng. **4**, 16–41 (2011)
28. W.E. Burcham, *Elements of Nuclear Physics* (Longman Group Limited, New York, 1981)
29. K. Bethge, *Kernphysik* (Springer, Berlin, Heidelberg, 1996)
30. L. Kissel, R.H. Pratt, S.C. Roy, Rayleigh scattering by neutral atoms, 100 eV to 10 MeV. Phys. Rev. A **22**, 1970–2004 (1980)
31. M.J. Berger et al., *NIST Standard Reference Database 8* (*XGAM*), in NIST, 2010, vol. 2022
32. J. Birkenstock, R.X. Fischer, T. Messner, BRASS, the Bremen Rietveld analysis and structure suite. Z. Kristallogr. **23**, 237–242 (2006)
33. G. Gottstein, *Physical Foundation of Materials Science* (Springer, Berlin, 2004)
34. B.D. Cullity, *Elements of X-ray Diffraction* (Addison-Wesley Publishing Company Inc., Reading, Mass, 1978)
35. J. Epp, X-ray diffraction (XRD) techniques for materials characterization, in *Materials Characterization Using Nondestructive Evaluation* (*NDE*) *Methods*, ed. by G. Hübschen, I. Altpeter, R. Tschuncky, H.-G. Herrmann, (Elsevier Ltd., 2016) pp. 81–124
36. J. Goldstein, D.E. Newbury, J.R. Michael, N.W.M., Ritchie, J.H.J. Scott, D.C. Joy, Scanning Electron Microscopy and X-ray Microanalysis, (Springer Nature, Berlin, Heidelberg, New York, 2018)
37. D.P. Woodruff, *Modern Techniques of Surface Science* (Cambridge University Press, 2016)
38. Berry, B. F., Voots, G. R., 1989. On-Site Verification of Alloy Materials with a New Field-Portable XRF Analyzer Based on a High Resolution HgI_2 Semiconductor X-ray Detector. In: J. Boogaard, G. M. van Dijk, (Eds.), 12th World Conference on Non-Destructive Testing. Elsevier, Amsterdam, pp. 737-742.
39. S. Piorek, J. Ojanpera, E. Piorek, J.R. Pastore, In-situ, positive materials identification at room and elevated temperatures with modern, portable, optical emission and X-ray fluorescence analyzers, in *Nondestructive Characterization of Materials*, ed. by R.E. Green, vol. VIII (Plenum Press, Springer Science + Business Media, LLC, 1998), pp. 461–468
40. J. Hsieh, *Computed Tomography: Principles, Design, Artifacts, and Recent Advances*, (SPIE Press, 2015)

41. A.C. Kak, M. Slaney, *Principles of Computerized Tomographic Imaging* (Society of Industrial and Applied Mathematics, University City Science Center, PA, 2001)
42. W.A. Kalender, in Computed Tomography, Fundamentals, System Technology, Image Quality, Applications, (Wiley, Hoboken, NJ, 2011)
43. T. Vontz, K. Goebbels, M. Maisl, H. Reiter, S. Hirsekorn, NDT for small defects by X-ray tomography, in *Review of Progress in Quantitative Nondestructive Evaluation*, vol. 7A, ed. by D.O. Thompson, D.E. Chimenti (Plenum Press, 1988), pp.389–397
44. R.J. Parmee, C.M. Collins, W.I. Milne, M.T. Cole, X-ray generation using carbon nanotubes. Nano Convergence **2** (2015)
45. S.R. Stock, X-ray microtomography of materials. Int. Mater. Rev. **44**, 141–164 (1999)
46. L. Vasarhelyi, Z. Konya, A. Kukovecz, R. Vajtai in Microcomputed tomography-based characterization of advanced materials: a review. Mater. Today Adv. **8** (2020)
47. J. Kastner, C. Heinzl, X-ray computed tomography for non-destructive testing and materials characterization, in *Integrated Imaging and Vision Techniques for Industrial Inspection, Advances in Computer Vision and Pattern Recognition*, ed. by L. Zheng, H. Ukida, P. Ramuhalli, K. Niel, (Springer, London, 2015)
48. T.O.J. Fuchs, R. Hanke, 3D X-ray tomography—basics and latest developments, in *Handbook of Advanced Non-Destructive Evaluation*, ed. by N. Ida, N. Meyendorf, (Springer International Publishing AG, Berlin, 2018)
49. S. Gondrom, J. Zhou, M. Maisl, H. Reiter, M. Kröning, W. Arnold, X-ray computed laminography: an approach of computed tomography for applications with limited access. Nucl. Eng. Des. **190**, 141–147 (1999)
50. P.J. Withers, M. Preuss, Fatigue and damage in structural materials studied by X-ray tomography. Annu. Rev. Mater. Res. **42**(42), 81–103 (2012)
51. J.E. Evans, H. Friedrich, Advanced tomography techniques for inorganic, organic, and biological materials. MRS Bull. **41**, 516–521 (2016)
52. F. Porsch, Automated seed testing by 3D X-ray computed tomography. Seed Sci. Technol. **48**, 73–81 (2020)
53. S. Fourmaux, E. Hallin, P.G. Arnison, J.C. Kieffer, Optimization of laser-based synchrotron X-ray for plant imaging. Appl. Phys. B **125** (2019)
54. S. Kneip, et al., X-ray phase contrast imaging of biological specimens with femtosecond pulses of betatron radiation from a compact laser plasma wakefield accelerator. Appl. Phys. Lett. **99** (2011)
55. J. Hiller, M. Maisl, L.M. Reindl, Physical characterization and performance evaluation of an X-ray micro-computed tomography system for dimensional metrology applications. Meas. Sci. Technol. **23**, 085404-1–85417 (2012)
56. G. Dehm, B.N. Jaya, R. Raghavan, C. Kirchlechner, Overview on micro-and nanomechanical testing: new insights in interface plasticity and fracture at small length scales. Acta Mater. **142**, 248–282 (2018)
57. S.M. Spearing, Materials issues in microelectromechanical systems (MEMS). Acta Mater. **48**, 179–196 (2000)
58. C.R.M. Grovenor, *Microelectronic Materials* (Routledge, New York, 1998)
59. R. Waser (ed.), *Nanoelectronics and Information Technology* (Wiley VCH, Weinheim, Germany, 2003)
60. R. Hanke, T. Fuchs, N. Uhlmann, X-ray based methods for non-destructive testing and material characterization. Nucl. Instrum. Methods Phys. Res. Sect. A **591**, 14–18 (2008)
61. W. Arnold, Nondestructive Testing of Electronic Components and Structural Ceramics, in *Reference Module in Materials Science and Materials Engineering*. ed. by S. Hashmi (Elsevier, Oxford, 2019), pp.1–4
62. I.P. Almeida et al., Dual-energy CT quantitative imaging: a comparison study between twin-beam and dual-source CT scanners. Med. Phys. **44**, 171–179 (2017)
63. H.E. Martz, S.M. Glenn, Dual-energy X-ray radiography and computed tomography. In: *Nondestructive Testing Handbook: Radiographic Testing* (*RT*), (Lawrence Liverpool National Laboratory, 2019)

64. X.M. Wu, Q. Wang, J.L. Ma, W. Zhang, P. Li, Z. Fang, A hyperspectral X-ray computed tomography system for enhanced material identification. Rev. Sci. Instrum. **88**, 083111–083121 (2017)
65. K.M. Champley et al., Method to extract system-independent material properties from dual-energy X-ray CT. IEEE Trans. Nucl. Sci. **66**, 674–686 (2019)
66. B.J. Heismann, J. Leppert, K. Stierstorfer, Density and atomic number measurements with spectral X-ray attenuation method. J. Appl. Phys. **94**, 2073–2079 (2003)
67. S. Osipov, E. Libin, S. Chakhlov, O. Osipov, A. Shtein, Parameter identification method for dual-energy X-ray imaging. NDTE Int. **76**, 38–42 (2015)
68. V. Rebuffel, J.M. Dinten, Dual-energy X-ray imaging: benefits and limits. Insight **49**, 589–594 (2007)
69. T. Hensler, M. Firsching, J.S.G. Bonilla, T. Woerlein, N. Uhlmann, K.E. Wirth, Non-invasive investigation of the cross-sectional solids distribution in CFB risers by X-ray computed tomography. Powder Technol. **297**, 247–258 (2016)
70. K. Wells, D.A. Bradley, A review of X-ray explosives detection techniques for checked baggage. Appl. Radiat. Isot. **70**, 1729–1746 (2012)
71. Z.R. Ying, R. Naidu, C.R. Crawford, Dual energy computed tomography for explosive detection. J. X-Ray Sci. Technol. **14**, 235–256 (2006)
72. B.R. Müller, A. Lange, M. Harwardt, M.P. Hentschel, Synchrotron-based micro-CT and refraction-enhanced micro-CT for non-destructive materials characterisation. Adv. Eng. Mater. **11**, 435–440 (2009)
73. A. Rack et al., High resolution synchrotron-based radiography and tomography using hard X-rays at the BAMline (BESSY II). Nucl. Instrum. Methods Phys. Res. Sect. A **586**, 327–344 (2008)
74. A. Lipson, S.G. Lipson, H. Lipson, *Optical Physics* (Cambridge University Press, Cambridge, UK, 2010)
75. P. Cloetens et al., Hard X-ray phase imaging using simple propagation of a coherent synchrotron radiation beam. J. Phys. D **32**, A145–A151 (1999)
76. A. Kupsch et al., Evaluating porosity in cordierite diesel particulate filter materials, part 1 X-ray refraction. J. Ceram. Sci. Technol. **4**, 169–175 (2013)
77. J. Banhart, H. Stanzick, L. Helfen, T. Baumbach, Metal foam evolution studied by synchrotron radioscopy. Appl. Phys. Lett. **78**, 1152–1154 (2001)
78. E.A.C. Panduro et al., Computed X-ray tomography study of carbonate precipitation in large portland cement pores. Cryst. Growth Des. **19**, 5850–5857 (2019)
79. J. Kirz, C. Jacobsen, M. Howells, Soft X-ray microscopes and their biological applications. Q. Rev. Biophys. **28**, 33–130 (1995)
80. B. Niemann, D. Rudolph, G. Schmahl, X-ray microscopy with synchrotron radiation. Appl. Opt. **15**, 1883–1884 (1976)
81. B. Kaulich, P. Thibault, A. Gianoncelli, M. Kiskinova, Transmission and emission X-ray microscopy: operation modes, contrast mechanisms and applications. J. Phys. Condens. Matter **23** (2011)
82. V.V. Lider, X-ray microscopy. Phys. Usp. **60**, 187–203 (2017)
83. G. Margaritondo, Y. Hwu, J.H. Je, Synchrotron light in medical and materials science radiology. Rivista del Nuovo Cimento **27**, 1–40 (2004)
84. B. Lengeler et al., Parabolic refractive X-ray lenses: a breakthrough in X-ray optics. Nucl. Instrum. Methods Phys. Res. Sect. A **467**, 944–950 (2001)
85. E. Hecht, *Optics* (Addison-Wesly Publishing Company, Reading, Mass, 2015)
86. M. Awaji et al., Zernike-type X-ray imaging microscopy at 25 keV with Fresnel zone plate optics. J. Synchrotron Radiat. **9**, 125–127 (2002)
87. G. Schmahl et al., X-ray microscopy studies. Optik **93**, 95–102 (1993)
88. C. Holzner, M. Feser, S. Vogt, B. Hornberger, S.B. Baines, C. Jacobsen, Zernike phase contrast in scanning microscopy with X-rays. Nat. Phys. **6**, 883–887 (2010)
89. E. Zschech, W.B. Yun, G. Schneider, High-resolution X-ray imaging -a powerful nondestructive technique for applications in semiconductor industry. Appl Phys A **92**, 423–429 (2008)

90. G.C. Yin et al., in Energy-tunable transmission X-ray microscope for differential contrast imaging with near 60 nm resolution tomography. Appl. Phys. Lett. **88**, 241115 (2006)
91. A.E. Hussein et al., Laser-wakefield accelerators for high-resolution X-ray imaging of complex microstructures. Sci. Rep. **9** (2019)
92. C. Fella, A. Balles, R. Hanke, A. Last, S. Zabler, Hybrid setup for micro-and nano-computed tomography in the hard X-ray range. Rev. Sci. Instrum. **88**(123702), 1–11 (2017)
93. L. Helfen et al., On the implementation of computed laminography using synchrotron radiation. Rev. Sci. Instrum. **82**, 063702–063711 (2011)
94. A.W. Hull, A new method for chemical analysis. J. Am. Chem. Soc. **41**, 1168–1175 (1919)
95. B.L. Averbach, M. Cohen, X-ray determination of retained austenite by integrated intensities. Trans. Am. Inst. Min. Metall. Eng. **176**, 401–415 (1948)
96. H.P. Klug, L.E. Alexander, *X-ray Diffraction Procedures: For Polycrystalline and Amorphous Materials* (Wiley, New York, 1974)
97. S. Mahadevan, G. Gnanaprakash, J. Philip, B.P.C. Rao, T. Jayakumar, X-ray diffraction-based characterization of magnetite nanoparticles in presence of goethite and correlation with magnetic properties. Phys. E. **39**, 20–25 (2007)
98. M.E. Fitzpatrick, A. Lodini, *Analysis of Residual Stress by Diffraction using Neutron and Synchrotron Radiation* (CRC Press, London, 2003)
99. V. Hauk, *Structural and Residual Stress Analysis by Nondestructive Methods* (Elsevier, Amsterdam, 1997)
100. B.B. He, *Two-Dimensional X-ray Diffraction* (Wiley, Hoboken, N.J., 2018)
101. E. Macherauch, Praktikum in der Werkstoffkunde, Friedr, (Vieweg & Sohn, Braunschweig, Germany, 1992)
102. C.E. Murray, I.C. Noyan, Applied and residual stress determination using X-ray Diffraction, in *Practical Residual Stress Measurement Methods*, ed. by G.S. Schajer, (John Wiley & Sons Ltd., Chichester, West Sussex, PO19 8SQ, United Kingdom, 2013)
103. I.C. Noyan, J.B. Cohen, *Residual Stress-Measurement by Diffraction and Interpretation* (Springer, New York, 1987)
104. S. Prevéy, X-ray diffraction residual stress techniques, in *Materials Characterization*, ed. by J.T. Bruno et al. (American Society for Metals, Metals Park, OH, 1986), pp. 380–392
105. U. Welzel, J. Ligot, P. Lamparter, A.C. Vermeulen, E.J. Mittemeijer, Stress analysis of polycrystalline thin films and surface regions by X-ray diffraction. J. Appl. Crystallogr. **38**, 1–29 (2005)
106. M.E. Fitzpatrick, A.T. Fry, P. Holdway, F.A. Kandil, J. Shackleton, L. Suominen, Determination of residual stresses by X-ray diffraction, in *Measurement Good Practice Guide*, National Physical Laboratory, Teddington TW11 0LW, United Kingdom, 2018
107. B.B. He, U. Preckwinckel, K.L. Smith, Fundamentals of two-dimensional X-ray difffraction (XRD^2), in *Advances in X-ray Analysis Denver X-ray Conference (DXC) on Applications of X-ray Analysis, JCPDS-International Centre for Diffraction Data, Denver, Co.*, vol. 43, pp. 273–280 (2000)
108. W. Pfeiffer, Stationary and Mobile X-ray Equipment, in *Structural and Residual Stress Analysis by Nondestructive Methods.* ed. by V. Hauk (Elsevier, Amsterdam, 1997), pp.116–128
109. H. Behnken, Some basic relations to the stress analysis using diffraction methods, in *Structural and Residual Stress Analysis by Nondestructive Methods.* ed. by V. Hauk (Elsevier, Amsterdam, 1997), pp.39–115
110. B. Eigenmann, E. Macherauch, X-ray investigation of stress states in materials—Part III. Materialwiss. Werkstofftech. **27**, 426–437 (1996)
111. B. Eigenmann, B. Scholtes, E. Macherauch, Determination of residual-stresses in ceramics and ceramic-metal composites by X-ray diffraction methods. Mater. Sci. Eng. A **118**, 1–17 (1989)
112. M.R. James, J.B. Cohen (eds.), *The Measurement of Residual Stress by X-ray Diffraction Techniques* (Academic Press Inc.; Published by Elsevier Inc., New York, 1980)
113. C.E. Murray, J.L. Jordan-Sweet, S.W. Bedell, E.T. Ryan, Stress determination through diffraction: establishing the link between Kroner and Voigt/Reuss limits. Powder Diffr. **30**, 99–103 (2015)

114. I.C. Noyan, T.C. Huang, B.R. York, Residual stress-strain analysis in thin films by X-ray diffraction. Crit. Rev. Solid State Mater. Sci. **20**, 125–177 (1995)
115. E. Kröner, Berechnung der elastischen Konstanten des Vielkristalls aus den Konstanten des Einkristalls. Z. Phys. **151**, 504–518 (1958)
116. E. Macherauch, H.W. Zoch, Röntgenographische Eigenspannungsbestimmung, in Praktikum in Werkstoffkunde, (Vieweg+Teubner, Springer Fachmedien, 2011)
117. E. Macherauch, P. Müller, Das $\sin^2(\psi)$ Verfahren in der röntgenographischen Spannungsmessung. Z. Angew. Phys. **13**, 305–312 (1961)
118. M. Gelfi, E. Bontempi, R. Roberti, L.E. Depero, X-ray diffraction Debye ring analysis for stress measurement (DRAST): a new method to evaluate residual stresses. Acta Mater. **52**, 583–589 (2004)
119. A. Kämpfe, B. Kämpfe, S. Goldenbogen, B. Eigenmann, E. Macherauch, D. Löhe, X-ray stress analysis on polycrystalline materials using two-dimensional detectors, in *Advances in X-ray Analysis, JCPDS-International Centre for Diffraction Data* vol. 43, ed. by T. Blanton, L. Brehm (2000), pp. 54–65
120. J. Ling, S.Y. Lee, Characterization of a portable X-ray device for residual stress measurements. Adv. X-ray Anal. **59**, 153–161 (2015)
121. B.L. Boyce, T.A. Furnish, H.A. Padilla, D. Van Campen, A. Mehta, Detecting rare, abnormally large grains by X-ray diffraction. J. Mater. Sci. **50**, 6719–6729 (2015)
122. T. Sasaki, Y. Hirose, K. Sasaki, S. Yasuka, Influence of image processing conditions of debye scherrer ring images in X-ray stress measurement using an imaging plate, in *Advances in X-ray Analysis, JCPDS-International Centre for Diffraction Data*, vol. 40, ed. by T. Blanton, L. Brehm (1997), pp. 588–594
123. T. Sasaki, Y. Kobayashi, X-ray Multiaxial Stress Analysis Using Two Debye Rings, in *JCPDS-International Centre for Diffraction Data*, vol. 52, ed. by T. Blanton, L. Brehm (2009), pp. 248–255
124. T. Sasaki, Y. Maruyama, H. Ohba, S. Ejiri, Two-dimensional imaging of Debye-Scherrer ring for tri-axial stress analysis of industrial materials. J. Instrum. **9** (2014)
125. J. Ramirez-Rico, S.Y. Lee, J.J. Ling, I.C. Noyan, Stress measurement using area detectors: a theoretical and experimental comparison of different methods in ferritic steel using a portable X-ray apparatus. J. Mater. Sci. **51**, 5343–5355 (2016)
126. S. Taira, K. Tanaka, T. Yamazaki, A method of X-ray microbeam measurement of local stress and its application to fatigue crack growth problems. J. Soc. Mater. Sci. Jpn **27**, 251–256 (1978)
127. D. Delbergue, D. Texier, M. Levesque, P. Bocher, Diffracting-grain identification from electron backscatter diffraction maps during residual stress measurements: a comparison between the $\sin^2$(and $\cos(\alpha)$ methods. J. Appl. Crystallogr. **52**, 828–843 (2019)
128. S.Y. Lee, J.J. Ling, S.H. Wang, J. Ramirez-Rico, Precision and accuracy of stress measurement with a portable X-ray machine using an area detector. J. Appl. Crystallogr. **50**, 131–144 (2017)
129. T. Miyazaki, Y. Maruyama, Y. Fujimoto, T. Sasaki, Improvement of X-ray stress measurement from a debye-scherrer ring by oscillation of the X-ray incident angle. Powder Diffr. **30**, 250–255 (2015)
130. S. Shukla, Rapid in-line residual stress analysis from a portable two-dimensional X-ray diffractometer. Measurement **157** (2020)
131. B. Scholtes, Röntgenographisches Verfahren, in *Handbuch für experimentelle Spannungsanalyse*. ed. by C. Rohrbach (VDI-Verlag, Düsseldorf, Germany, 1989), pp.435–464
132. H. Ruppersberg, I. Detemple, J. Krier, Evaluation of strongly nonlinear surface-stress fields $\text{sigma}_{XX}(z)$ and $\text{sigma}_{YY}(z)$ from diffraction analysis. Phys. Status Solidi A **116**, 681–687 (1989)
133. H. Ruppersberg, I. Detemple, Evaluation of the stress-field in a ground steel plate from energy-dispersive X-ray diffraction experiments. Mater. Sci. Eng. A **161**, 41–44 (1993)
134. S. Hartmann, H. Ruppersberg, Thermoelastic strain of the lattice plane spacings in white cast iron. Mater. Sci. Eng.A **325**, 414–423 (2002)

135. H. Ruppersberg, I. Detemple, J. Krier, Sigma$_{xx}$(z) and sigma$_{yy}$(z) stress-fields calculated from diffraction experiments performed with synchrotron radiation in the omega-mode and PSI-mode techniques. Z. Kristallogr. **195**, 189–203 (1991)
136. H. Ruppersberg, Stress fields in the surface region of pearlite. Mater. Sci. Eng. A **224**, 61–68 (1997)
137. M. Voigt, *Lehrbuch der Kristallphysik* (Teubner, Berlin, 1929)
138. A. Reuss, Berechnung der Fließgrenze von Mischkristallen auf Grund der Plastizitätsbedingung für Einkristalle. Z. Angew. Math. Mech. **9**, 49–58 (1929)
139. C.E. Murray, Equivalence of kroner and weighted voigt-reuss models for X-ray stress determination. J. Appl. Phys. **113** (2013)
140. B. Eigenmann, E. Macherauch, X-ray Investigation of stress states in materials—Part II. Materialwiss. Werkstofftech. **26**, 199–216 (1995)
141. V.G. Navas, O. Gonzalo, I. Bengoetxea, Effect of cutting parameters in the surface residual stresses generated by turning in AISI 4340 steel. Int. J. Mach. Tools Manuf. **61**, 48–57 (2012)
142. A. Gupta, Determination of residual stresses for helical compression spring method. Mater. Today Proc. **25**, 654–658 (2020)
143. T.M. Pollock, S. Tin, Nickel-based superalloys for advanced turbine engines: chemistry, microstructure, and properties. J. Propul. Power **22**, 361–374 (2006)
144. A.K. Suri, Material development for India's nuclear power programme. Sadhana Acad Proc. Eng. Sci. **38**, 859–895 (2013)
145. B.K. Shah, A. Kumar, M. Bandyopadhyay, A.K. Bandyopadhyay, P.G. Kulkarni, Qualification of beta heat treatment of uranium fuel rods by ultrasonics. Insight **41**, 707–709 (1999)
146. C.S. Barrett, T.B. Massalski, *Structure of Metals* (Pergamon Press, Oxford, New York, Oxford UK, 1980)
147. S. Suwas, N.P. Gurao, Crystallographic texture in materials. J. Indian Inst. Sci. **88**, 151–177 (2008)
148. I.L. Dillamore, J.G. Roberts, A.C. Bush, Occurrence of shear bands in heavily rolled cubic metals. Met. Sci. **13**, 73–77 (1979)
149. L.A.I. Kestens, H. Pirgazi, Texture formation in metal alloys with cubic crystal structures. Mater. Sci. Technol. **32**, 1303–1315 (2016)
150. H.J. Bunge, 3-dimensional texture analysis. Int. Mater. Rev. **32**, 265–291 (1987)
151. A. Volodarskaja, V. Vodarek, J. Holesinsky, S. Miklusova, O. Zacek, Analysis of microstructure and microtexture in grain-oriented electrical steel (GOES) during manufacturing process. Metalurgija. **54**, 615–618 (2015)
152. C.T. Mouni et al., Influence of prior deformation temperature on strain induced martensite formation and its effect on the tensile strengthening behaviour of type 304 SS studied by XRDLPA. Mater. Sci. Eng. A **826** (2021)
153. L.G. Schulz, A direct method of determining preferred orientation of a flat reflection sample using a geiger counter X-ray spectrometer. J. Appl. Phys. **49**, 1030–1033 (1949)
154. H.J. Bunge, *Texture Analysis in Materials Science* (Butterworths, London, 1982)
155. R.W. Cahn, Measurement and control of texture, in *Processing of Metals and Alloys*, ed. by R.W. Cahn, (VCH Verlagsgesellschaft mbH, Weinheim, Weinheim, 1991)
156. M. Heatherly, W.B. Hutchinson, *An Introduction to Texture in Metals* (The Institution of Metallurgists, London, 1979)
157. H.J. Bunge, *Mathematische Methoden der Texturanalyse* (Akademie-Verlag, Berlin, 1969)
158. H.J. Bunge, H.R. Wenk, J. Pannetier, Neutron-diffraction texture analysis using a 2-theta-position sensitive detector. Textures Microstruct. **5**, 153–170 (1982)
159. S. Kalpakjian, R.S. Schmid, S. Werner, Werkstofftechnik, in *Manufacturing Processes for Engineering Materials*, (Pearson Education Germany, Munich, 2011)
160. W.T. Lankford, S.C. Snyder, J.A. Bausher, New criteria for predicting the press performance of deep drawing sheets. Trans. Am. Soc. Met. **42**, 1197–1205 (1950)
161. C.M. Vlad, Verfahren zur Ermittlung der Anisotropie-Kennzahlen für das Tiefziehverhaltens kohlenstoffarmer Feinbleche. Materialprüfung. **14**, 179–182 (1972)

162. M. Spies, M. Schneider, Nondestructive Analysis of the Deep-Drawing Behaviour of Rolling Sheets with Ultrasonic Techniques, in *Nondestructive Characterization of Materials*, ed. by P. Höller, V. Hauk, G. Dobmann, C. Ruud, R. Green (Springer, 1989), pp. 296–302
163. M. Spies, E. Schneider, Non-destructive analysis of textures in rolled sheets by ultrasonic techniques. Textures Microstruct. **12**, 219–231 (1990)
164. K. Goebbels, H.-J. Salzburger, Zerstörungsfreies Messen der plastischen Anisotropie von Blechen mit Ultraschall. Mater. **25**, 279–281 (1983)
165. V.N. Serebryany, Plastic anisotropy prediction by ultrasonic texture data. Textures Microstruct. **25**, 223–228 (1996)
166. O. Cassier, C. Donadille, B. Bacroix, Lankford Coefficient Evaluation in Steel Sheets by an Ultrasonic Method, in *Nondestructive Characterization of Materials*. ed. by P. Höller, V. Hauk, G. Dobmann, C.O. Ruud, R.E. Green (Springer, Berlin, Saarbrücken, Germany, 1989), pp.303–311
167. M. Borsutzki, C. Thoma, W. Bleck, W. Theiner, Online determination of the material properties of cold-rolled steel strip. Stahl Eisen **113**, 93–99 (1993)
168. A. Moreau et al., On-line measurement of texture, thickness and plastic strain ratio using laser-ultrasound resonance spectroscopy. Ultrason. **40**, 1047–1056 (2002)
169. F. Roters, P. Eisenlohr, L. Hantcherli, D.D. Tjahjanto, T.R. Bieler, D. Raabe, Overview of constitutive laws, kinematics, homogenization and multiscale methods in crystal plasticity finite-element modeling: theory, experiments, applications. Acta Mater. **58**, 1152–1211 (2010)
170. Z.J. Gao, J.Y. Li, Y.D. Wang, The crystallographic textures and anisotropy in 2507 super duplex stainless steel. Steel Res. Int. **90** (2019)
171. C. Genzel, Line broadening by non-oriented micros RS, in *Structural and Residual Stress Analysis by Nondestructive Methods*. ed. by V. Hauk (Elsevier, Amsterdam, 1997), pp.435–456
172. P. Scherrer, Bestimmung der Größe und der inneren Struktur von Kolloidteilchen mittels Röntgenstrahlen. Nachrichten von der Gesellschaft der Wissenschaften, Göttingen, Mathematisch-Physikalische Klasse. **2**, 98–100 (1918)
173. J.I. Langford, A.J.C. Wilson, Scherrer after 60 years -survey and some new results in determination of crystallite size. J. Appl. Crystallogr. **11**, 102–113 (1978)
174. C.E. Krill, R. Birringer, Estimating grain-size distributions in nanocrystalline materials from X-ray diffraction profile analysis. Philos. Mag. A **77**, 621–640 (1998)
175. X. Zhu, R. Birringer, U. Herr, H. Gleiter, X-ray diffraction studies of the structure of nanometer-sized crystalline materials. Phys. Rev. B **35**, 9085–9090 (1987)
176. S.A. McDonald et al., Non-destructive mapping of grain orientations in 3D by laboratory X-ray microscopy. Sci. Rep. **5** (2015)
177. A. Kumar, R.R. Adharapurapu, J.W. Jones, T.M. Pollock, In situ damage assessment in a cast magnesium alloy during very high cycle fatigue. Scr. Mater. **64**, 65–68 (2011)
178. Y.H. Zhao, X.Z. Liao, Z. Jin, R.Z. Valiev, Y.T. Zhu, Microstructures and mechanical properties of ultrafine grained 7075 Al alloy processed by ECAP and their evolutions during annealing. Acta Mater. **52**, 4589–4599 (2004)
179. B.E. Warren, B.L. Averbach, The separation of cold-work distortion and particle size broadening in X-ray patterns. J. Appl. Phys. **23**, 497–497 (1952)
180. G.K. Williamson, W.H. Hall, X-ray Line broadening from filed aluminium and wolfram. Acta Metall. **1**, 22–31 (1953)
181. A. Viswanath, B.P.C. Rao, S. Mahadevan, P. Parameswaran, T. Jayakumar, B. Raj, Nondestructive assessment of tensile properties of cold worked AISI type 304 stainless steel using nonlinear ultrasonic technique. J. Mater. Process. Technol. **211**, 538–544 (2011)
182. S.K. Rai, A. Kumar, V. Shankar, T. Jayakumar, B.S. Rao, K., Raj, B., Characterization of microstructures in Inconel 625 using X-ray diffraction peak broadening and lattice parameter measurements. Scr. Mater. **51**, 59–63 (2004)
183. S. Mahadevan, T. Jayakumar, B.P.C. Rao, A. Kumar, K.V. Rajkumar, B. Raj, X-ray diffraction profile analysis for characterizing isothermal aging behavior of M250 grade maraging steel. Metall. Mater. Trans. A **39A**, 1978–1984 (2008)

184. L. Crow, Neutron Detectors for Imaging, in *Neutron Imaging and Applications: A Reference for the Imaging Community*. ed. by I.S. Anderson, R. McGreevy, H.Z. Bilheux (Springer, New York, 2009), pp.47–66
185. W. Reimers, A.R. Pyzalla, A.K. Schreyer, H. Clemens (eds.), *Neutrons and Synchrotron Radiation in Engineering Materials Science: From Fundamentals to Material and Component Characterization* (Wiley-VCH Verlag, Weinheim, Germany, 2008)
186. A. Schreyer, Physical Properties of Photons and Neutrons, in *Neutron and Synchrotron Radiation in Engineering Materials Science: From Fundamentals to Material and Component Characterization*. ed. by W. Reimers, A.R. Pyzalla, A. Schreyer, H. Clemens (Wiley-VHC Verlag, Weinheim, Germany, 2008), pp.79–89
187. N. Chankow, Neutron radiography, in *Nondestructive Testing Methods and New Applications*, ed. by M. Omar, (InTechOpen Lt., London, 2012)
188. L. Josic, A. Steuwer, E. Lehmann, Energy selective neutron radiography in material research. Appl. Phys. A **99**, 515–522 (2010)
189. J.R. Granada, Total scattering cross section of solids for cold and epithermal neutrons. Z. Naturforschung Section A J. Phys. Sci. **39a**, 1160–1167 (1984)
190. K.V. Anisovich et al., Neutron Radiography, in *Non-Destructive Testing*. ed. by V.V. Kluyev (Publishing House Spektr, Moscow, 2010), pp.992–999
191. J.R. Santisteban et al., Strain imaging by Bragg edge neutron transmission. Nucl. Instrum. Methods Phys. Res. Sect. A **481**, 765–768 (2002)
192. A.S. Tremsin, J. Rakovan, T. Shinohara, W. Kockelmann, A.S. Losko, S.C. Vogel, in Non-destructive study of bulk crystallinity and elemental composition of natural gold single crystal samples by energy-resolved neutron imaging. Sci. Rep. **7** (2017)
193. A. El-Ghany El Abd, J.J. Milczarek, Neutron radiography study of water absorption in porous building materials: anomalous diffusion analysis. J. Phys. D **37**, 2305–2313 (2004)
194. L.G. Tumlinson, H.Y. Liu, W.K. Silk, J.W. Hopmans, Thermal neutron computed tomography of soil water and plant roots. Soil Sci. Soc. Am. J. **72**, 1234–1242 (2008)
195. N. Raghu, V. Anandaraj, K.V. Kasiviswanathan, P. Kalyanasundaram, Neutron Radiographic Inspection of Industrial Components using Kamini Neutron Source Facility, in *Neutron and X-ray Scattering 2007*, vol. 989, ed. by A. Ikram, A. Purwanto, A. Zulfia, S. Hendrana, Z. Nurachman (American Institute of Physics, 2008), pp.202–205
196. N. Kardjilov, I. Manke, A. Hilger, M. Strobl, J. Banhart, Neutron imaging in materials science. Mater. Today **14**, 248–256 (2011)
197. A. Peiter, Neutronentechnik, in *Handbuch der Spannungsmesspraxis*, ed. by A. Peiter, (Vieweg & Sohn, Braunschweig, 1992), pp. 157–172
198. L. Pintschovius, Neutron Diffraction Methods, in *Structural and Residual Stress Analysis by Nondestructive Methods*. ed. by V. Hauk (Elsevier, Amsterdam, 1997), pp.495–521
199. P.J. Withers, Mapping residual and internal stress in materials by neutron diffraction. C.R. Phys. **8**, 806–820 (2007)
200. W. Woo, Z. Feng, X.L. Wang, S.A. David, Neutron diffraction measurements of residual stresses in friction stir welding: a review. Sci. Technol. Weld. Joining **16**, 23–32 (2011)
201. P. Staron, Stress Analysis by Angle Dispersive Neutron Diffraction, in *Neutron and Synchrotron Radiation in Engineering Materials Science: From Fundamentals to Material and Component Characterization*. ed. by W. Reimers, A.R. Pyzalla, A. Schreyer, H. Clemens (Wiley-VCH Verlag, Weinheim, Germany, 2008), pp.137–153
202. J.R. Santisteban, Stress Analysis by Energy-Dispersive Neutron Diffraction, in *Neutrons and Synchrotron Radiation in Engineering Materials Science: From Fundamentals to Material and Component Characterization*. ed. by W. Reimers, A.R. Pyzalla, A. Schreyer, H. Clemens (Wiley-VCH Verlag, Weinheim, Germany, 2008), pp.155–175
203. Y.V. Taran, M.R. Daymond, D. Eifler, T. Nebel, J. Schreiber, Investigation of mechanical features of low cycle fatigue specimens of austenitic steel AISI type 321 under applied load by neutron diffraction stress analysis. Mater. Sci. Technol. **21**, 35–45 (2005)
204. Y.V. Taran, J. Schreiber, A.M. Balagurov, U. Stuhr, H. Kockelmann, V.B. Zlokazov, Triaxial residual stresses in composite tube from austenitic stainless steel with welded ferritic steel cladding. Z. Kristallogr. **26**, 355–360 (2007)

205. A. Steuwer, M.J. Peel, P.J. Withers, Dissimilar friction stir welds in AA5083-AA6082: The effect of process parameters on residual stress. Mater. Sci. Eng. A **441**, 187–196 (2006)
206. D. Thibault, P. Bocher, M. Thomas, M. Gharghouri, M. Cote, Residual stress characterization in low transformation temperature 13%Cr-4%Ni stainless steel weld by neutron diffraction and the contour method. Mater. Sci. Eng. A **527**, 6205–6210 (2010)
207. T. Gnaupel-Herold, Techniques for neutron stress determination with high spatial resolution. J. Nondestr. Eval. **28**, 149–155 (2009)
208. H.A. Crostack, W. Reimers, U. Selvadurai, Surveillance of materials degradation by means of diffraction methods, in *Nondestructive Characterization of Materials*, ed. by P. Höller, V. Hauk, G. Dobmann, C. Ruud, R.E. Green, (Springer, Saarbrücken, 1989)
209. H.J. Bunge, Inhomogeneous textures. Z. Met. **73**, 483–488 (1982)
210. H.G. Brokmeier, W.M. Gan, C. Randau, M. Voller, J. Rebelo-Kornmeier, M. Hofmann, Texture analysis at neutron diffractometer STRESS-SPEC. Nucl. Instrum. Methods Phys. Res. Sect. A **642**, 87–92 (2011)
211. H.A. Crostack, W. Reimers, G. Eckold, *Analysis of Strain Hindering in Polycrystalline Materials*. Proceedings 8th Riso International Symposium on Metallurgy and Materials Science, Roskilde, DK, 1987, pp. 291–297
212. S.F. Bordin, S. Limandri, J.M. Ranalli, G. Castellano, EBSD spatial resolution for detecting sigma phase in steels. Ultramicroscopy **171**, 177–185 (2016)
213. M.N. Alam, M. Blackman, D.W. Pashley, High-angle kikuchi patterns. Proc. R. Soc. Lond. A **221**, 224–242 (1954)
214. J.A. Venables, C.J. Harland, Electron backscatterung patterns—new technique for obtaining crytallographic information in scanning electron microscope. Phil. Mag. **27**, 1193–1200 (1973)
215. S.I. Wright, B.L. Adams, Automatic analysis of electron backscatter diffraction patterns. Metall. Trans. A **23**, 759–767 (1992)
216. S.I. Wright, M.M. Nowell, J.F. Bingert, A comparison of textures measured using X-ray and electron backscatter diffraction. Metall. Mater. Trans. A **38A**, 1845–1855 (2007)
217. M.R. Barnett, M.D. Nave, C.J. Bettles, Deformation microstructures and textures of some cold rolled Mg alloys. Mater. Sci. Eng. A **386**, 205–211 (2004)
218. A. Kumar, T.M. Pollock, Mapping of femtosecond laser-induced collateral damage by electron backscatter diffraction. J. Appl. Phys. **110** (2011)
219. D.J. Prior et al., The application of electron backscatter diffraction and orientation contrast imaging in the SEM to textural problems in rocks. Am. Miner. **84**, 1741–1759 (1999)
220. F.J. Humphreys, Review grain and subgrain characterisation by electron backscatter diffraction. J. Mater. Sci. **36**, 3833–3854 (2001)
221. J. Cizek, Characterization of lattice defects in metallic materials by positron annihilation spectroscopy: a review. J. Mater. Sci. Technol. **34**, 577–598 (2018)
222. M.J. Puska, Abinitio calculations of positron-annihilation rates in solids. J. Phys. Condens. Matter. **3**, 3455–3469 (1991)
223. M.J. Puska, R.M. Nieminen, Theory of positrons in solids and on solid surfaces. Rev. Mod. Phys. **66**, 841–897 (1994)
224. M. Eldrup, B.N. Singh, Studies of defects and defect agglomerates by positron annihilation spectroscopy. J. Nucl. Mater. **251**, 132–138 (1997)
225. F.A. Selima, Positron annihilation spectroscopy of defects in nuclear and irradiated materials -a review. Mater. Charact. **174** (2021)
226. P.G. Coleman, Experimental techniques in positron spectroscopy, in *Principles and Applications of Positron and Positronium Chemistry*, ed. by Y.C. Jean, P.E. Mallon, D.M. Schrader, (World Scientific Publishing Co. Pt. Ltd., Singapore, 2003), pp. 37–72
227. R. Krause-Rehberg, H.S. Leipner, *Positron Annihilation in Semiconductors: Defect Studies* (Springer, Heidelberg, 1999)
228. C.F. Coleman, Positron-annihilation -potential new ndt technique. Part 1. Introduction and Technique. NDT Int. **10**, 227–234 (1977)

229. X. Hu, T. Koyanagi, Y. Katoh, B.D. Wirth, Positron annihilation spectroscopy investigation of vacancy defects in neutron-irradiated 3C-SiC. Phys. Rev. B **95**, 104103–1–104103–11 (2017)
230. R. Krause-Rehberg et al., Defects in plastically deformed semiconductors studied by positron annihilation: silicon and germanium. Phys. Rev. B **47**, 13266–13276 (1993)
231. R. Rajaraman, G. Amarendra, C.S. Sundar, Defect evolution in steels: insights from positron studies. Phys. Status Solidi C **6**, 2285–2290 (2009)
232. K.V. Rajkumar et al., Investigation of microstructural changes in M250 grade maraging steel using positron annihilation. Phil. Mag. **89**, 1597–1610 (2009)
233. Y. Uematsu, T. Kakiuchi, K. Hattori, N. Uesugi, F. Nakao, Non-destructive evaluation of fatigue damage and fatigue crack initiation in type 316 stainless steel by positron annihilation line-shape and lifetime analyses. Fatigue Fract. Eng. Mater. Struct. **40**, 1143–1153 (2017)
234. J. Krawczyk, W. Bogdanowicz, A. Hanc-Kuczkowska, A. Tondos, J. Sieniawski, Influence of heat treatment on defect structures in single-crystalline blade roots studied by X-ray topography and positron annihilation lifetime spectroscopy. Metall. Mater. Trans. A **49A**, 4353–4361 (2018)
235. H.S. Leipner, C.G. Hubner, T.E.M. Staab, R. Krause-Rehberg, Open-volume defects in plastically deformed semiconductors. Mater. Sci. Forum **363–365**, 61–63 (2001)
236. D. Giebel, J. Kansy, A new version of LT program for positron lifetime spectra analysis. Mater. Sci. Forum **666**, 138–141 (2011)
237. J.V. Olsen, P. Kirkegaard, N.J. Pedersen, M. Eldrup, PALSfit: a new program for the evaluation of positron lifetime spectra. Phys. Status Solidi C **4**, 4004–4006 (2007)
238. L. Resch et al., Precipitation processes in Al-Mg-Si extending down to initial clustering revealed by the complementary techniques of positron lifetime spectroscopy and dilatometry. J. Mater. Sci. **53**, 14657–14665 (2018)

Chapter 3
Non-destructive Materials Characterization by Electromagnetic Techniques

Abstract In this chapter, electro-magnetic (EM) and micro-magnetic (MM) techniques for applications in non-destructive material characterization (NDMC) are discussed. While EM techniques can be used for all electrically conducting and ferromagnetic materials, MM techniques are applicable only to ferromagnetic materials. There are many commonalities between the EM and MM techniques in terms of the sensors' primary excitation and reception of secondary magnetic fields. After explaining the basic concepts of conductivity, electromagnetism, and magnetic properties of materials, applications of these for the characterization of microstructures and stresses are presented. The effects of conductivity and permeability of materials on the amplitude and phase of eddy current signals are analyzed for use in NDMC. Furthermore, the physical principles underlying various magnetic hysteresis loop parameters and MM parameters are discussed, followed by their applications in the field of metallurgical engineering and material science. The analogy of the dislocation movement under stress fields, which determines the mechanical properties, and the domain-wall movement under magnetic fields governing the magnetic properties provide the means for NDMC.

3.1 Basic Concepts of NDMC Based on Electromagnetism

Electromagnetic techniques applied in Non-destructive Materials Characterization can be divided into five categories based on the frequency employed:

1. Use of static electric and magnetic fields exploiting time-independent phenomena, e.g. for conductivity measurements and for magnetic memory methods.
2. Use of slowly time-varying quasistatic electromagnetic phenomena exploiting alternating currents (ac) at frequencies from some mHz to some kHz, e.g. for magnetic hysteresis loop and low-frequency magnetic Barkhausen emission (MBE) measurements.
3. Use of electromagnetic phenomena in the kHz to the MHz radio-frequency (rf) range, e.g. in MBE and eddy current techniques.

W. Arnold et al., *Non-destructive Materials Characterization and Evaluation*, Springer Series in Materials Science 329,
https://doi.org/10.1007/978-3-662-66489-6_3

4. Use of electromagnetic wave propagation phenomena at frequencies from some MHz up to THz, for example in composite materials.
5. Electromagnetic waves of frequencies greater than 10^{16} Hz in X-rays and γ-ray measurements, see Chap. 2 of this book.

The techniques described in point 4 are only peripherally discussed in this book because they are more relevant for defect detection in non-destructive testing. The techniques described in points 1–3 are predominant in NDMC dealing with various electrical, magnetic or electromagnetic interactions and will be treated in detail in this chapter. The techniques described in the last point are discussed in Chap. 2.

The electromagnetic interactions used in NDMC occur primarily by the magnetic field of coils generated with AC currents. For the detection, specially designed probes are used. The wide variety of transmitter and receiving devices such as coils, Hall probes, tape recorder heads, giant magneto-resistors, Förster probes, piezoelectric, and electromagnetic acoustic transducers (EMATS, see Chap. 1, Sect. 1.5.4) allows a manifold of measurement schemes and their engineering realizations. The spatial resolution is given by the size of the exciting and receiving coils and by the interaction volume determined by their evanescent waves. In contrast to ultrasonics, coupling media are in general not required due to the inductive coupling of the measurement devices.

Two material properties govern electromagnetic NDMC. These are the electric conductivity σ [1/Ωm = S/m], which is equal to the inverse resistivity $1/\rho_{el}$, and the magnetic permeability $\mu = \mu_r \mu_0$, where $\mu_0 = 4\pi \times 10^{-7}$ [Vs/Am] is the vacuum permeability and μ_r is the relative permeability. We first explain the basics of these material parameters in two short sections based on elements of solid-state physics. In this context, we refer to Förster [1] who gave an excellent short summary on the history, possibilities, and requirements for NDMC using electromagnetic techniques. An account on all aspects of electromagnetic testing principles and their theoretical and material science foundations are given in [2].

3.1.1 Electrical Conductivity

The carriers of conductivity in metals are electrons. Their behavior can only be understood due to their properties as matter waves. If they were completely free, they would propagate in crystals without scattering by the atomic nuclei. However, there are always some crystal defects, which leads to scattering and hence to a reduction of the mean free path of the electrons. Before we discuss the scattering of electrons, let us first discuss the conduction process and the consequences of the material wave properties of the electrons.

In a metal the electrons occupy quantum states. Because electrons are Fermions, only two electrons with a spin up and a spin down can occupy one level (remember that Fermions cannot occupy a given state and having the same quantum numbers, see Fig. 3.1a). For one free electron/atom $\approx 6 \times 10^{23}$ levels must be filled per mol.

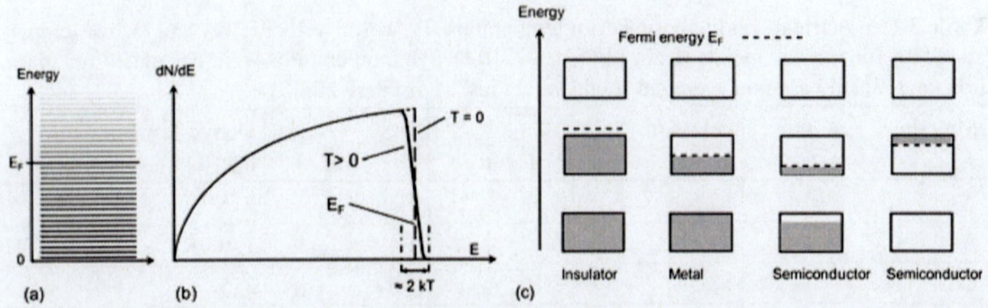

Fig. 3.1 **a** Energy levels for free electrons in a metal. The electrons fill the levels until the Fermi energy. Only two electrons with opposite spins can occupy a given state. Thermal excitation results in the occupation and de-occupation of the levels above and below the Fermi energy with a width of $\approx 2\, k_B T$ according to the Fermi–Dirac statistics, as indicated in **a** by the changing grey level from dark to light grey of the states and in **b** according to the density of states as a function of energy E; **c** The electronic states are arranged in energy bands which are caused by the periodic boundary conditions for the electrons moving in the potential of the atoms. For further details see [3–5]

They extend up to an energy $E_f \approx 5$ eV which is called the Fermi energy. The equivalent temperature T_f is several 10 000 K and is called the Fermi temperature. Only a fraction of the electrons, namely $kT/kT_f = T/T_f$, can actually participate in the conduction process due to Fermi statistics, see Fig. 3.1b, and therefore $T/T_f \approx 10^{-2}$ [3–5]. For example for Fe $T_f \approx 130\,000$ K and for Cu $T_f \approx 81\,600$ K (Table 3.1). The average velocity of the electrons v_{av} at the Fermi level is given by

$$\frac{mv_{\text{av}}^2}{2} \cong kT_f \tag{3.1}$$

which entails a velocity of the order 10^6 m/s, a surprisingly large value. More accurately, for Fe and Cu, the Fermi velocities are respectively 1.98×10^6 m/s and 1.57×10^6 m/s.

Furthermore, due to the matter-wave properties of the electrons in the periodic potential of the atoms, the boundary conditions result in energy bands which the electrons can occupy. If the Fermi energy falls roughly in the midst of a band, so that the electrons find free states when they are excited, the solid is a metal. This band is called the conduction band. The electrons behave then essentially as if they were free. If the Fermi level is such that no states can be excited, the material is an insulator. If only a few occupied states are in the conduction band or if the conduction band is almost filled, we have semiconductors or semi-metals, see Fig. 3.1c. Further details of band theory, for example the role of indirect band gaps in electrical conductivity, can be found in the classic textbooks of solid state physics [3, 4, 6].

Electrons undergo scattering by lattice defects, by phonons (lattice vibrations), and by electron–electron scattering, and other processes. Let us discuss the effect of scattering processes in Ohm's law. The differential form of Ohm's law states that the current density $\boldsymbol{j}$ is given by

Table 3.1 Electrical conductivity, Fermi temperature T_f, Fermi velocities v_f and Debye temperatures θ_D for various metals (from [3, 4, 10–12]). Debye temperatures were calculated for brass, iron, and nickel based on averaged sound velocities, as discussed in [8]

Material	σ [MS/m]	T_f [10^4 K]	v_f [10^6 m/s]	Debye Temperature θ_D [K]
Ag	62.5	6.38	1.39	215
Au	43.5	6.42	1.40	170
Cu	58	8.16	1.57	315
Al	36	13.6	2.03	394
Mg	23.2	8.23	1.58	318
Co	17	13.8	1.47	436
Zn	16.7	11.0	1.83	234
Brass ($Cu_{60}Zn_{40}$)	15	12	1.25	315
Pb	2.8	11.0	1.83	95
Ni	14	9.4	1	462
Fe	6–10	13	1.98	472

$$j = \sigma \boldsymbol{E} \tag{3.2}$$

where $\boldsymbol{E}$ is the electric field which drives the current and σ is the electrical conductivity of the material. The electrical conductivity may be anisotropic and σ becomes then a tensor. For the purpose of this book, we neglect this. The conductivity is determined by the number of electrons participating in the conduction process and by the time τ between two scattering events:

$$\sigma = \frac{ne^2\tau}{m} = \frac{1}{\rho} \tag{3.3}$$

where ρ is the specific resistance. Equation (3.3) is easy to interpret. One expects that the charge transported is proportional to the charge density ne. The factor e/m enters because the acceleration of the electron is proportional to the charge and inversely proportional to its mass. The time τ is the so-called collision time. In this time the electron is accelerated and after a collision event, the process starts again. The mean free path of the electron between two collisions is given by:

$$l = v_f \tau \tag{3.4}$$

The conductivity of copper is about 6×10⁵ S/cm (Table 3.1). About every Cu atom contributes one electron to the conduction process, as it has one electron *in the outer shell ($1s^2$, $2s^2$, $2p^6$, $3s^2$, $3p^6$, $3d^{10}$, $4s^1$). Therefore, the number of electrons per volume, n, in Cu equals to Avogadro's number multiplied by the density of Cu and divided by the molecular weight, i.e. $n = (8.92/63.55) \times 6.022 \times 10^{23}$ [$g/cm^3/g$] $\approx 8.45 \times 10^{22}$ cm^{-3}. Inserting n into (3.3) yields $\tau \approx 2 \times 10^{-14}$ s. The Fermi velocity for copper is $v_f = 1.57 \times 10^6$ m/s (Table 3.1) and hence the mean free path of an electron between two collisions is $l \approx 40$ nm according to (3.4).*

At very low temperatures, usually at He-temperature (≈ 4 K), and for $T << \theta_D$, where θ_D is the Debye temperature, the collisions of electrons with phonons (lattice vibrations) and electrons are frozen out and only the resistance due to collisions with lattice defects and impurities remain. For example, the mean free path can become as large as 0.3 cm and even larger for copper at He-temperatures. Measuring the specific resistance of a metal at very low temperature is thus a simple and well-known technique to obtain an estimate of the density of intrinsic defects and impurities. The total specific resistance of a metal can be written as

$$\rho = \rho_L + \rho_i = \rho_L + \rho_{defects} + \rho_{impurities} \tag{3.5}$$

where ρ_L is the resistance caused by the thermal lattice vibrations or phonons and electrons and

$$\rho_i = \rho_{defects} + \rho_{impurities} \tag{3.6}$$

is the intrinsic resistance due to lattice defects, $\rho_{defects}$, and impurities, $\rho_{impurities}$. Equation (3.5) is also known as Matthiessen's rule. Defects and impurities modify the scattering process for electrons either by modification of the local potential and/or strain, see for further details [3, 7–9].

The phonon's contribution to electron scattering and hence to the specific resistance for temperatures $T >> \theta_D$ is given by

$$\rho_L \propto T \tag{3.7}$$

where θ_D is the Debye temperature. The Debye temperature is a measure of how many phonon modes are present in a crystalline lattice. It can be calculated by an average of the longitudinal and shear sound velocities. The proportionality of (3.7) to the temperature for $T >> \theta_D$ follows from the probability of scattering of an electron at a local lattice deformation caused by phonons. For $T \ll \theta_D$:

$$\rho_L \propto T^5 \tag{3.8}$$

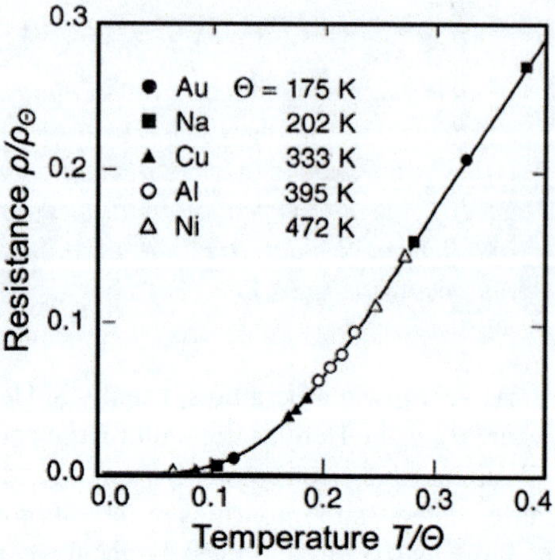

Fig. 3.2 Normalized electrical resistivity ρ_L/ρ_θ for several metals with simple Fermi surfaces plotted as a function of normalized temperature T/θ_D where θ_D is the Debye temperature (Debye temperatures may vary from different sources due to material variations (from [7], with permission from Springer

For this temperature regime the phonons have long wavelengths and their number is proportional to T^3. The long wavelengths entail that the electrons can only be scattered by a small angle due to momentum conservation. Small angle scattering, however, contributes only little to the specific resistance [4]. Figure 3.2 shows the temperature-dependent part of electrical conductivity caused by phonon scattering, (3.8).

Scattering of electrons at defects can be made visible by scanning tunneling microscopy (STM) and by atomic force microscopy, enabling one to image eddy currents, see Figs. 3.3 and 3.4. In the case of the STM image, the spatial resolution is sufficient to image the local density of states at the Fermi level and its change by a defect which acts as a repulsive potential causing a phase shift of the electronic wave function [13]. The spatial resolution of the eddy current image is approximately 50 nm and hence the contrast is determined by an average conductivity σ and by a local average scattering time τ [14], see 3.3.

3.1.2 Skin Effect

If an electromagnetic wave impinges on a metal surface with the conductivity σ, it penetrates a certain depth δ. This penetration depth of the electromagnetic wave can be calculated using Maxwell equations [15, 16]:

$$rot\boldsymbol{B} = \mu_r\mu_0\boldsymbol{j} + \frac{\partial \boldsymbol{D}}{\partial t} \tag{3.9}$$

$$rot\boldsymbol{E} = -\frac{\partial \boldsymbol{B}}{\partial t} \tag{3.10}$$

Fig. 3.3 Electrons occupying surface states on the close-packed surfaces of noble metals form a two-dimensional nearly free electron gas. These states can be probed using the scanning tunnelling microscope (STM) providing a unique opportunity to study the local properties of electrons in low-dimensional systems. The figure shows the direct observation of standing-wave patterns of electrons in the local density of states of a Cu(111) surface using a Scanning Tunneling Microscope at low temperatures. The spatial oscillations are quantum–mechanical interference patterns caused by scattering of the two-dimensional electron gas at the step edges and at the point defects. The image sized 50×50 nm^2 is obtained at a constant current. Three monatomic steps and approximately 50 point defects are visible which scatter the electrons. Spatial oscillations representing the local density of states with a periodicity of ≈ 1.5 nm near the edges are clearly visible. The vertical scale is enlarged in order to display the spatial oscillations more clearly (from [13], with permission from Springer Nature)

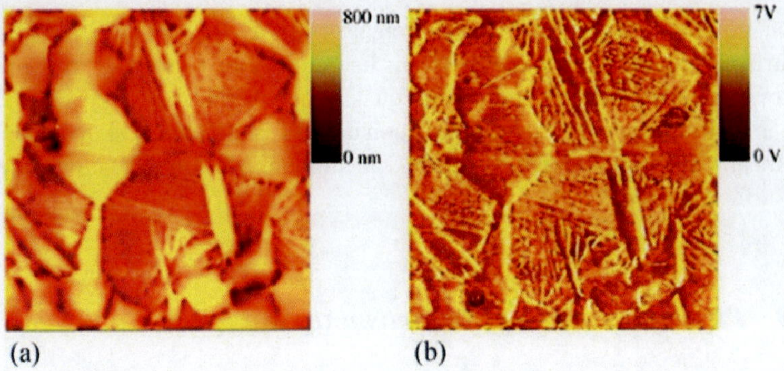

Fig. 3.4 Electrical conductivity image based on eddy current techniques. An electromagnetic coil is used to generate eddy currents in an electrically conducting material. The eddy currents generated in the conducting sample are detected and measured with a magnetic tip attached to a flexible cantilever of an atomic force microscope: **a** Topography image and **b** image of relative electrical conductivity variations in a polycrystalline dual phase titanium alloy (Ti-6Al-4 V). The contrast in the image stems from the scattering of electrons at the microstructure, which determines the local electrical conductivity (3.3), and hence the eddy current force between the magnetic tip and the sample. Here, the scan size was 50×50 μm^2. The frequency of the cantilever oscillations was 92 kHz (from [14], with the permission of AIP Publishing)

Here, $\boldsymbol{B}$ is the magnetic induction, $\boldsymbol{D}$ is the dielectric displacement, and $\boldsymbol{E}$ is the electric field with the relation $\boldsymbol{D} = \varepsilon_r \varepsilon_0\ \boldsymbol{E}$. It is assumed that the frequency is low so that $\omega\tau << 1$ holds. In this case the displacement current $\partial \boldsymbol{D} / \partial t$ can be neglected in (3.9). Furthermore, μ_r and μ_0 are the relative and absolute permeability. The two equations lead to:

$$rot \times rot\boldsymbol{E} = -rot\frac{\partial \boldsymbol{B}}{\partial t} = \frac{\partial}{\partial t}(rot\boldsymbol{B}) = \mu_r \mu_0 \sigma \frac{\partial \boldsymbol{E}}{\partial t} \tag{3.11}$$

or

$$\boldsymbol{\Delta E} = \mu_r \mu_0 \sigma \frac{\partial \boldsymbol{E}}{\partial t} \tag{3.12}$$

This is a wave equation for the electric field $\boldsymbol{E}$ propagating in the metal in the low-frequency regime. The wave is overdamped due to the term $\mu_r \mu_0 \partial \boldsymbol{E} / \partial t$ in (3.11). The solution for $\boldsymbol{E}$ is

$$\boldsymbol{E}(z, t) = \boldsymbol{x} E_0 e^{-\omega t} e^{-ikz} \tag{3.13}$$

where $\boldsymbol{x}$ is the unit vector in x-direction. Furthermore $k = (1 + i)/\delta$ where

$$\delta = 1 \Big/ \sqrt{\pi f \sigma \mu_r \mu_0} \tag{3.14}$$

Equation (3.13) means that the electromagnetic wave decays exponentially from the surface into the conductive material. Its wavevector k is equal to the inverse penetration length δ, and its real part is equal to the imaginary part, i.e. (3.13) represents an overdamped wave. The penetration depth δ is also called skin depth (3.14). This means that electromagnetic NDMC techniques yield information only from near-surface regions.

3.1.3 Physical Origin of Magnetism in Materials

The magnetism of materials is caused by the magnetic moments of their electrons. Whether these lead to a macroscopic moment depends to a large extent on the crystal structure, on the material's temperature, on the alloying elements, and on a number of other factors, see for example [17]. Magnetism can only be explained by quantum mechanical principles as discussed in many textbooks for example in [3, 4, 6]. Here, we have to restrict ourselves to the basic ideas of magnetic phenomena which are relevant for NDMC, but we will discuss in more detail the phenomena related to micro-magnetism.

Any material develops a magnetic moment M per volume if brought into a magnetic field H. For isotropic materials, the factor of proportionality is called the susceptibility:

$$\chi = \frac{\boldsymbol{M}}{\boldsymbol{H}} \tag{3.15}$$

The permeability is defined as:

$$\mu = \frac{\boldsymbol{B}}{\boldsymbol{H}} \tag{3.16}$$

Because $\boldsymbol{B}$ and $\boldsymbol{M}$ are nonlinear functions of $\boldsymbol{H}$ in magnetic materials, it is appropriate to use also the differential susceptibility and permeability

$$\chi' = \frac{d\boldsymbol{M}}{d\boldsymbol{H}} \tag{3.17}$$

$$\mu' = \frac{d\boldsymbol{B}}{d\boldsymbol{H}} \tag{3.18}$$

In the SI unit system, one uses the terms relative and absolute permeability

$$\mu_r = \frac{\mu}{\mu_0} \tag{3.19}$$

where μ_O is the absolute permeability of the vacuum $\mu_O = 4\pi \times 10^{-7}$ [Henry/m = Vs/Am]. The relative permeability of free space is $\mu_r = 1$. Finally, the relation

$$\mu_r = 1 + \chi \tag{3.20}$$

always holds. In the case that the total magnetic moment of the outer electrons of a material is zero, it is diamagnetic. Examples of diamagnetic materials are copper, silver, gold, bismuth, and beryllium. In an external magnetic field, the action of the electron currents is such that they create a magnetic field that is opposing the original field according to Lenz's law and therefore $\mu < 1$ and hence $\chi < 0$. The amplitude of χ is of the order of 10^{-5}–10^{-6}. A detailed calculation shows that

$$\chi = -\frac{\mu_0 NZe^2}{6m}\langle r^2\rangle \tag{3.21}$$

Here, N is the number of atoms/volume, Z is the atomic number, and m is the electron mass. Furthermore, $\langle r^2 \rangle$ is the mean squared distance of the electrons from the field axis through the nucleus. It represents the charge distribution in the plane of the orbit perpendicular to the field axis, which can be calculated by quantum mechanical principles.

In the case that there are permanent magnetic moments, the substance becomes paramagnetic in an external field. The magnetic moments become partially oriented. The degree of orientation is determined by the ratio of the potential energy of the dipoles in the magnetic field relative to the thermal energy opposing the orientation, $\boldsymbol{p_m B}/kT$, where $\boldsymbol{p_m}$ is the dipole moment. Both the classical and the quantum mechanical calculations give the same result for the susceptibility:

$$\chi = \mu_0 N p_m^2 / 3kT \tag{3.22}$$

which is Curie's law. Paramagnetic materials are aluminum, platinum and manganese. In these materials $\chi \approx 10^{-3} - 10^{-5}$ and always positive.

Ferromagnets are iron, cobalt, nickel and several rare earth metals and their alloys. Their susceptibility $\chi \approx 50$ to $\approx 10^4$, and hence their permeability is much larger than 1, see Table 3.2. There are many industrial products and constructions where ferromagnetic materials are employed that require a thorough safety check. Here, NDMC techniques exploiting ferromagnetic properties play a prominent role. Therefore, the occurrence of ferromagnetism is discussed in more detail.

But let us first return to some basic principles. According to Ampère's law

$$\oint \boldsymbol{B}d\boldsymbol{s} = \mu_0 I \tag{3.23}$$

magnetic dipoles have their origin in the flow of electric currents. It is well-known that a current $\boldsymbol{I}$ flowing through a plane loop produces a magnetic field which is described at large distances as a magnetic dipole field. Its magnitude on the z-axis is given by [15]

$$\boldsymbol{B} = \frac{\mu_0 \boldsymbol{IA}}{2\pi r^3} \tag{3.24}$$

where $\boldsymbol{A} = \pi R^2 \boldsymbol{e_z}$ is the area vector, r the distance from the current loop, and R is the loop radius. The dipole moment of the loop is given by:

$$\boldsymbol{p_m} = \boldsymbol{I}\pi R^2 \tag{3.25}$$

The dipole orientation is perpendicular to the plane of the loop. Employing this relation, the magnetic dipole moment of an electron on a circular orbit of radius R with the angular frequency ω_0 produces a current:

$$I = -e\omega_0 / 2\pi \tag{3.26}$$

and the angular moment of the circulating electron is

$$\boldsymbol{L} = m_e(\boldsymbol{R} \times \boldsymbol{v}) = m_e R^2 \omega_0 \tag{3.27}$$

and hence

$$\boldsymbol{p_m} = \frac{-e}{2m_e} \boldsymbol{L} \tag{3.28}$$

Table 3.2 Magnetic properties of some ferromagnetic elements and some soft and hard magnetic iron alloys. These values are approximate because they depend on the microstructure, heat treatment, composition (alloying elements), and many other parameters. Materials with a large coercivity H_c, larger than $\approx$ 10 kA/m are called magnetically hard materials and those with smaller H_c magnetically soft materials. The initial and the maximum permeability are respectively designated μ_i and μ_{max}. Note that there are also variations of the coercivity if there is a large variation in the permeabilities, which we do not discuss further. Data were assembled from various graphs in [20]. Further data can be found in the references of (Jiles, 2015); see also http://hyperphysics.phy-astr.gsu.edu/hbase/Tables/magprop.html#c2

Material	*Saturation induction* B_s [T]	*Coercivity field* H_c [kA/m]	*Curie Temperature* T_c [K]	μ_i	μ_{max}
Fe	2.15	0.080	1041	150	5000
Co	1.44	0.796	1394	70	245
Ni	0.48	0.056	633	300	2500
Fe_3C (cementite)	1.2	0.068	483	120	350 – 1,400
Fe-Si alloy (4.25% Si)	1.57	0.024	1000	500	7000
Fe–Ni (50% Ni)	1.6	0.031	813	100 – 30,000	2,000 125,000
Ni–Fe (80% Ni)	1.3	0.004	873	800 – 8000	12,00 – 80,000
Iron-carbon steel (quenched) (99% Fe; 1%C)	1.8	4.1	996	50	5000
Steel (98.1% Fe; 0.9% C; 1% Mn)	1	4	1023	120	2000
Fe-Co Steel (36% Co; 5.75% Cr)	0.96	18	1083	200–1,000	5,000 – 70,000
Alnico 5 (50% Fe; 24% Co; 15% Ni; 8% Al; 3% Cu)	1	56	1123		

The cause of magnetism are the angular moments of the electrons which lead to ring currents (3.26), which in turn produce magnetic fields. Because the angular moment of the electrons is quantized, the same holds for their magnetic moment. Applying this, and knowing that their orbital moments are given in units of the Planck constant ħ, $|\boldsymbol{L}| = \sqrt{l(l+1)}\hbar$ (l is an integer number), one obtains

$$\boldsymbol{p}_m = -\frac{e}{2m_e}\boldsymbol{L} = -l\sqrt{l(l+1)}\frac{e\hbar}{2m_e} \tag{3.29}$$

Here, a few remarks on the use of quantum numbers in the theory of the electron orbits in atoms are appropriate. A two page summary of these lines can be found in [16] and a detailed theory of atomic physics in many textbooks on quantum mechanics, see for example [18]:

(a) The principal quantum number n determines the energy of electrons in an atomic orbit. It can accept only the integer values $n = 1, 2, 3, ...$. The corresponding shells are called the *K, L, M, N, …*
(b) The angular momentum of the electron orbit is determined by the quantum number l, which is restricted to the values $l = 0, 1, 2,.., (n - 1)$. The total angular momentum associated with a given value of l is

$$\hbar[l(l+1)]^{1/2} \tag{3.30}$$

Electrons associated with states $l = 0, 1, 2, 3, 4,..$ are called *s, p, d, f, g,* …electrons, respectively. This means that electrons in an s state always have zero angular momentum and thus a vanishing magnetic moment.
(c) The possible components of the angular momentum along any specified direction (such as the direction of an external magnetic field $\boldsymbol{H}$) are determined by the magnetic quantum number m, where m is restricted to the set of values $m_l = l, (l\text{-}1), \ldots 0.0, \ldots -(l-1), -l$. For example, a p electron has the possible components of the angular momentum along the direction of a magnetic field $\hbar$, 0, - $\hbar$. Consequently, the possible magnetic moment components along the direction of an applied magnetic field are

$$-\frac{e\hbar}{2m_e},\ 0,\ \frac{e\hbar}{2m_e} \tag{3.31}$$

The quantity $\mu_B = e\hbar/2m_e = 9.274 \times 10^{-24}$ [Am^2] is called the Bohr magneton.
(d) Finally, the electron itself has an angular momentum known as the spin. The possible angular momentum components, $\pm\ \hbar/2$, along an external field direction is given by the spin quantum number $s = \pm\ 1/2$.

In ferromagnetic materials magnetic moments are spontaneously oriented in domains below the temperature T_c, which is the Curie temperature. When applying a magnetic field below T_c, the material can be completely magnetized. In this ferromagnetic state, the susceptibility χ as a function of temperature T is given by the Curie–Weiss law:

$$\chi(T) = \frac{C}{T - T_c} \tag{3.32}$$

Examples of ferromagnetic elements are iron, $T_c = 1041$ K, nickel, $T_c = 633$ K, and cobalt, $T_c = 1394$ K. Several of the rare earth metals also exhibit ferromagnetism.

In the ferromagnetic state below the Curie temperature, the magnetic moments are aligned parallel within domains. Weiss assumed that there is an interaction field (which is now called the Weiss field), which causes the alignment. This field must be of the order of

$$\mu_B\, B \approx kT_c \tag{3.33}$$

For $T_c \approx 1000$ K, this yields $B \approx 10^3$ T corresponding to 10^7 Gauss which is a rather large field. Remember the Earth magnetic field is $\approx 50\ \mu T$ or 0.5 Gauss. As the temperature of a ferromagnet is increased, the energy increases while the interaction energy remains unaffected. At the critical temperature T_c, the thermal energy overcomes the aligning of the interaction field and the magnetic order disappears. But which interaction causes the ordering below T_c?

It is tempting to assign the ordering to the interaction between the magnetic dipole moments in the solid. From the discussion above, these moments should be of the order of 1 Bohr's magneton. The field of such a magnetic dipole at its neighbor is of the order of $\mu_0\mu_B/a^3 \approx 0.5\ \text{T} \mathrel{\hat{=}} 5$ kG ($a = 0.286$ nm was used for the lattice constant of iron) about four orders of magnitude smaller than Weiss's interaction field. Therefore, a magnetic dipole–dipole interaction must be ruled out as the origin of the ordering in the ferromagnetic state.

If two ions in a crystal are sufficiently far apart, so that their quantum mechanical eigenfunctions do not overlap, their electrostatic interaction is rather weak. If, however, the wavefunctions overlap, there is an additional term called exchange energy *J,* which is due to Pauli's exclusion principle that the total wavefunction of the combined system must be antisymmetric with respect to interchanging the electrons. Symmetric and antisymmetric wave functions correspond to parallel and antiparallel spins. Therefore, the exchange energy can be considered as an interaction between the spins. A positive J leads to parallel spin orientation and hence to ferromagnetic ordering. A negative J leads to an antiparallel alignment. This can be cast in a formal way into an equation:

$$H = -2J_i \boldsymbol{S}_i \cdot \boldsymbol{S}_j \tag{3.34}$$

which reminds one of a dipole–dipole interaction. However, the interaction is of electrostatic origin. A detailed discussion of the exchange interaction can be found in [3, 17]. The latter reference was written more for a material scientist's back-ground.

As said above, the spins are oriented parallel in the ferromagnetic state in larger volumes. This state is also called collective of structural magnetism. The most important cases are shown in Fig. 3.5.

1. Ferromagnetic structure: The spin moments of neighboring atoms are oriented parallel. Examples are Fe, Co, Ni, Gd, EuS, and many alloys.
2. Antiferromagnetic structures: The spin moments of neighboring atoms are oriented antiparallel. Therefore, the magnetic moment is zero. Examples are Mn, Cr, and FeO_2.

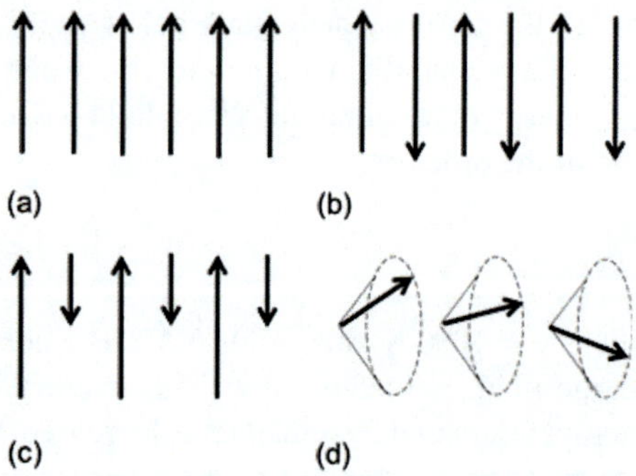

Fig. 3.5 Spin structures: **a** Ferromagnetic; **b** Antiferromagnetic; **c** Ferrimagnetic; **d** Spiral structure

3. Ferrimagnetic structures: The antiparallel spin moments are of different magnitudes. Examples are ferrites like Fe_3O_4, garnets, sulfides, and selenides.
4. Spiral structures: The orientations of the spin moments rotate by a certain angle from atom to atom. Examples are Tb, Dy, and Ho.

The reason of this behavior can again be found in Pauli's principle. The exchange interaction depends on the angle between the spins, and as a result it is positive for ferromagnetic materials leading to parallel orientation of the spins. It is negative for antiferromagnetic materials, which leads to an antiparallel orientation of the spins (Fig. 3.6).

Finally, the exchange interaction J_i also depends on the crystal orientation due to the spin–orbit coupling, which leads to a deviation of the spherical charge symmetry. Without going into details, the anisotropy can be cast into an effective magnetic field H_{an} which acts in addition to the exchange interaction. It is of the form

$$H_{an} = \frac{2K_i}{\mu_0 M_s} \tag{3.35}$$

The constants K_i are a measure of the effect of anisotropy and depend on the crystal symmetry and orientation of the H-field in crystals [19], and on composition for alloys. They can be positive or negative. Absolute values for K_i for single crystals

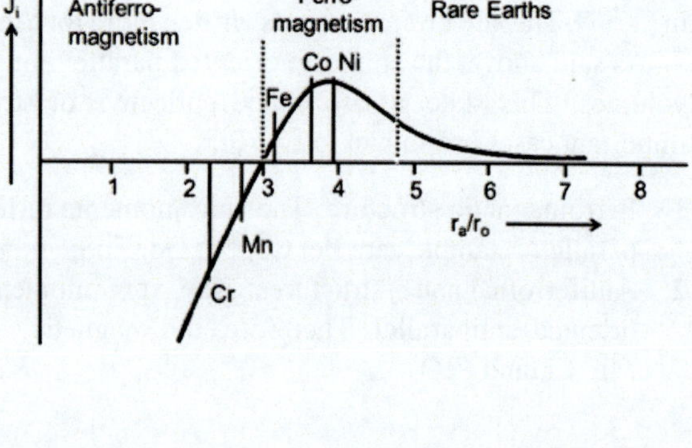

Fig. 3.6 The exchange integral J_i depends on the ratio of the atomic distance r_a and the electron orbital radius r_0

and compounds can be found in [20]. Here, we restrict ourselves to the first order K_I of the K_i in a series expansion. Typical values are $K_I \approx 10^3$ [J/m^3] for Fe–Ni alloys. This results in an anisotropy field of ≈ 1.3 [kA/m] for a saturation induction of 1.6 [T] (Table 3.2).

3.1.4 Hysteresis and Domain Walls of Magnetic Materials

The behavior of ferromagnetic materials depends not only on the atomic structure but also on their mechanical and thermal treatments. This is the physical basis for a whole class of NDMC techniques. When magnetized, ferromagnetic materials show strong hysteresis after an initial curve, see Fig. 3.7a. The main parameters characterizing a hysteresis curve are the coercivity H_c, the remanence B_r, the saturation magnetization M_s, and hysteretic losses, part of them are listed in Table 3.2. Material with a large coercivity H_c are called magnetically hard materials and those with small H_c magnetically soft materials. The flux density B reaches the saturation value B_s for large $H \rightarrow H_s$, which is the field needed to drive the magnetization into saturation, and then no further increase of B is possible, see Fig. 3.7a. In any magnetic material, the flux density B is given by (3.36):

$$\boldsymbol{B} = \mu_0 \mu_r(H)\boldsymbol{H} \tag{3.36}$$

As can be seen from Fig. 3.7, the permeability μ_r is itself a function of H. A typical curve for $\mu_r(H)$ is shown in Fig. 3.7. In order to characterize the hysteresis as well as the $\mu(H)$ curves, additional permeability parameters are used in the scientific literature, in particular for the purpose of NDMC. These are values of the incremental permeability $\mu_\Delta = (1/\mu_0)d\boldsymbol{B}/d\boldsymbol{H}$ at certain values of H. The slope of the initial curve, also called virgin curve, $\mu_i = (1/\mu_0)d\boldsymbol{B}/d\boldsymbol{H}$ for $H \rightarrow 0$, is called the initial permeability. The maximum slope of this curve is called μ_{max}. In the course of the

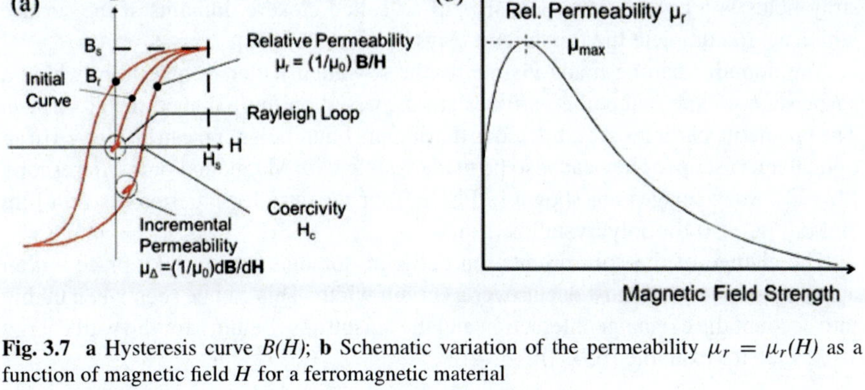

Fig. 3.7 **a** Hysteresis curve $B(H)$; **b** Schematic variation of the permeability $\mu_r = \mu_r(H)$ as a function of magnetic field H for a ferromagnetic material

book, other permeability values are defined when required. Table 3.2 shows that the permeability can vary over nearly 6 orders of magnitude for different ferromagnetic materials.

For small magnetic field values, the hysteresis curve follows a reversible loop which is described by the so-called Rayleigh law [17]:

$$\boldsymbol{B}(\boldsymbol{H}) = \mu(0)\boldsymbol{H} + \nu \boldsymbol{H}^2 \tag{3.37}$$

The term $\mu(0)$ $\boldsymbol{H}$ represents the reversible change in the magnetic induction, whereas the term ν $\boldsymbol{H}^2$ represents the irreversible part.

Finally, the area which the hysteresis loop *B(H)* encompasses, is directly proportional to the energy necessary to magnetize the material. The corresponding magnetic energy density w_m is given by:

$$w_m = \boldsymbol{B}\,\boldsymbol{H} \tag{3.38}$$

Like metallic materials in general, ferromagnetic engineering materials consist of polycrystals with a more or less random distribution of grain orientations. Besides the grains, there are also the magnetic domains. In general, the magnetic domains are not oriented along the grain boundaries or the grain structure, but have their own sub-structure. The domains are oriented in such a way that the macroscopic magnetization is as small as possible. In each domain, the magnetization is oriented preferably into easy directions. In iron these are the < 100 > directions, and in nickel the < 111 > and < 110 > directions. They reflect the magnetic anisotropy caused by the spin–orbit coupling together with the crystal structure.

Let us first discuss the behavior of single crystals. If one divides a monodomain crystal (Fig. 3.8a) into two halves of opposite magnetization, a part of the magnetic field lines will be shortened without having opposite lines (Fig. 3.8b). As a result, the demagnetization factor of the crystal will be reduced further and further if this process is continued. This increases, however, the number of boundary layers, called domain walls (Figs. 3.8c). The domain size, which results from considerations of minimizing the magnetostatic energy, is of the order of 10 μm. Completely free of stray fields is a crystal which consists of so-called closure domains at the surface which act to complete the flux circuit (Figs. 3.8d and e) [21].

The domains can be made visible by the so-called Bitter technique in which a suspension of fine magnetite particles are dispersed on the polished metal surface. The magnetite particles are attracted to the domain boundaries and can be viewed in an optical microscope. They can also be made visible with Magnetic Force Microscopy [22, 23]. Such images are shown in Fig. 3.9 for the case for a permalloy thin film and in Fig. 3.10 for polycrystalline iron.

The change of the spin orientation between domains cannot take place within atomic distances, but must occur over a certain width. This can be seen when taking into account the exchange interaction and the anisotropy yielding for the wall energy W_{dom} (see for example [6, 17]):

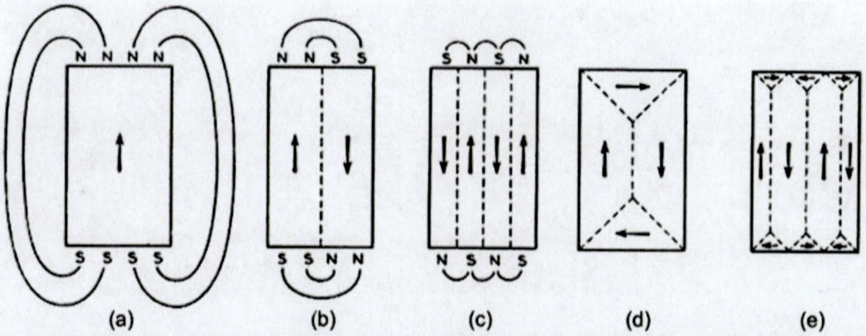

Fig. 3.8 Decreasing demagnetization of the magnetic material with increasing number of domains (a, b, c). Closed magnetic flux by 90° walls (d, e) and zero magnetic energy (reprinted from [21], with permission from the American Physical Society)

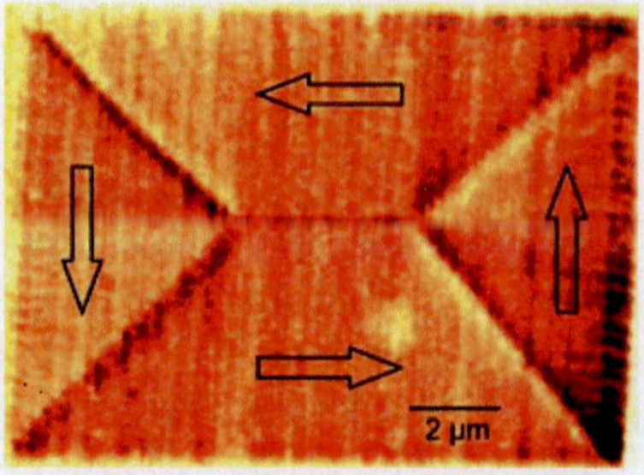

Fig. 3.9 Flux-closure domain pattern observed in a permalloy thin-film structure. The $15 \times 10\,\mu m^2$ magnetic force microscopy image shows the intersection of four 90° Bloch walls with one 180° wall. Apart from the overall wall topology, even finer structures of the walls become visible due to distinct differences in their stray fields. These fine structures have their origin in the underlying global flux-closure behavior. The stray fields are also caused by rough edges and defects (white spot) (from U. Hartmann, Department of Physics, Saarland University, private communication, 2022, with kind permission)

$$W_{dom} = W_{ex} + W_{an} = \frac{\pi^2 J_i S^2}{Na^2} + K_1 Na \tag{3.39}$$

where W_{ex} and W_{an} are the contributions to the domain wall energy due to the exchange energy and the anisotropy energy, and N is the number of atoms with interatomic spacing a in the domain wall of width δ_w. Differentiating 3.39 with respect to $Na = \delta_w$ and setting the resulting equation to zero, one obtains for the domain wall thickness δ_w corresponding to the minimum wall energy:

$$Na = \delta_{\mathrm{w}} = \left(\frac{\pi^2 J_i S^2}{K_1 a}\right)^{0.5} \tag{3.40}$$

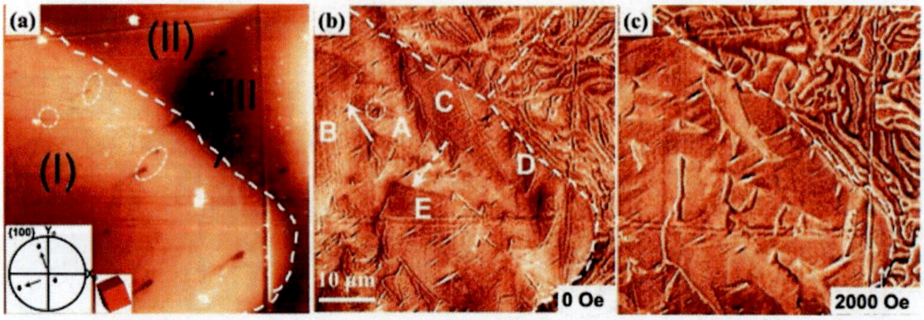

Fig. 3.10 **a** Contact mode AFM topography showing three different grains in a polycrystalline iron specimen and corresponding magnetic force microscopy phase maps in **b** demagnetized condition and **c** in presence of an in-plane magnetic field of 2000 Oe (≙ 159.2 kA/m) applied in the vertically upward direction. Various micro-magnetization phenomena such as reversible and irreversible domain wall movements, expansion and contraction of domains, Barkhausen jumps, generation of a spike domain, and bowing of a pinned domain wall (E) could be observed. The pole figure map and 3D cube orientation of grain (I) in **a** indicate the presence of two easy magnetization directions < 100 > lying on the plane, thus domains aligned in two perpendicular directions are expected (from [22], with permission from Elsevier)

where $J_i S^2$ is the exchange interaction, K_1 is the anisotropy energy, and a is the lattice constant. For iron $K_1 \approx 4.2 \times 10^3 \mathrm{J/m^3}$ with $a = 2.86\,\text{Å}$ [20]. The energy of the exchange interaction can be estimated from the Curie temperature [21] to be $J_i S^2 \approx 0.15 kT_C \approx 2.1 \times 10^{-21}\mathrm{J}$. With these values, one obtains $\delta_w \approx 131$ nm for the wall thickness. The number of atoms across the wall in a cross-section is $N = \delta_w / a \approx 460$ for iron, comprising 100 – 1000 atoms for other ferromagnetic materials. Within the wall, the spin direction changes gently by small increments from one direction to the other in the plane of the wall, see Fig. 3.11. Inserting 3.40 into 3.39, yields for the wall energy:

$$W_{dom} = 2\left(\frac{\pi^2 J_i S^2 K_1}{a}\right)^{0.5} \tag{3.41}$$

resulting in $W_{dom} \approx [1.1\ \mathrm{mJ/m^2}]$ with the numbers listed above for iron.

The domains in a thin ferromagnetic film can extend over its full width. In this case Bloch walls, which have their magnetization normal to the plane of the material cause a larger demagnetization energy in comparison to a wall in which the spins rotate within the plane of the film. These walls are called Néel walls. They do not occur in bulk specimens because they generate a high demagnetization energy within the volume of the domain wall. It is only in films that this energy becomes lower than the demagnetization energy of the wall [17] (Fig. 3.12).

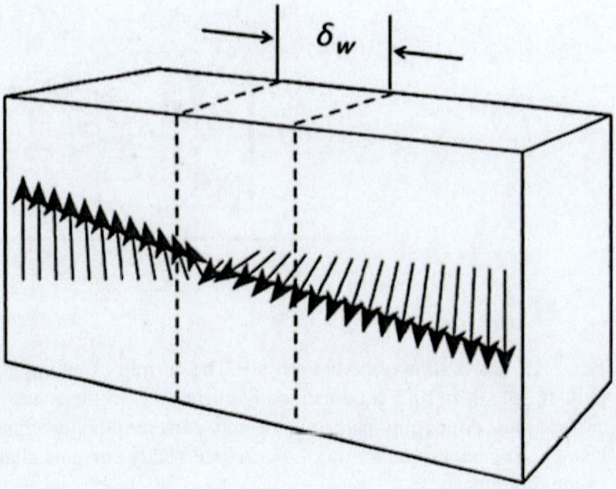

Fig. 3.11 Change of spin orientation within a domain wall. The thickness of the wall, also called a Bloch wall, is about $N \approx 100 - 1000$ lattice constants for various ferromagnetic materials. For iron, the domain wall width $\delta_\omega \approx 400$ Å and $N \approx 460$. In modified form from [21], with permission from the American Physical Society

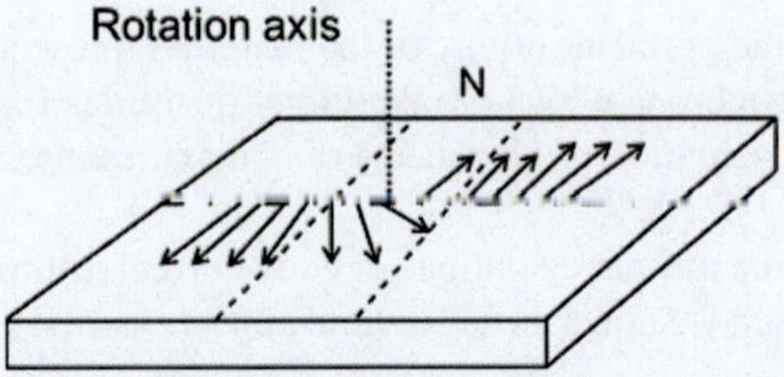

Fig. 3.12 Néel wall in a thin ferromagnetic film. Here, the orientation of the spins changes from one direction by 180° to the opposite direction in the surface plane, which is energetically favored once the film thickness is below a certain critical value. Reproduced from [133], with permission from the AIP Publishing

3.2 NDMC by Eddy Current Techniques

With eddy current techniques (ECT), detection and sizing of defects and material characterization are performed. The used frequency and the sensor design for the eddy current measurements are the parameters, which must be optimized in order to obtain a high reliability. In multi-frequency eddy current (MFEC) testing technique data are processed, which are obtained by applying more than one EC frequency. In this way, disturbing influences such as sensor lift-off, sensor tilt, edge effects or material property discontinuities can be compensated [25–27].

3.2.1 Basics of Eddy Current Testing for NDMC

In eddy current testing, the real and imaginary parts of the complex impedance Z of test coils are measured as two independent quantities from which one derives the

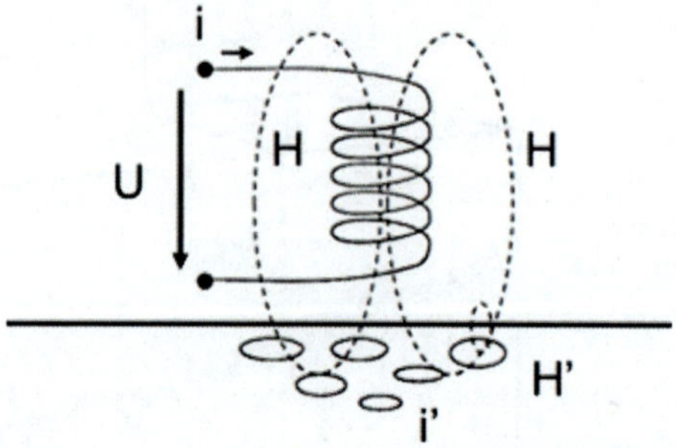

Fig. 3.13 Formation of eddy currents: The current *i* flowing through the coil generates a magnetic field *H,* which in turn produces eddy currents *i'* in the material. When moving the coil over the surface of a conducting material, the eddy currents change when encountering a defect or a material change. The back-interaction of the induced eddy currents changes the coil's impedance, which is eventually measured

conductivity σ and the permeability μ_r of the material. These are in turn influenced by those material properties which are the target quantities for NDMC. A detailed introduction and presentation of all aspects of eddy current testing can be found in [28–30].

We will first outline the basic relations of eddy current testing. Let us assume that we have a coil in which we inject an AC current *i* by an external voltage *U* (Fig. 3.13):

$$U = U_0 e^{j(\omega t + \beta_u)} \tag{3.42}$$

$$i = i_0 e^{j(\omega t + \beta_i)} \tag{3.43}$$

One defines in analogy to the Ohmic resistance in a DC-system the complex impedance *Z* of the coil as:

$$Z(X + jY) = \frac{U}{i} = \frac{U_0}{i_0} e^{j(\beta_u - \beta_i)} = \frac{U_0}{i_0} e^{j\beta} \tag{3.44}$$

where $\omega = 2\pi f = 2\pi/T$ is the angular frequency, f is the frequency, and T is the time period of the excitation voltage and the ensuing current, respectively. Furthermore, $\beta = \beta_u - \beta_i$, where β_u and β_i are the phase angles of the voltage and the current. The real part of the coil impedance (resistance) is $\mathrm{Re}(Z) = X$ and $\mathrm{Im}(Z) = jY$ is its imaginary part. There is a phase difference between the applied voltage and the current due to the coil's imaginary part of the impedance $jY = j\omega L$. The current in the coil produces a magnetic field ***H(t)***, which one may calculate using Ampère's law or the appropriate Maxwell equation. If the coil is brought in close proximity to a conducting object (Fig. 3.13), ***H(t)*** induces eddy currents *i'* in the object according to Faraday's law of induction. These currents are accompanied by secondary magnetic fields ***H'*** that are oriented opposite to the exciting field. Therefore, the coil is exposed to a modified field and the resulting coil impedance Z_1 will differ from Z_0 (Fig. 3.14).

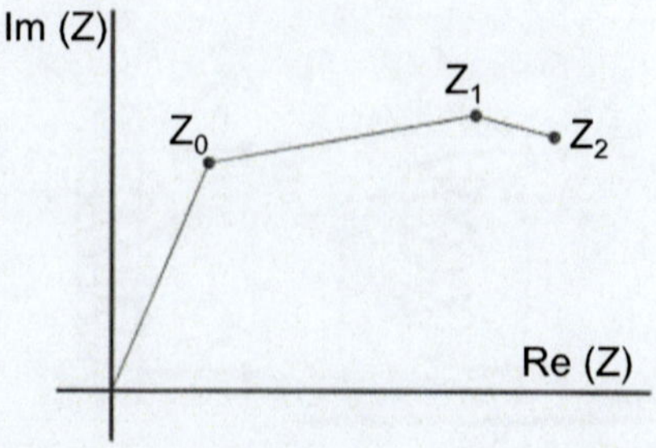

Fig. 3.14 In the complex plane, Z_0 is the impedance of the coil far away from any material. The impedance Z_1 represents the situation when the coil is in close proximity with an intact material. Variations of surface and near-surface conditions due to defects, varying heat treatments, changes in geometry, etc. lead to changes $\Delta Z = Z_2 - Z_1$ in the complex impedance plane for the coil. After calibration, ΔZ is measured for the purpose of NDMC

The parameters which determine the change of the impedance from Z_0 to Z_1 are the electric conductivity σ, the magnetic permeability μ of the object, the EC frequency *f*, and the coil-object distance. The latter is called the lift-off effect.

Let us call Z_1 the impedance of the pick-up coil when in close proximity with a defect-free material. Moving now the coil over a component made of the material with variations of surface and near-surface conditions, due to heat treatment, changes in surface geometry, defects within the skin depth etc. will lead to variations ΔZ of the complex impedance of the coil, i.e. $\Delta Z = Z_2 - Z_1$ (Fig. 3.14). The numerous effects which influence the impedance, such as changes of conductivity $\Delta\sigma$, changes of permeability $\Delta\mu$, lift-off or filling factor effects, or ΔZ due to geometrical changes and defects like cracks, makes it necessary to find additional measuring quantities in order to enable the separation of the different influencing factors.

In the case of an eddy current measuring circuit with a voltage source of high Ohmic internal resistance, the current and its phase becomes independent of the coil impedance. In this case, the real and imaginary parts of U are found by measuring the voltage at $t = 0$ and $t = T/4$, where $U_{real} = U_0 \cos(\beta_u)$ and $U_{img} = -U_0 \sin(\beta_u)$, respectively.

Only strong enough changes of σ and/or μ in an interaction volume determined by the coil's design and the employed frequency leads to distinguishable impedance changes with sufficient signal-to-noise ratio (Fig. 3.15a), in particular for small loops in the hysteresis curve (Fig. 3.7a). Alternatively, one can make the interaction volume larger by using encircling coils (Fig. 3.15b). In this case, the filling factor plays the role of the lift-off effect. In case the test sample is a rod, the filling factor η is defined as the ratio of the rod's cross-section to the effective cross-section of the coil.

Since 1970 multi-frequency eddy current (MFEC) techniques have been developed [30–33]. By using several frequencies different influencing parameters can be separated from each other facilitating the interpretation of eddy current signals.

In transient ECNDE, eddy currents are induced in the test-piece by a step-like or pulsed excitation of the coil, rather than by a sinusoidal current excitation. This

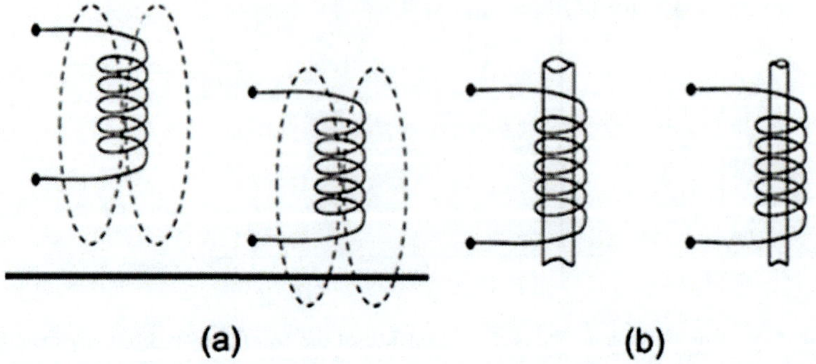

Fig. 3.15 **a** Change of the impedance of a pick-up coil by ΔZ when approaching a component to be tested (indicated by the solid line). The change depends not only on σ and/or μ_r of the material, but also on the distance of the coil which is called the lift-off effect; **b** For encircling coils, the filling factor plays the same role as the lift-off effect

leads to a spectrum of frequencies and hence to an increased information content [34, 35]. It significantly reduces the electronics required in comparison to multiple-frequency excitation for providing direct access to the wanted parameters. The full impedance locus curve (Fig. 3.17) can be measured by pulsed EC [26]. Knowing the coil's transfer function allows EC pulse shaping, which effectively produces a homogeneously weighted frequency spectrum in order to optimize the signal content for the wanted parameter underlying σ or μ.

In [36], the governing equations of eddy current inspection using scanning pick-up and encircling coils are solved. For a scanning pick-up coil, a vector potential $A_\varphi(r, z, r_0, z_0)$ is defined from which the magnetic induction is derived by $\boldsymbol{B} = \text{rot}\,\boldsymbol{A}$. Due to the rotational symmetry of the coil, see Fig. 3.16, $\boldsymbol{A}$ possesses only a φ-component which is a function of r and z, where r is the radial coordinate and z is coordinate above the sample. For a periodic current of small amplitude i, and hence small magnetic field strength, which entails a reversible permeability, $\boldsymbol{A}$ can be derived from the solution of the differential equation for a single current loop:

$$\left(\frac{d^2}{dr^2} + \frac{1}{r}\frac{d}{dr} + \frac{d^2}{dz^2} - \frac{1}{r^2} - j\omega\sigma\mu\right)\boldsymbol{A} + i\mu\delta(r - r_0)\delta(z - z_0) = 0 \qquad (3.45)$$

It is assumed that the test material is isotropic, homogeneous, and linear. The cross-section Q of the coil has the dimension $\delta(r\text{-}r_0)$, and $\delta(z\text{-}z_0)$. Equation 3.45 is solved for the four spatial volumes 1, 2, 3, and 4 by taking into account the proper boundary conditions (Fig. 3.16). The appearance of μ in the second part of 3.45 entails that the impedance curve for magnetic materials is always outwards of those for $\mu = 1$, see Fig. 3.17. Using simplifying assumptions, 3.45 was solved analytically and numerically by [32], but also by many other groups, see [28] and the references contained therein.

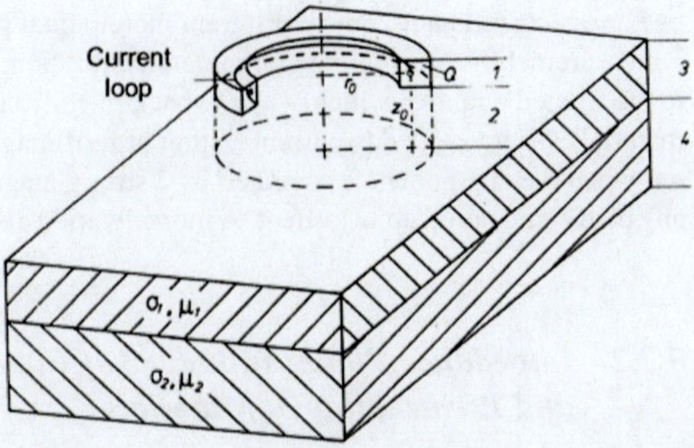

Fig. 3.16 Parameters which enter eddy current testing using a pick-up coil

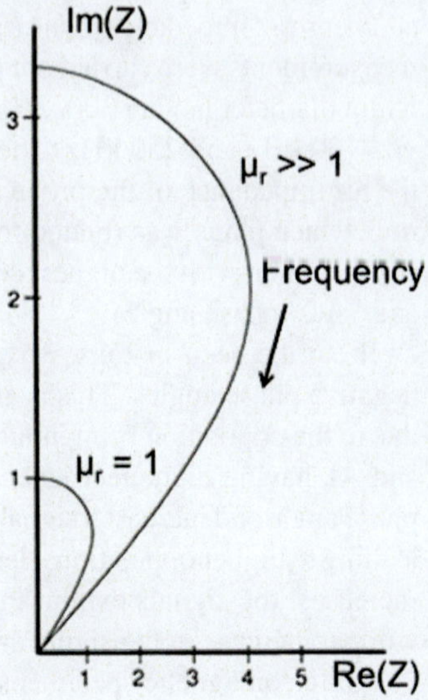

Fig. 3.17 Impedance plane for a pick-up coil. The two curves depict the impedance of the coil as a function of the parameter frequency. The inner curve is valid for a non-magnetic material ($\mu_r =$ 1) and the outer curve for a ferromagnetic material ($\mu_r >$ > 1). For a given impedance curve, any point can be selected by either varying the frequency f or the conductivity σ. This is the so-called equivalency principle in EC testing

A given working point on the impedance curve (Fig. 3.17) can be realized by different parameter combinations of ω, σ, *and* μ. For materials with large permeability μ, the impedance curve approaches for high frequencies the curve for μ $= 1$. If the conductivity σ and its variability $\Delta\sigma$ standing for material property changes are small, eddy current testing is then based on incremental permeability μ_Δ measurements, and its correlation to the wanted material properties.

If in eddy current testing of ferromagnetic materials the currents injected into the test coils are small, the *B/H* loops in the hysteresis curve are small as well and close to $H = 0$ (Fig. 3.7a). The *H*-fields are thus far lower than the coercivity field H_c and $B_s/\mu_r\mu_0$, i.e. the field leading to saturation of the flux density B_s. This holds for a totally demagnetized component. If the material of a component has experienced prior magnetization, however, then the loops in the remanence are between the induction fields $B = 0$ and $B = B_r$, the maximum remanence. If the operating point on the hysteresis curve is not known, different incremental permeabilities may result from a measurement rendering the assignment/calibration of the data difficult in respect to the wanted quantity, which is a drawback of eddy current testing of ferromagnetic materials. Therefore, eddy current testing of ferromagnetic materials is often carried out when the component is saturated by a strong magnetic field in order to suppress any disturbing permeability effect. A more detailed discussion follows in Sect. 3.2.5.

3.2.2 *Impedance Plane: Influence of Conductivity and Permeability of Materials*

The effects of electrical conductivity and magnetic permeability of specimens on the eddy current impedance plane response can be visualized in Fig. 3.18. The impedance measurements were carried out using a commercial LCR meter. An absolute probe of 5 mm diameter having 300 μH inductance was used at two different test frequencies viz. 100 kHz and 250 kHz. The impedance values were normalized with respect to the air impedance of the probe (shifted to the origin of the impedance plot) and the impedance plane was rotated to align the impedance change with respect to air for the specimen with the highest conductivity (99.9% Cu) along the negative imaginary axis (-90° phase angle).

It can be seen in Figs. 3.18a and b that all the non-magnetic specimens have negative phase angles. This is attributed to the reduction in the probe magnetic field due to the opposition from induced eddy current in the specimen. Specimens, like Cu and Al, having high electrical conductivity values exhibit higher values of negative phase angle and increased signal amplitude, due to induction of the high eddy currents leading to higher opposition. Hence, as the electrical conductivity of the test specimen increases, the signals exhibit in the impedance plane a larger negative phase angle with an increase in the signal amplitude.

All ferromagnetic specimens exhibit a positive phase angle due to the increase in the inductive reactance as compared to air. Here, the opposition to the probe magnetic field from the induced eddy current is being over ridden by the large contribution to the magnetic field by the magnetic domains present in the specimen. Hence for ferromagnetic materials, the phase angle increases in positive direction as the magnetic permeability increases for a given electrical conductivity, i.e. the signals move up the impedance plane with higher amplitudes.

Further, for both non-magnetic and ferromagnetic specimens, an increase in test frequency leads to the induction of more eddy currents in the specimen, which in turn increases the opposition to the probe magnetic field and to a reduction of the inductive part of the probe impedance. Hence, as the test frequency is increased, the signals move down the impedance plane exhibiting more negative phase angle (as seen in Fig. 3.18c), with higher amplitude for non-magnetic materials.

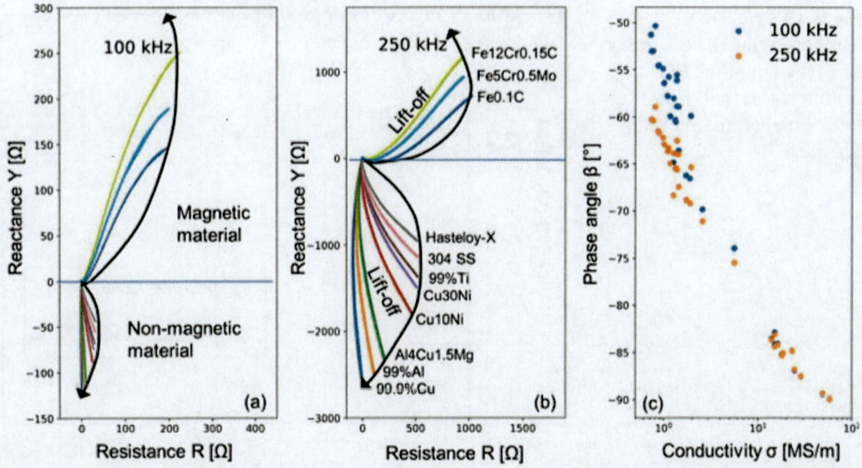

Fig. 3.18 Eddy current impedance plane showing the effect of conductivity and permeability of metals and alloys at **a** 100 kHz and **b** 250 kHz. The colors designate different materials and the arrows increasing permeability μ and conductivity σ; **c** shows variations of the phase angle β with conductivity σ for 35 different non-ferromagnetic metals and alloys, including those shown in **a** and **b**. The effect of the test frequency on the phase angle can be observed in **c** for materials with lower conductivity, as the impedance planes at corresponding frequencies are rotated to bring the impedance change with respect to air for the specimen with the highest conductivity (99.9% Cu) along the negative imaginary axis (−90° phase angle). Nominal conductivity values for the alloys are obtained from [37]

These results demonstrate the basis of eddy current based NDMC. Any change in material, leading to a decrease in conductivity viz. alloying, cold work, deformation or radiation damage, is expected to decrease the impedance and phase angle (less negative) of a non-magnetic material. Similarly, the changes leading to an increase in conductivity viz. annealing and precipitation of secondary phases upon heat treatment, lead to more negative phase angles. In case of a ferromagnetic material such as a ferritic/martensitic steel, hardening due to martensite formation is expected to decrease the amplitude and phase angle due to decreased permeability and conductivity.

3.2.3 Quality Assurance of Materials by Eddy Current Techniques

Fava and Ruch [38] designed and characterized rectangular planar coils for eddy current testing applied for crack detection in Zircaloy-4 plates. The eddy current testing system was calibrated with test materials with different electrical conductivities in order to differentiate the hydrogen content in Zircaloys. A frequency range between 0.4 and 2.0 MHz was selected for the inspection tasks. An advantage of the

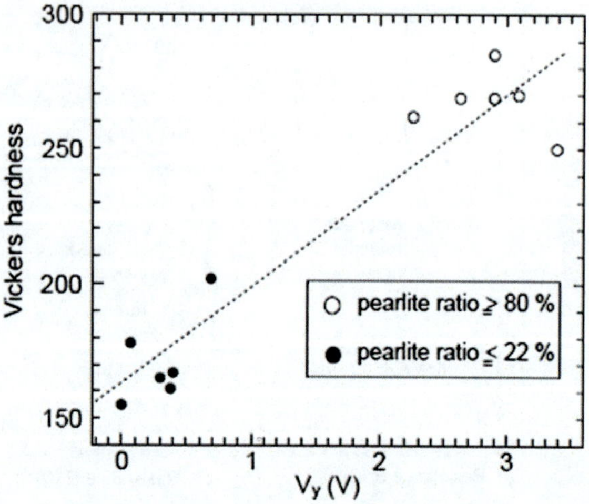

Fig. 3.19 Vickers's hardness of ductile cast irons as a function of the EC instrument' signal (from [39], with permission from Elsevier)

rectangular probes was the elimination of rim effects in the detection of edge cracks, retaining the high sensitivity to crazing, and other shallow imperfections. With one coil design, hydrogen contents of 4.9, 7.3 and 12.1 atomic percent (+ -10%) could be distinguished.

Uchimoto et al. [39] proposed a method for the evaluation of the matrix structures of cast iron by means of eddy current measurements consisting of a differential probe with two coils. Experimental results of gray and ductile cast irons show that EC signals were sensitive to the difference in the matrix microstructure. It was shown that the relationship between the percentage of pearlite in ductile cast iron and EC signals can be fitted by a functional dependence containing the Vickers hardness (Fig. 3.19). Likewise, Konoplyuk et al. [40] demonstrated a good correlation of the hardness and the tensile properties of ductile cast iron with empirically determined parameters of eddy current measurements.

Dobmann et al. [41] developed an EC system in order to detect areas of higher hardness (so-called hardness spots) on steel plates in a production setting. Hardness spots are local areas with increased mechanical hardness on the surface of semi-finished or end products in steel manufacturing. The cause of these spots was attributed to effects in the casting or rolling process.

If there is an increase in local hardness in the microstructure, e.g. due to the occurrence of martensite or bainite, as it can be the case of hardness spots on heavy plates, one can exploit the relation between hardness and permeability. The harder the material, the lower is usually its magnetic permeability, see Sect. 3.3.3. Furthermore, the electrical conductivity is reduced since the local harder microstructure has a higher dislocation density, and the dislocations are scattering centers for the conduction electrons in the eddy currents.

A two-step procedure was undertaken: (1) Excitation of a yoke coil with an AC current at a frequency of 60 kHz in order to measure the impedance relative to the Z_0 value where the coil was far apart from the plate surface (see Fig. 3.14), and

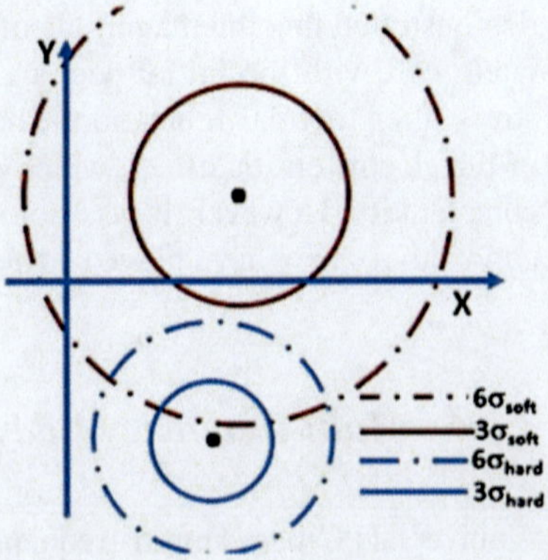

Fig. 3.20 Impedance value distributions for hardness spots (in blue) in the softer microstructure (in red). The σ-values denote the deviations from the mean values (from [41], with permission from SIMP Publishing)

a secondary pulsed excitation triggered with a preset delay in order to generate a small loop in the hysteresis curve at a definite operating point (see Sect. 3.2.1, and Fig. 3.7a). The latter served to measure the incremental permeability μ_{Δ}. These data were calibrated with destructive tests, which yielded the hardness in so-called Leeb rebound hardness tests. In these tests, an indenter is accelerated against the spots whose hardness is to be determined. The final and the initial relative velocity of the indenter is measured. Their ratio is the so-called coefficient of restitution (COR). COR $\times$ 1000 is then the Leeb hardness. Calibration tests were made at hardness spots, and on plates with and without scale (rolling skin).

The Leeb hardness can be converted into Vickers hardness by separate calibration tests. The measured impedance values in the impedance plane are schematically shown in Fig. 3.20. The softer microstructure lies in the upper part of the impedance plane due to higher permeability, as discussed in Sect. 3.2.2.

Rajkumar et al. [42] studied the effect of aging of solution annealed M250 grade maraging steel on the microstructure, room temperature hardness, and eddy current parameters. Specimens of M250 maraging steel were subjected to solution annealing at 1093 K (821 °C) for 1 h followed by aging at 755 K (483 °C) for various durations in the range of 0.25–100 h. The hardness was found to increase continuously with aging due to precipitation of intermetallics reaching a maximum, which decreased thereafter due to the reversion of martensite to austenite. The aging treatment led to a decrease in dislocation density and quenched-in point defects during initial aging, increase in volume fraction of intermetallic precipitates at intermediate duration, and a systematic increase in reverted austenite at longer duration beyond 40 h. These microstructural changes altered the electrical resistivity and magnetic permeability which, in turn, influenced the induced voltage in the eddy current coil. Two eddy current parameters, viz. magnitude and phase angle of the induced voltage, were measured for heat treated specimens. The eddy current parameters revealed a clear correlation to the microstructural changes in M250 maraging steel specimens.

Shot-peened nickel-base superalloys exhibit an increase in eddy current conductivity with increasing frequency. This was exploited for non-destructive measurements of subsurface residual stresses by Blodgett and Nagy [43]. Experimental results demonstrated that the magnitude of the increase in eddy current conductivity correlated well with the initial peening intensity as well as with the remnant residual stress after thermal relaxation. The stress dependence was explained on the base of the electro-elastic effect, which describes the stress dependence of the electrical conductivity. However, it was not excluded that paramagnetic contributions to the conductivity may have played a role as well.

3.2.4 Multi-frequency Eddy Current Techniques

A universal PC-based multi-frequency eddy current system (MFEC) was developed at Fraunhofer IZFP [31]. The system consists of a hardware apparatus controlled by a PC for data acquisition, visualization, data storage, and documentation of the results. The basic concept is to control the system functions by software and to shield the hardware in order to minimize electronic noise. The signal processing software is based on a regression analysis similar to the 3MA approach, as discussed in Sect. 3.3.9. Here, the real and/or imaginary parts of the EC coil impedance at different frequencies are the measuring quantities. The implemented algorithms can be interpreted as numerical filters applied to solve different inspection tasks. The objective of the signal processing is to differentiate between disturbing signals and the target values. The filter procedure encompasses the calibration of the inspection system requiring calibration samples, representing the relevant influencing parameters similar to the actual inspection tasks. The developed sensors covered EC frequencies of 50, 280, and 600 kHz in order to detect and size surface-breaking flaws, to determine surface waviness, and to characterize the δ-ferrite content and its distribution. In addition, a low-frequency (LF) sensor for EC frequencies of 0.500, 2.8, and 5 kHz was developed in order to measure clad thicknesses, sub-cladding flaws, and to detect and size subsurface volumetric and planar flaws.

Hidden corrosion is a typical defect, which can occur in adhesive lap-joints of layered aluminum aircraft structures. Using a suitable sensor, Disqué and Becker [44] evaluated adhesive failures in the fuselage skin using the MFEC technique, as shown in Figs. 3.21a and b. After applying a filtering algorithm, the corrosion indications in the different layers were separated. The results of detection of the corrosion in the third layer, covered and shielded by the corrosion in the first layer, were made visible in the impedance curves, which served as data for visualization, like C-scans. The data shown in Fig. 3.21b represent an example of how the suppression of lift-off effects is accomplished by evaluating Re(Z) and Im(Z) using multiple frequencies.

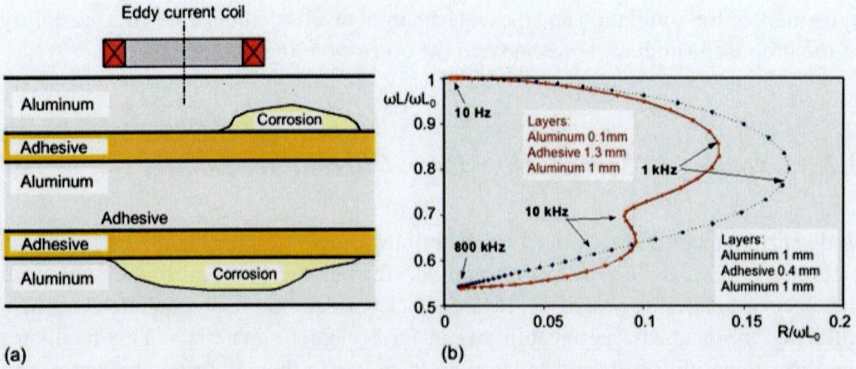

Fig. 3.21 **a** Three-layer laminated aluminum structure inspected by the multi-frequency eddy current method; **b** Impedance plane from which the parameters for detection of corrosion are derived (from [44], with permission from Wiley)

3.2.5 Hysteresis in Ferromagnetic Materials and Eddy Current Techniques

In the case of ferromagnetic materials, the magnetic field of the test coil may be adjusted so that the nonlinearity of the hysteresis curve, i.e. the generation of odd harmonics in the tangential field becomes noticeable (see also Sect. 3.3.9). When the magnetic field excited at the frequency f_1 in the test coil is close to H_c of the material being examined, the fifth harmonic becomes dominant. The third and seventh harmonics are noticeable near the saturation field strength [45, 46]. This behavior can be used to find correlations of the different harmonics with material properties like hardness and stress as a function of depth into the material by varying the excitation frequency and using (3.14), i.e. exploiting the skin effect.

Chan et al. [47] presented a nonlinear eddy current inspection technique for assessing the case hardening profile based on the assumption that the magnetic characteristics of the hardened region are different from those of the host material. An electromagnetic excitation sensor array was used to both apply sinusoidal excitations to the component and measure the nonlinear response at multiple excitation frequencies at different spatial locations, exploiting the varying penetration regions due to the skin depth. Each response signal obtained from the component under test was compared with that from a reference component subjected to the same excitation. Finally, two pattern recognition algorithms were used to process selected characteristics of the difference signal in order to determine the case depth profile of the component.

In order to explore and to optimize nonlinear eddy current testing systems, a forward numerical analysis code for nonlinear ECT signal simulation was carried out by Xie et al. [48]. To this end, an edge element numerical code of the transient vector potential A was developed. The dependence of the third harmonic component of the response signals on the initial permeability was investigated. The qualitative agreement of the simulation and the experimental results demonstrated the feasibility of the proposed simulation method and the corresponding numerical code.

3.2.6 Sorting of Materials by Eddy Current Techniques

Eddy current techniques are used for material sorting. Alloying elements change the electrical conductivity σ of ferrous and non-ferrous metals because they influence the electron scattering processes, see Sect. 3.1.1. In addition, alloying elements also influence the magnetic permeability μ in ferromagnetic materials. This facilitates the discrimination of different materials from each other. Both parameters σ and μ determine the impedance Z of an eddy current (EC) coil (see Sect. 3.2.2). For calibration, it is essential that as many as possible "accept" and "reject" test specimen are available in order to define the threshold criteria for selection, that is to say more data are better. Furthermore, a "master" test piece, often placed in a second coil, is very helpful for comparison, and it is mandatory that calibrations are regularly repeated. All disturbing effects like lift-off or filling-factor variations have to be taken into account. These procedures apply for single-frequency, multi-frequency, and for pulsed EC testing. Acceptance criteria are often based on experience besides objective, generally accepted parameters. This is discussed in much detail in [27]. There are now additional tools for calibration like computer based algorithms, envisaged already quite early [49], see also the preceding discussions.

Single frequency EC testing delivers the two parameters of interest σ and μ in one impedance measurement. For a given operation point Z_I and allowing for some scatter in the complex impedance plane, a yes/no decision is possible with regard to the material's suitability. With multi-frequency EC testing, disturbing effects can be better suppressed and a more specific sorting is possible. Overdetermination ensures the decision parameters for a given material and/or defines several decision windows for several classes of materials. Like in multi-frequency EC testing, with pulsed EC testing one uses the additional information content due to the various frequencies in order to suppress disturbing influences like geometric tolerances for evaluating the materials sorting parameters.

There are some general experiences concerning the sorting of materials using EC technologies. The technique works better with encircling coils than with pick-up probes as sensors because the interaction volume is larger due to the skin effect. For high frequencies, a good resolution for all μ values is provided, but the separation between σ and μ is reduced, and the different interaction depths result in different measurement volumes. It is desirable that the sorting of a steel material is possible with only the two parameters σ and μ in a so-called σ-μ meter. There were quite a lot of efforts to find the limits of the σ-μ meter concept. Following earlier work by Becker et al. [50], Karpen et al. [51] examined the limits for a σ-μ meter by numerical modeling as well as by experimental verification. The electrical conductivities σ of the calibration materials, representing alloyed steels, varied between 1.3 and 3.5 [MS/m]. They were measured with a potential probe. The relative permeabilities μ_r

of the test materials varied between 10 and 55. The flux was injected into the test materials by a core-up sensor with a ferrite concentrator. The exciting field strength was about 100 A/m at a frequency of 100 Hz allowing one to describe the hysteresis curve by a Rayleigh loop (Fig. 3.7a). The resolution obtained was $\Delta\sigma/\sigma < 9\%$. and $\Delta\mu/\mu < 15\%$.

A multiparameter eddy current technique for on-line material sorting of rolled steel bars has been developed by Sakomoto [52] exploiting the dependence of the resistivity ρ_{el} of the steel on its chemical composition:

$$\begin{aligned}\rho_{el}\left[\Omega mm^2/m\right] &= 1/\sigma_{el}\\ &= 0.1 + 0.05C + 0.135Si + 0.053Mn\\ &\quad + 0.11P + 0.10S + 0.054Cr \end{aligned} \tag{3.46}$$

where the quantities are the weight percentage of the different elements C (ranging from 0.09 to 0.59 wt%), Si (ranging from 0.22 to 1.53 wt%), Mn (ranging from 0.45 to 0.88 wt%), Cr (ranging from 0.03 to 1.10 wt%), and Mo (ranging from 0.01 to 0.16 wt%). The content of S and P was the same for all examined steels. The accuracy of the correlation obtained was $\Delta\sigma = \pm 0.01$ [Ωmm^2/m] $= 10^{-8}$ [Ωm], see Fig. 3.22. Assuming a conductivity of 5 [MS/m] for the steel monitored, yielded a relative accuracy of 5% for the resistance (solid lines in Fig. 3.22).

The magnetic permeability depends not only on chemical composition but also on residual stress and the thermomechanical treatment, here the reduction rate of the steel bars. Sakamoto [52] experienced that the initial permeability μ_i and the incremental permeability μ_Δ measured at small magnetic fields led to a correlation with the chemical composition. The carbon content, however, could be determined by measuring the incremental permeability μ_Δ close to the magnetic saturation state using the correlation function:

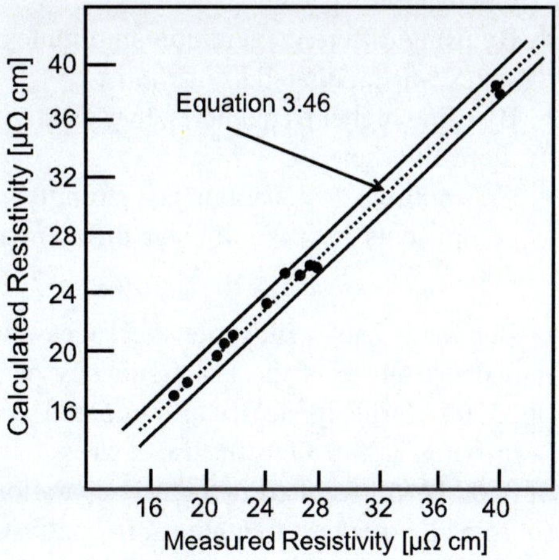

Fig. 3.22 Correlation between electrical resistivity and chemical composition of rolled steel bars. The dashed line represents (3.46). The solid lines are each 1 [µΩcm] apart from the dashed line and show the scattering of the data (figure designed according to the data in [52])

$$\mu_\Delta = 0.1 + 0.5C - 0.05Si + 0.17Cr \tag{3.47}$$

To suppress disturbing influences by diameter variations (up to 3%) of the material in the encircling coil and by temperature drifts, a two-frequency method was used in order to obtain signals with better significance. Signal averaging was also used in order to suppress small fluctuations of the electric and magnetic properties. A difference of 0.05% could be discriminated for the C content. The multiparameter method allowed to measure resistivity and incremental permeability independently.

If the disturbing influences like geometry, remanence, surface roughness, or external stray fields, could not be eliminated by the two-frequency technique, pulsed EC with a wider frequency spectrum led to the suppression of the additional disturbing parameters.

For semifinished steel bars, tubes, and crankshafts, a 100% automated process control line employing multifrequency eddy current systems was reported by Nehring [53]. Using circumferential probes as well as pick-up coils, they were designed to suppress disturbing parameters due to the geometries of the components, residual magnetism, and local fluctuations of the permeability and the conductivity of the materials. These systems were extended for measuring the hardness depth in inductive hardened flanges, steel strips and magnetic inserts in fuel pump cases [54].

Sorting high speed (HS) steels versus HS Cobalt (HSCO) steels by the MFEC method [55] started with a series of components with known quality parameters. The frequency and amplitude parameters were optimized in order to best separate the two steel qualities. Sorting with reference to different heat treatments of forged components, like bars, connecting rods, and lever rods made of 41Cr4 steel, was successfully realized with pulsed EC techniques [26]. A circumferential coil was used for the tests.

- The eddy current impedance values were required to use a number of measuring parameters which enhanced the selectivity for sorting the as-delivered, tempered, and hardened parts.
- By using different excitation amplitudes, the separation of the different material states was facilitated.
- By using higher frequencies, the selectivity of the different material states became more significant.
- The optimal test parameters strongly depended on the geometry of the test components because of their different filling factor in the circumferential test coil.

Figures 3.23a–c illustrate the successful separation of heat treatments in the impedance plane at the test frequency of 3 Hz at higher magnetic field strength of 3500 [A/m] (Fig. 3.23b) as compared to those at 220 [A/m] (Fig. 3.23a) and 350 [A/m] (Fig. 3.23c). Only the use of an excitation frequency of 3 Hz and a field strength of 3500 [A/m] resulted in a clear separation of the hardened condition (designated 6) from all other heat treatments. In contrast, Figs. 3.23d–f show how the impedance data clustered closely for the test components bar, connecting rod, and lever rod when using only a singular test frequency of 100 Hz with a single field strength of

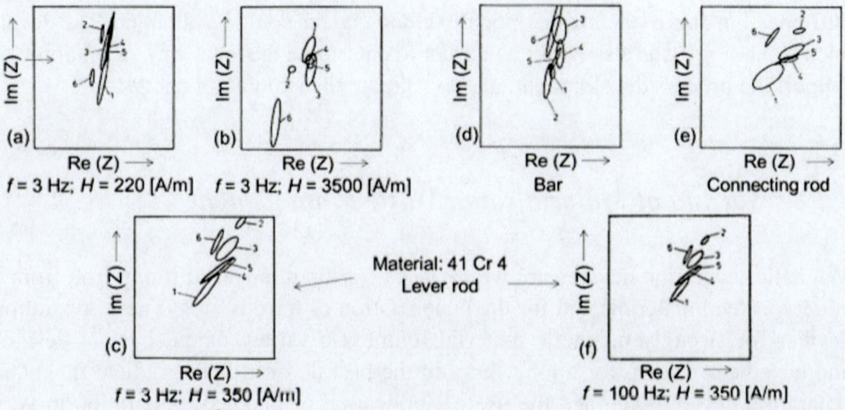

Fig. 3.23 Schematic representation of the impedance plane for material sorting for parts having the geometries of a bar, connecting rod, and lever rod made out of the steel 41Cr4. The numbers mean: **1** forged, undefined cooling rate; **2** forged with defined cooling rate (pearlitic structure); **3** forged and normalized; **4** forged and conventionally tempered; **5** forged/tempered by forging temperature; and **6** forged and hardened (from [26], with permission from Springer Nature)

350 [A/m], rendering their separation difficult. Different component geometries thus always require renewed optimization of the test parameters.

3.2.7 *Quality Monitoring of Composite Materials by Microwave Eddy Current Techniques*

The development of eddy current technology with frequency ranges up to 100 MHz made it possible to extend its application to less conductive multi-layered materials like carbon-fiber reinforced plastics (CFRP) [56, 57]. At such high frequencies, the displacement currents can no longer be neglected (see Sect. 3.1.2). Both permittivity and conductivity of the CFRP structure influence the complex impedance plane. The electrical conductivity contains information about the texture of fibers like their orientations, gaps, and undulations. Due to the capacitance of the carbon rovings, the measurement of the permittivity allows the determination of local curing defects of the CFRP structure such as hot spots, thermal impacts, polymer degradations, and the homogeneity of the fiber distribution.

Furthermore, small deviations in the fiber angles during preforming have a significant effect on the mechanical properties of the final composite. Bardl et al. [58] scanned draped fabrics with a robot-guided high-frequency eddy current system operated at a frequency of 6 MHz in order to obtain C-scan conductivity and permittivity images. Fourier transforming these images, the fiber orientations not only of the upper, but also of the lower, optically hidden layers could be obtained. The developed technology was also applied in order to determine the local yarn orientation, to support the process development, and to validate the simulation results.

3.2.8 Sorting of Materials and Waste Management

Magnetic separation devices are widely used for the removal of tramp iron from a variety of feed materials and for the beneficiation of ferrous ores. These separation devices for strongly magnetic materials employ a variety of mechanical designs and have been known for a long time. In the past decades high-gradient magnetic separation devices extended the useful application of magnetic separation to very weakly magnetic materials having small particle sizes. Potential applications of these new devices are in pollution control, chemical processing, and the beneficiation of nonferrous low-grade ores. The principle of operation of magnetic separation devices is the interaction between magnetic forces and the gravitational, the hydrodynamic, and the interparticle forces within the separator. For a review describing the principles of these separators see [59]. Here, we discuss an operation principle which is called eddy current separation and which fits well into the scheme of this book.

Eddy current separation (ECS) is used for sorting nonferrous metals based on conductivity, density, and geometry. A time-varying magnetic field induces electrical currents in conductive particles, which produce a magnet dipole moment $\boldsymbol{m}$ in a particle [60]:

$$\boldsymbol{m}=\frac{2\pi B_0 a^3}{\mu_0}\left(1-3\frac{qa\cosh(qa)-\sinh(qa)}{(qa^2)\sinh(qa)}\right)z \tag{3.48}$$

Here, $\boldsymbol{B} = B_0\cos(\omega t)z$ is the time-varying magnetic flux density, a is the particle radius and $q = \sqrt{j\omega\mu_0\sigma}$. The force acting on the magnetic dipole moment by the field $\boldsymbol{B}$ is given by:

$$\boldsymbol{F} = \nabla(\boldsymbol{m}\cdot\boldsymbol{B}) \tag{3.49}$$

Equation (3.49) shows that a homogeneous field produces no net force. However, if the magnetic field possesses a linear gradient α, for example $\boldsymbol{B} = (B_0 + \alpha x)z$, a net force will be generated in $\boldsymbol{x}$-direction whose time averaged value $\boldsymbol{F}_{avg}$ is given by [60]:

$$\boldsymbol{F}_{avg} = \frac{3\pi\alpha B_0 a^3}{\mu_0}\left(\frac{1}{3}-\frac{1}{q_r}\frac{\sinh q_r-\sin q_r}{\cosh q_r-\cos q_r}\right)\boldsymbol{x} \tag{3.50}$$

Here, $q_r = a\sqrt{2\omega\mu_0\sigma} = a/2\delta$ is the ratio of the particle radius to twice the skin depth δ (3.14). The magnitude of the forces is shown in Fig. 3.24.

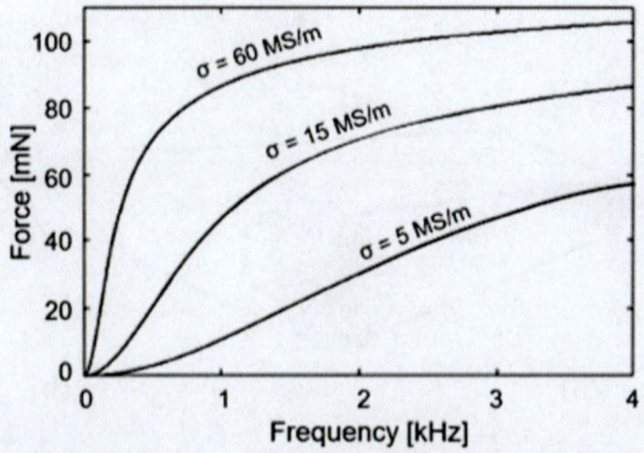

Fig. 3.24 Forces acting on metal spheres with varying conductivity σ. The particles have a radius $a = 10$ mm when excited by a sinusoidal magnetic field intensity of $B_0 = 50$ mT at a field gradient of $\alpha = -1.0$ T/m (from [61], with permission from Elsevier)

Eddy current separation is remarkably efficient and environmentally friendly, making it essential for the metal recycling industry. Although several different designs of eddy current separators have been examined, the most common in use today remains a belt-driven rotary drum design. The engineering tasks to be accomplished are as complex as for material sorting, see for example [61] and [62].

Ruan and Xu [63] extended the modeling of ECS in order to separate aluminum flakes from plastic parts in crushed waste toner cartridges. A rotating drum wrapped by permanent magnets produces an alternating magnetic field (Fig. 3.25). Like in the calculations leading to (3.48 to 3.50), when conductive metal flakes pass through a magnetic field, eddy currents will be generated in the flakes whose magnetic fields are opposite to the direction of the magnetic field generated by the magnet rotor. Ruan and Xu [63] considered additional factors such as the magnet area facing the particle, the horizontal cross-section of the flake, and an orientation factor for circular, rectangular, and triangular aluminum flakes. The theoretical results agreed with the results of the test experiments. The repulsive forces F_r act on the metal flakes separating them in the material flow and thus achieving separation. The repulsive forces are related to the strength of the magnetic field of the rotors, the conductivity, density, area, and the shapes of the aluminum flakes of the shredded toner carts.

O'Toole et al. [64] demonstrated a sorting system based on multi-frequency eddy current spectroscopy, an extension of the eddy current separation method discussed above. By a specially designed coil-array system, the scrap parts were sorted according to their real and imaginary impedance $Z(\omega) = X + iY$ (3.44) by an automated classification system trained with test parts. The frequencies used were 2, 4, 8, 16, and 64 kHz. The developed system was applied for non-ferrous metals such as copper, aluminum, and brass found in end-of-life waste. Two features were selected: (1) a high-frequency component, where the skin depth was negligible compared to the size of the parts, in order to model the geometry of the metal fragments, and (2) a low-frequency component, where the skin-depth is significant, where both geometry and conductivity of the fragments entered. A simple but effective classifier which used also geometric features was created which turned out to be well-suited for industrial needs.

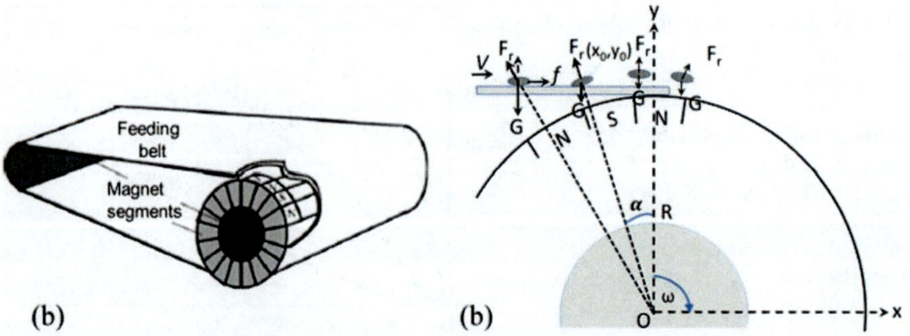

Fig. 3.25 **a** Eddy current separator with belt and alternating magnet sections; **b** Flakes with induced currents and repulsive forces F_r. The magnetic flux intensity at the drum surface was 0.24 T, the area of a magnet section $\approx 10^{-3}$ m^2, $R \approx 0.1$ m, and the relative permeability of the magnets $\mu = 5000$ (in modified form from [63], with permission from Elsevier)

3.3 NDMC by Micromagnetics

3.3.1 Basic Ideas of Micromagnetism for NDMC

The process of magnetization of a magnetic material, called "Micromagnetics", is strongly influenced by its microstructure. The parameters associated with the hysteresis curve are influenced by the same microstructural features, which determine the material's strength. They are also sensitive to external loading and residual stresses, to processing routes and to material degradation processes. These properties open the possibility for NDMC. Parallel to applications in electrical and materials engineering, there were many efforts over the past decades in order to understand the properties of magnetic materials as an emerging field of solid-state physics and quantum mechanics. Some of them are presented in this book in order to introduce the underlying principles, however, for an in-depth discussion we refer partially to the original literature and partially to more recent books summarizing this field [17, 19, 21, 24, 65–70].

An external magnetic field moves the Bloch walls with increasing field strength. At small field strengths, the motion of the domain walls is reversible. Upon further increase of the field, irreversible motions of the domains set in. All domains that have an acute angle with the field direction grow at the cost of the domains with obtuse angles. A further increase of the external field increases the magnetization only if threshold values are surpassed that depend on the orientation of the domains relative to the field, and on the presence of microstructural defects such as dislocations, segregates, dissolved atoms, vacancies, precipitates, grain and phase boundaries. These defects or inhomogeneities determine the coercivity, the shape of the hysteresis curve, and the permeability.

Mechanical properties of crystalline materials are governed by the movement of dislocations upon application of external loading. This is analogous to the movement of domain walls upon the application of an external magnetic field. The motion of both dislocations and domain walls is influenced by microstructural features including dislocation density, grain size and secondary phases. This forms the basis for the measurement of micromagnetic parameters to determine the material's strength parameters indirectly, as discussed in Sect. 3.3.3.

3.3.2 *Motion of Bloch Walls and Micromagnetism*

Bloch walls are pinned in local minima of the potential energy landscape determined by the defect structure. When applying an increasing magnetic field, either a quasi-DC or an AC field or a combination of both, the motion of the domain walls is first reversible but proceeds then in discrete jumps which are called "Barkhausen jumps" [71]. The domain's magnetization changes its orientation as a whole, into a new direction with the smallest possible angle to the field direction. Furthermore, magnetostriction, which is always accompanied with ferromagnetism, leads to a change in domain size. With further increasing field strength, the domains become finally oriented parallel to the field, reaching saturation. These processes manifest themselves in different parts of the hysteresis curve. The following three parts are shown in Fig. 3.26:

(i) Reversible boundary displacements at low fields
(ii) Irreversible 180° displacement encompassing a major portion of magnetization at intermediate field values and
(iii) Domain rotation by irreversible 90° domain at high fields

The wall energy has the dimension energy/area (3.41), which is equivalent to a force/length or surface tension, and that hints that a domain wall is responsive to mechanical stresses. If one applies an external magnetic field, there is a change in

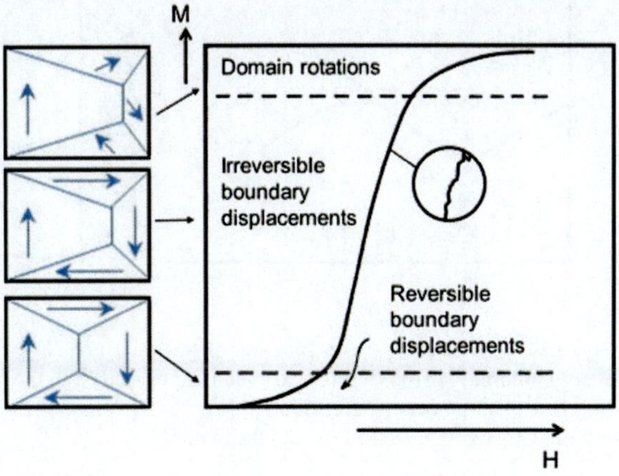

Fig. 3.26 Magnetization curve M of a ferromagnetic material by an external field H starting from the virgin state. The motion of the domains occurs in small irregular steps which are called Barkhausen jumps (reprinted from [21] in modified form, with permission from the American Physical Society)

the energy density E

$$E = -\mu_0 \boldsymbol{M}_S \cdot \boldsymbol{H} \tag{3.51}$$

of a domain which leads to stresses acting on the domain walls. They will start to set the domain walls into motion, if they are not impeded by obstacles. These are internal stresses, inclusions, grain boundaries, and dislocations. The gain in wall energy dW when moving a 180° domain wall of unit area by a distance dx may be written as

$$dW(x + dx) - dW(x) = dW = 2\mu_0 \boldsymbol{M}_S \cdot \boldsymbol{H} dx \tag{3.52}$$

The factor 2 stems from the fact that the magnetization changes by 180° within the wall. Upon applying an external field $\boldsymbol{H}$, the domains start to move in a reversible way as long as the field is smaller than H_0, see Fig. 3.27a. For larger fields than H_0, the energy density of the wall becomes larger than the local stress gradient and the potential barrier can be surmounted leading to a Barkhausen jump. This field, corresponding to the coercivity, is given by:

$$H_0 = \frac{1}{2\mu_0 M_S}\left(\frac{dW}{dx}\right)_{\max} \tag{3.53}$$

The motion of a domain wall generates eddy currents which in turn lead to their viscous damping by the material's resistance and to heating of the component under test. The motion of the domain walls as well as the ensuing spatial and temporal distribution of eddy currents contribute to the hysteresis losses in electrical engineering applications [19] and are to be minimized in transformer sheets. The viscous damping of the domain wall motion also contributes to internal friction in ultrasonic attenuation measurement in the MHz range [72, 73], and has been measured in polycrystalline white cast iron [74].

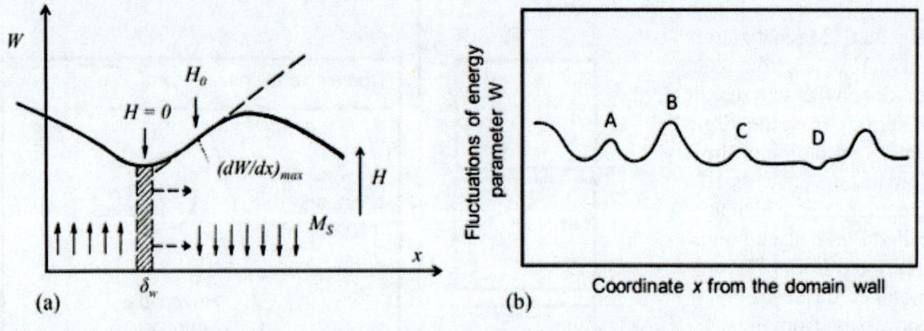

Fig. 3.27 **a** Wall energy as a function of position. For $H = 0$, the wall rest in a local potential energy minimum of $W(x)$. If the external field exceeds H_0, the wall can surmount the barrier and a Barkhausen jump occurs, according to [67, 75]; **b** Potential energy landscape for a domain wall at different positions $A, B, C, D, \ldots$in the material

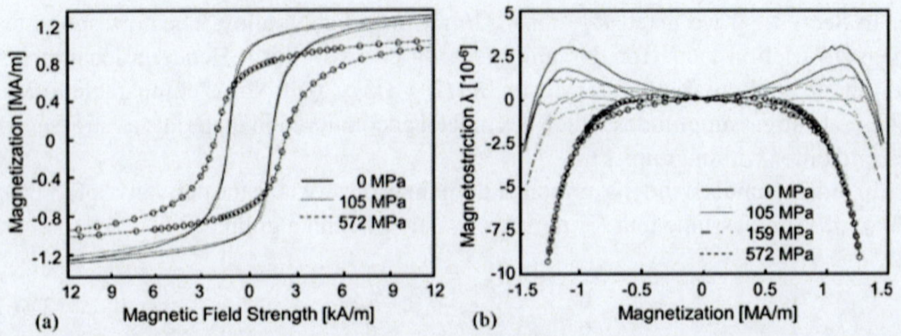

Fig. 3.28 **a** Hysteresis curves *M(H)* of a ferromagnetic iron-carbon (0.86% C) heat-treated sample showing the sensitivity to mechanical stresses. The shape of the hysteresis curve, the coercivity as well as the saturation magnetization depend on mechanical stresses. A loop for the sample is shown for zero stress (0 MPa), for conditions of maximum remnant magnetization (105 MPa), and when the sample was subjected to stresses well beyond the yield point (572 MPa); **b** Corresponding magnetostriction coefficient λ (in modified form from [76], with permission from Elsevier)

Hysteresis curves *B*(H) of ferromagnetic materials are sensitive to mechanical stresses, as can be seen schematically in Fig. 3.28 for unidirectional tensile stress applied parallel to the magnetization on a ferromagnetic iron-carbon (0.86% C) heat-treated sample. In materials with positive magnetostriction, the magnetization is increased by tension and the material expands when magnetized. In materials with negative magnetostriction, the magnetization is decreased by tension and the material contracts when magnetized. Shape changes minimize the energy density. These effects are caused microscopically by the reorientation of domains by rotationally processes and by the irreversible motion of Bloch-walls. The 180° Bloch walls are stress sensitive only because of their coupling to 90° walls.

The stress dependence of the magnetization curve is caused by inhomogeneities like stresses of the second and third kind (see also Chaps. 1 and 2, Sects. 1.11.1 and 2.4.3 Stress Measurements, respectively), by magnetic or non-magnetic inclusions, and by dislocations. They hinder the motion of the domain walls and can be viewed as a contribution to the anisotropy factor:

$$K_\sigma = \frac{3}{2}\lambda\sigma(x) \tag{3.54}$$

Here, λ is the magnetostriction coefficient and $\sigma(x)$ is the spatially varying stress. Complementing the anisotropy constant K_1 for the wall energy with the stress energy, i.e. by adding K_σ of (3.54) to K_1 in (3.41), yields:

$$W_{dom} = 2\left(\frac{\pi^2 J_i S^2 \left(K_1 + 3\lambda\sigma(x)/2\right)}{a}\right)^{0.5} \tag{3.55}$$

In Sect. 3.1.3 we listed $K_1 \approx 10^3$ [J/m^3] for a Fe–Ni alloy. The corresponding magnetostriction value for this alloy is $\lambda \approx 2 \times 10^{-5}$ [20]. Hence, a local stress amplitude of about 50 MPa results in $3\lambda\sigma/2 \approx 1.5 \times 10^3$ [N/m^2] comparable to K_1. These are stress amplitudes which are indeed encountered in materials as stresses of second order within grains [77].

In order to understand the motion of domain walls, we take the derivative of (3.55), dW_{dom}/dx, and assume that K_1 remains constant within a grain:

$$\frac{dW_{dom}}{dx} = 3\delta_w \lambda_s \frac{d\sigma}{dx} \tag{3.56}$$

The domain wall thickness δ_w enters in (3.56) by taking into account (3.41). Equation (3.56) means that stress gradients lead to a change in energy density of the domain, which leads to a bowing out of the domain wall and eventually to an irreversible expansion of the domain wall. Considering the limiting cases that the extent L of the stress gradients are smaller or larger than the domain wall thickness δ_w, Kersten [67] derived the expression for all δ_w/L ratios:

$$H_c = \frac{2}{3}\frac{\lambda_S \Delta\sigma}{\mu_0 M_S}\frac{2\delta_w/L}{1+2(\delta_w/L)^2} \tag{3.57}$$

where $\Delta\sigma$ is the stress amplitude. The situation is shown schematically in Fig. 3.29a for the cases $L > \delta_w$ and $L < \delta_w$.

Both, magnetic and non-magnetic inclusions reduce the energy for domain walls when they are intersected. Hence the walls are attracted to the inclusions which impede their motion. Kersten [67, 75] derived the expression

$$H_c = \alpha^{2/3}\frac{(3\lambda\sigma/2 + K_i)}{\mu_0 M_S}\frac{2\delta/d}{1+2(\delta/d)^2} \tag{3.58}$$

for the coercive field. Here, $\alpha = \pi d^3/6s^3$ is the average volume share of the spherical inclusions of diameter d and distance s (Fig. 3.29b). There is a pre-factor in this equation which is of the order of 1 depending on whether 90° or 180° walls are considered. Also, the shape of the inclusions plays a role. The equation is based on the consideration that the inclusions lead to a change of the energy density of the magnitude $(3\lambda\sigma/2 + K_i)$, and hence to a local stress. Kersten provided data on the validity of (3.57 and 3.58), but cautioned that these equations are rather an order of magnitude estimate than a rigorous theory. In Jiles [17] one finds further corresponding data, also in comparison with the theory of Néel taking into account the magnetostatic energy of the defects [78], see Fig. 3.30, see also Sect. 3.3.9.

Using magnetic force microscopy Abuthahir and Kumar [22] visualized bowing of a domain wall with the application of external magnetic field on the surface of an iron specimen. The domain wall dynamics was presented in form of the traces of the domain wall corresponding to increasing field from zero to 160 kA/m (Figs. 3.10b

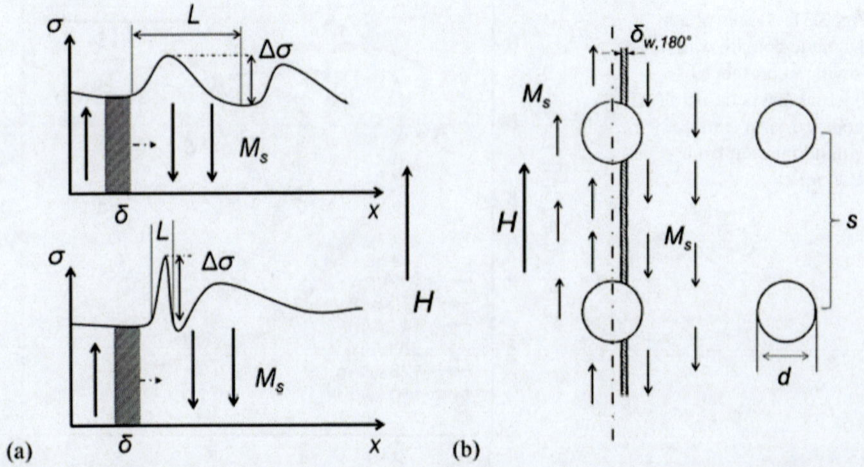

Fig. 3.29 **a** Stress variations $\Delta\sigma(x)$ in a micro-inhomogeneous material which determine the local wall energy and hence the magnetic field which is required to displace domain walls of thickness δ_w (hashed areas); **b** Schematic figure of the variation of the wall energy in a micro-inhomogeneous material due to non-magnetic inclusions (circles) with diameter d and distance s, which act as pinning centers because they reduce the wall energy and increase the local magnetic field which is necessary to displace the domain walls (**a** redesigned according to [67]), and **b** in modified form from [75], with permission from Springer Nature)

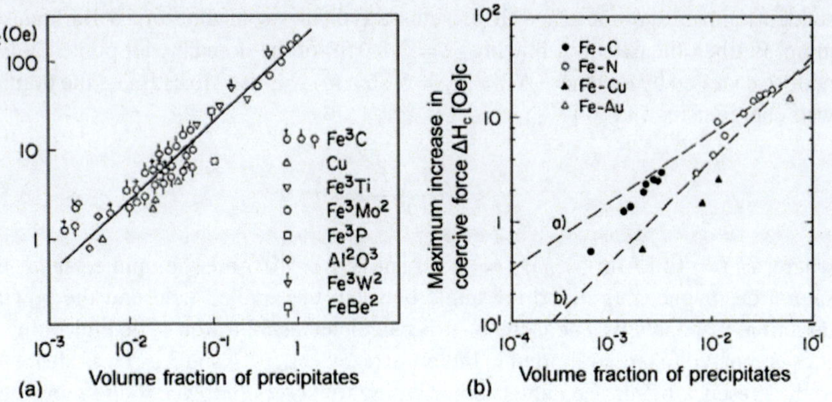

Fig. 3.30 **a** Coercivity of various steels as a function of the total volume fraction of inclusions. Experimental result of Kersten and calculations by Néel (solid line); **b** Increase in coercivity of iron caused by precipitates of interstitials or substitutional solutes. Theoretical calculations are represented by the two dashed lines: **a** due to Kersten and **b** due to Néel [78] (from [17], with permission from Francis & Taylor)

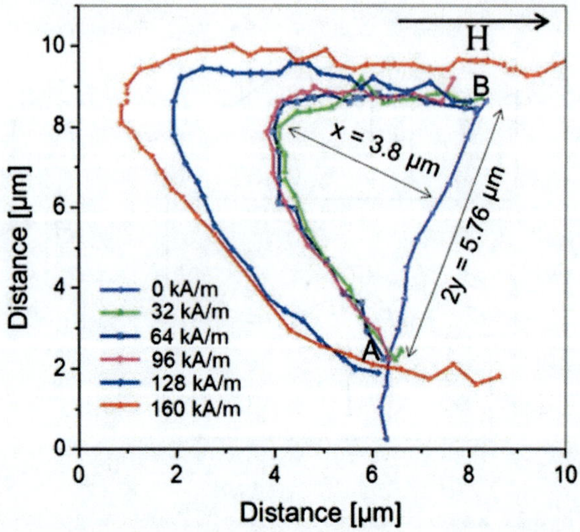

Fig. 3.31 Traces of a magnetic domain wall bowing in presence of external magnetic field (in modified form from [22], with permission from Elsevier

and c, (3.31)). The domain wall was pinned at two points A and B. With increase in the applied magnetic field to 32 kA/m, bowing of the domain wall was observed. No change in the domain wall was observed with further increase in the applied field up to 96 kA/m followed by a sudden jump on application of 128 kA/m field. The sudden jump in the domain wall presents a typical visualization of a Barkhausen jump. Further, the extent of bowing ($x = 3.8$ µm) of the domain wall pinned at two points separated by distance (AB = $2y = 5.76$ µm) was used to estimate the domain wall energy using (3.59) [79]:

$$\frac{x}{H} = \frac{y^2 \mu_0 M_s}{2} \cos(\theta) \tag{3.59}$$

where, H (= 32 kA/m), $\mu_0 M_s$ (= 2 T) and θ (= 40°) are the applied field, the saturation magnetization and the angle between the applied field and the bowing direction, respectively. The value of γ is calculated as 52 mJ/m^2. The high value of γ as compared to those reported in literature (see Sect. 3.1.4 and 3.41) was attributed to the presence of surface defects contributing to larger hindrance to the movement of surface domain walls. An analogue experiment has been carried out by Batista et al. [80] ascribing the bowing-out of the domain wall to average internal stresses $\sigma_{av} = 89$ MPa.

3.3.3 Material Properties and Magnetic Hysteresis Loop

The basics of a magnetic hysteresis loop and its important parameters are described in Sect. 3.1.4. The magnetic hysteresis (B-H) loops for two material conditions, let us call them x and y, with different mechanical hardness are schematically presented in Fig. 3.32a. For most material conditions of a ferromagnetic material, it can be observed that a mechanically harder material is also magnetically harder (y in Fig. 3.32a). This is attributed to the fact that the pinning points for the movement of dislocations, which cause increase in mechanical hardness, are often also the pinning points for the movement of magnetic domain walls, which means increasing magnetic hardness. For example, some means of increasing the mechanical hardness of a steel are by increasing its carbon content, quenching it from above the austenizing temperature in order to form a hard martensite phase, cold working or deformation in order to increase the dislocation density, and by refinement of the grain size. All these factors also lead to an increase of the magnetic hardness of the steel, i.e. the shape of the magnetic hysteresis loop changes from condition x to condition y, as shown in Fig. 3.32a (red hysteresis curve). A magnetically harder material exhibits a larger coercivity ($H_c^y > H_c^x$), a lower saturation magnetization ($B_s^y < B_s^x$), a lower remnant magnetization ($B_r^y < B_r^x$), and a lower permeability than a magnetically softer material. A typical example of the correlation between the mechanical and the magnetic hardness is shown in Fig. 3.32b.

Modified 9Cr-1Mo ferritic steel (T91/P91) specimens were subjected to a series of heat treatments consisting of soaking for 5 min at the selected temperatures, starting from the α-phase region (1073 K) to the $\gamma + \delta$ phase region (1623 K), followed by oil quenching [81]. As the soaking temperature increased above the Ac_1 temperature, the hardness was found to increase with increasing temperature. This was attributed to the increase in the amount of a harder phase (martensite) formed with an increase in the soaking temperature in the $\alpha + \gamma$ phase region. The hardness continued to increase even beyond the Ac_3 temperature due to the dissolution of the increased amount of NbC and V_4C_3, which increased the solid-solution strengthening as well as the formation of martensite with finer laths on quenching, due to the depression of the martensite starting temperature in the presence of a higher amount of solute elements. Above about 1373 K, in the absence of NbC and V_4C_3, the grain size increased rapidly, leading to a lower hardness with a further increase in the soaking temperature. Beyond 1473 K, the formation of soft δ ferrite further decreased the hardness, even though it refined the grain size. It can be seen in Fig. 3.32b that the coercivity exhibited the same trend with increasing soaking temperature. This clearly demonstrates the potential of measuring the basic magnetic hysteresis loop parameters for microstructural characterization of ferromagnetic steels.

3.3.4 *Material Properties and Magnetic Barkhausen Emissions*

When a ferromagnetic material is subjected to a time-varying magnetic field (Fig. 3.33a), the response of the material are discrete changes in flux density during magnetization, which are caused by the motion of the domain walls surmounting

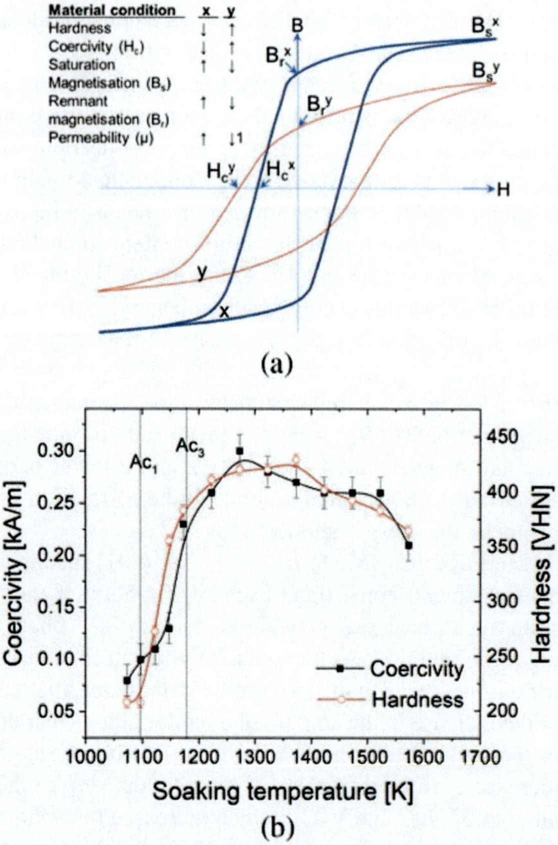

Fig. 3.32 **a** Schematic magnetic hysteresis (B-H) loops for two material conditions with different hardness. The hysteresis curve in red holds for the harder material; **b** Variations in coercivity and hardness for modified 9Cr-1Mo ferritic steel specimens heat-treated at different soaking temperatures, followed by quenching in oil

individual potential minima (Fig. 3.27a) of the potential landscape (Fig. 3.27b). The discrete flux changes induce electrical pulses in a pick-up coil. The maximum signal is obtained when the magnetic field is close to or at the coercivity H_c. These phenomena are called Magnetic Barkhausen Emissions (MBE) or Barkhausen Noise (MBN) (Fig. 3.33b). It is sensitive to microstructural variations and strains in the material, see Sect. 3.3.5.

From the discussion in Sect. 3.3.4, it follows that MBE could serve as a NDMC technique for the microstructural and mechanical property characterization of steels. MBE are sensitive to various parameters which affect the domain configurations and the domain-wall pinning sites, which in turn are strongly influenced by the grain size and composition, by the presence of ferrite, pearlite and martensite phases, surface conditions, hardness, residual stress, fatigue, creep and other causes of damage, and also by the action of external factors such as applied stress. There are many studies by different groups, which examined Barkhausen emission in order to exploit it for non-destructive materials characterization [82–88], also in combination with resistivity measurements [89].

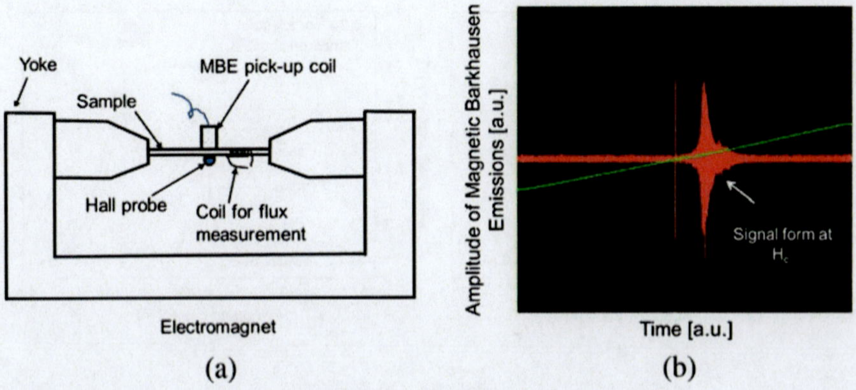

Fig. 3.33 **a** Scheme of experimental set-up to measure magnetic Barkhausen emissions. A sample is magnetized and the ensuing Barkhausen noise is picked up by a coil; **b** Barkhausen noise plotted as a function of time because the triangularly modulated magnetic field (**b**, green line) is driven by a corresponding electronic circuitry. The tangential magnetic field at the sample surface is measured by a Hall probe giving the magnetic field strength which is then plotted or digitized together with the generated emissions. A commercially available set-up is shown in Fig. 3.47

Jiles [90] reviewed the theory of the Barkhausen emissions in magnetic materials starting with the equations that are commonly used when describing MBE. Barkhausen emissions can be described in the presence of hysteresis using a hysteretic-stochastic process model. It contains information about the microscopic magnetization processes, i.e. from domain wall motions and domain rotations. There are two limiting cases: flexible domain wall motions and rigid domain wall motions. Both have reversible and irreversible components. The same holds for the domain rotations, depending on the anisotropy and the magnitude of the angle of rotation. The stochastic processes of the Barkhausen effect are taken into account, based on the work of Bertotti [19], by postulating that the domain walls move in a randomly fluctuating potential which varies from point to point in the magnetic material causing the emitted stochastic Barkhausen noise, when local minima are surmounted by the domain walls, see Fig. 3.27b. The random fluctuations in the potential reflect the distribution of the parameters listed above in a materials microstructure [21]. Jiles [90], then generalizes the stochastic model by including hysteresis, yielding equations for the amplitude of Barkhausen emissions under simplified conditions.

Moorthy et al. [91, 92] examined MBE in order to characterize the microstructures of tempered 0.2 wt% C steel, 2.25Cr-1Mo steel, and 9Cr-1Mo steel samples. They proposed a two-stage process of irreversible domain-wall movements during magnetization by considering the lath or grain boundaries and second phase precipitates as the two major obstacles to domain-wall movement. The domain walls overcome these two major obstacles over a range of critical field strengths with mean values that are characteristic of the obstacles. If these two mean values are close to each other, then a single peak, sometimes associated with slope changes, appears in the MBE behavior. If the mean values are widely separated, however, then the two-peak

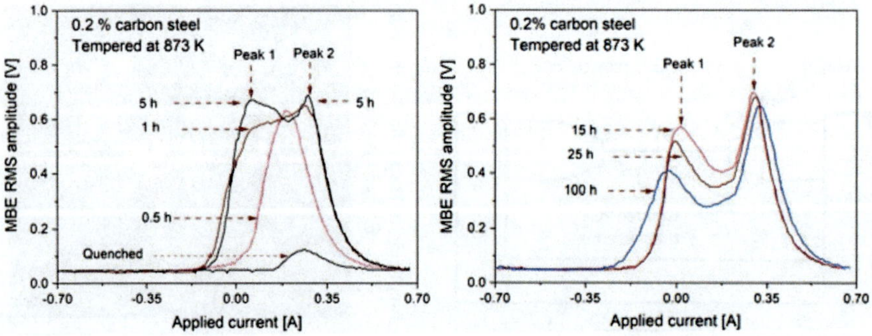

(a) (b)

Fig. 3.34 Magnetic Barkhausen RMS amplitude as a function of the current applied to the magnetizing yoke for half the magnetization cycle for carbon steel samples: **a** quenched and tempered at 873 K for 0.5. 1, and 5 h, and **b** tempered at 873 K for 15, 25 and 100 h. From [91], with permission from Francis & Taylor and courtesy of IGCAR, Kalpakkam

MBE signals appear to indicate the influence of these two major obstacles separately. Based on this, Moorthy et al [91] explained the influence of the dissolution of martensite and/or bainite and the precipitation and growth of the second phase precipitates with different sizes and morphologies on the MBE. Furthermore, they conjectured that the presence of two or more different types of carbide in the Cr–Mo steels significantly reduces the MBE compared with the growth of single carbides in the carbon steel, thus establishing the viability of the MBE technique for identifying different stages of tempering in ferritic steels (3.34).

3.3.5 *Applications of Micromagnetism for Microstructure and Grain Size Measurements*

As discussed above, the microstructural characteristics of steels have a strong effect on magnetic properties because they influence domain wall pinning and domain wall motion, which in turn is responsible for the hysteretic behavior of magnetic properties by affecting coercivity, remanence, differential magnetic permeability, and hysteresis loss. One important feature is the average grain diameter d. Grain boundaries present an obstacle to domain wall motion and act as pinning centers for the domain walls. With increasing grain size, the total length of the grain boundary decreases, and hence the number of pinning centers hindering the domain wall motion decreases. Since the coercivity H_c reflects the amount and strength of pinning, H_c decreases as the grain size increases. The exact dependence of H_c on the grain size has been a subject of discussion in many articles as pointed out by Sablik [93]. However, all references cited in Sablik reported an inverse relationship with grain size.

Using the mathematical model of hysteresis curves by Jiles and Atherton [79], Sablik [93] predicted that the coercivity $\boldsymbol{H_c}$ increases with inverse grain size ($1/d$), whereas the differential permeability at $\boldsymbol{H_c}$ and the remanence $\boldsymbol{B_r}$ decreases with inverse grain size. The hysteresis loss W_H first increases with inverse grain size, but then decreases and, depending on the anisotropy K_u, peaks at a particular inverse grain size. For the coercive field, the initial linear increase on $1/d$ was predicted to become nonlinear. Furthermore, at some small grain size, the coercive field peaks and starts to decline.

Sakamoto et al. [94] measured MBE in two different microstructures of carbon steels consisting of ferrite grains with or without dispersed cementite particles. The Barkhausen emissions were characterized by their RMS amplitude, which is the sum of the induced electrical pulses caused by the magnetic flux change in each micro-region. The MBE pulses were approximated by Gaussian pulses:

$$V(t) = \frac{\Delta\Phi}{\sqrt{2}\sigma} \exp\left(-(t - t_0)^2/2\sigma^2\right) \tag{3.60}$$

where $\Delta\Phi$ is the quantity of the magnetic flux change in a micro-region, σ is the standard deviation of the pulses, and at $t = t_o$ the Gaussian pulse is at its maximum. The induced electrical pulses in the test coil are obtained by the summation of the noise signals generated by successive Gaussian pulses. Figure 3.35a and b display the experimental measured values. In the case of the ferritic structure, the measured RMS value of the MBE was found to be proportional to $\propto d_g^{-0.5}$, where d_g is the grain size (Fig. 3.35a). Taking into account the microstructural features for the reversed domain motions in the ferrite grains and the speed of the domain wall motion, the authors also found theoretically that the RMS values of the MBE is $\propto d_g^{-0.5}$ as observed.

The situation is different for the ferrite grains with dispersed cementite particles. The motion of the reversed domains is hindered by the cementite particles by pinning, and as a result an RMS value $\propto d_p^2$ is obtained theoretically and experimentally, where d_p is the diameter of the cementite particles except for data with $d_p > 1$ μm (Fig. 3.35b).

In order to detect coarse-grained zones with grain sizes up to 300 μm in hot-rolled sheets made of low-carbon aluminum killed-steel, an on-line system was installed [95]. A measuring head with a constant lift-off of 3 mm was used in order to detect the Barkhausen emission signals at an excitation frequency of 5 Hz in a picking line at speeds up to 200 m/min. They were normalized to the maximum of the magnetic Barkhausen noise M_{max} and to the width $\Delta M(H)$ of the noise amplitude distribution, averaging over five curves at $f_e = 5$ Hz. Due to the sheet's anisotropy, a set-up arrangement parallel to the rolling direction was used. Thresholds were set for $M_{\max}$ and $\Delta M(H)$, where lowest M_{max} and the highest $\Delta M(H)$ represent the acceptable quality thresholds. These threshold settings in order to detect coarse-grained zones were based on practical experience. It turned out that when monitoring the coils in the production line, M_{max} was the more sensitive parameter for detecting coarse-grained zones in comparison to $\Delta M(H)$.

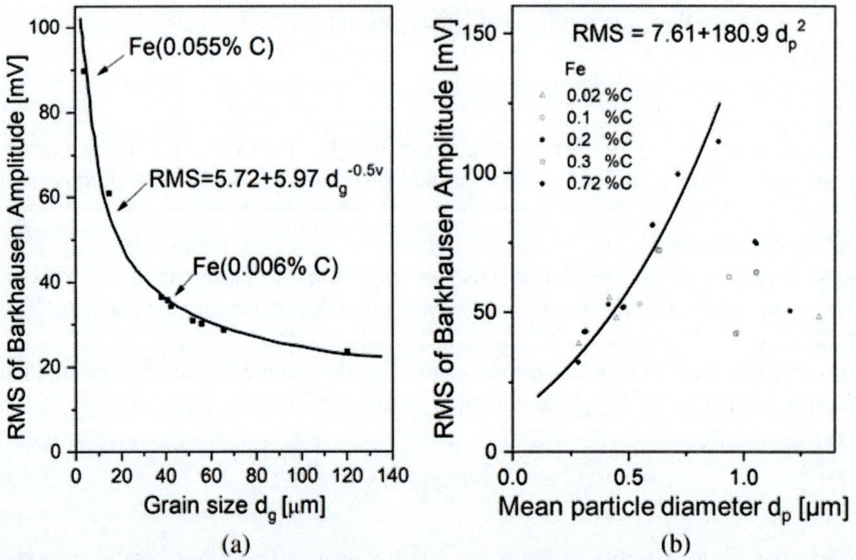

Fig. 3.35 **a** RMS values of the MBE as a function of ferrite grain size; **b** RMS values of the MBE as a function of cementite particle diameter d_p in carbon steel. Data for mean particle diameters larger than 1 μm did not fit the curve $\propto d_p^2$ (reprinted from [94]), with permission from IEEE Magnetic Society

Santa-aho et al. [96] reviewed the design and modeling of MBE probes. As stated by the authors, MBE measurements will only be consistent and comparable in cases where the magnetization distribution is similar in each measurement. The flux density control in the component to be tested is important for achieving this goal. The challenges to be faced in magnetic field simulations are related to magnetic saturation, which requires a sufficiently fine mesh, in order to achieve convergence.

3.3.6 *Applications of Micromagnetism for Material Strength Parameter Measurements*

Mitra et al. [97] studied in AISI 304ss steels the austenitic transformation to the ferromagnetic bcc-martensite phase during plastic deformation. The magnetic characterization was carried out in samples having different percentages of martensite in 304ss obtained by plastic deformation. Among the magnetic properties, the hysteresis loop and the Barkhausen emissions were studied using a surface probe. A linear increase in remanence with the deformation was observed. The coercivity also increased with the percentage of martensite, but tended toward saturation with an increase of martensite. A large number of small amplitude Barkhausen emissions signals at very low volume fraction of martensite indicated that magnetization rotation took place within a small

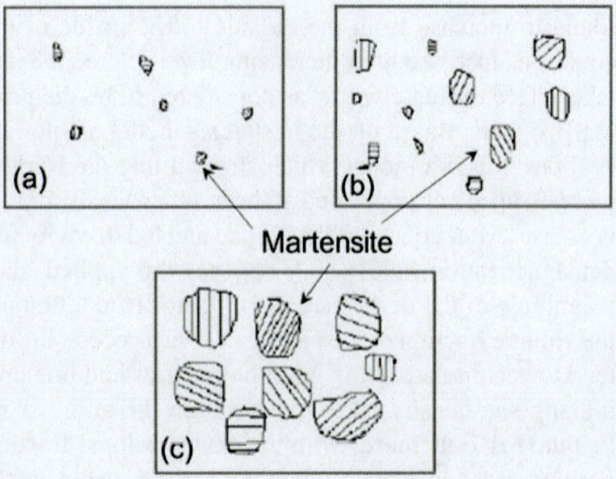

Fig. 3.36 **a** At the initial stage of plastic deformation or strain, the effect of strain causes the formation of magnetic martensite particles; (**b**, **c**) With further strain, the martensite fraction increases. When the martensite clusters are wide apart, each magnetic moment interacts only with the other moments within the martensite cluster (in modified form from [97], with permission from Springer Nature)

region. Mitra et al. [97] proposed that the nucleation and the growth of martensite clusters takes place during plastic deformation. The intra-cluster exchange interaction is predominant at low volume fraction of martensite, and hence the coercivity was found to increase with the radius and exchange interaction between the magnetic moments within the martensite cluster (Fig. 3.36). It was further assumed that the saturation magnetization of the martensite is a constant, and because the matrix is austenite, it was also assumed that the saturation magnetization of the matrix is zero. With the growth of the cluster size, inter-cluster interaction takes place and as a consequence the coercivity H_c depends on the separation between the clusters, which is described by the factor $(1 - \exp(-r/r_0))$ with a characteristic scaling length r_0:

$$H_c = \frac{1}{\mu_0 M_s} K_1 \exp(-k_B T / \langle E_{ex} \rangle)(1 - \exp(-r/r_0)) \tag{3.61}$$

Here, $\langle E_{ex} \rangle$ is the exchange energy between two magnetic moments, T is the temperature, and k_b is the Boltzmann constant. Finally, measurements of the angular variation of Barkhausen emission signals indicated that their recording could be used for the determination of rolling texture in steels in production lines.

Magnetic Barkhausen emissions were measured in a number of specimens of Cr–Mo steel in order to study the effects of creep by Mitra et al. [98]. The steel was subjected to a temperature of 550 °C and pressure of 17 MPa for 180 000 h. It was then subjected to a temperature of 520 °C and a pressure of 17 MPa for a further 15,000 h. The wall thickness of the pipe was 100 mm. Test specimens were prepared from a section of the pipe, which were cut into several specimens that were representative of different regions from the inside to the outside wall. Scanning electron micrographs showed that all the specimens were at an early stage of creep. The number density of cavities in specimens which were taken from the outer surface of the pipe was lower than in specimens taken from the inner surface, indicating lower creep on the outer surface. The RMS and the peak-to-peak voltages of MBE were found to increase from the outside to the inside of the pipe, indicating that these emissions increase with increasing level of creep. Segregation of impurities, which takes place during creep, was considered to be the primary cause of the increase in MBE signals. Based on the hysteresis model of Jiles and Atherton [79], Mitra et al. [98] developed a model which showed that the Barkhausen signal strength should increase up to a critical value of the number density of cavities, beyond which it should decrease. After creep has developed and led to cavity formation, they produce a local demagnetization field which opposes the applied magnetic field. As a result, the magnitude of the demagnetization factor is to a first approximation proportional to the volume fraction of the cavities. In the process, the dislocations and the impurities tend to accumulate at the grain boundaries and pile up there, and a reduction of the pinning site density takes place within the grain. A numerical simulation showed that the MBE increases with the cavity volume fraction due to the reduction of the peening sites and decreases beyond a critical value. As the specimens measured in this study were in the initial stage of creep only, an increase in Barkhausen voltage was observed. In the experimental measurements an angular variation in RMS voltage was also observed which was explained by the orientation of the easy magnetization axis due to the development of long-range residual stresses in a direction different from the pipe axis.

A micromagnetic technique was used for the on-line determination of the yield strength, tensile strength, and fracture strain for hot steel plates in the 5 to 45 mm thickness range at temperatures up to 400° C [99]. The sensor was a yoke arrangement (Fig. 3.37a). By an automated electronic control unit, the hysteresis curve was swept through with a frequency of $f_e = 0.083$ Hz up to saturation in order to determine the maximum incremental permeability $\mu_\Delta(H) = (1/\mu_0)d\boldsymbol{B}/d\boldsymbol{H}$ at a frequency of $f_\Delta = 70$ Hz (Fig. 3.37b and c). These parameters were selected in order not to influence the manufacturing process, i.e. the speed of the production line, and in order to obtain information also from the depth of the steel plates. For the set-up used, after calibrations on cut plates, on-line tests with measuring cycle times of 25 s led to an acceptable agreement between the non-destructively and the destructively determined values for the parameters tensile strength, yield point, and elongation (Fig. 3.38). The NDMC quantity used for calibration was the so-called magnetic index MI, which was defined by Komine and Nishifuji [99] as the variations of the imaginary part of the probe coil impedance.

For cold-worked ferritic stainless-steel strips, a correlation was found between the magnetic remanence $\boldsymbol{B}$ and the tensile strength [100], which was exploited in a rolling mill. A magnetizing unit was coupled to the strip, together with a Hall sensor. The Hall voltage is a measure of the apparent remanence which is used as the quantity yielding the tensile strength after calibration. The remanence B_r reacted sensitively to disturbances during the heat treatment process after cold working (Fig. 3.39). Bussière [101] reported in a review article the correlations between B_r, the Rockwell hardness *HRc,* and the coercivity H_c of ferritic chromium stainless steel strips. The correlations were the basis of an in-line monitoring device.

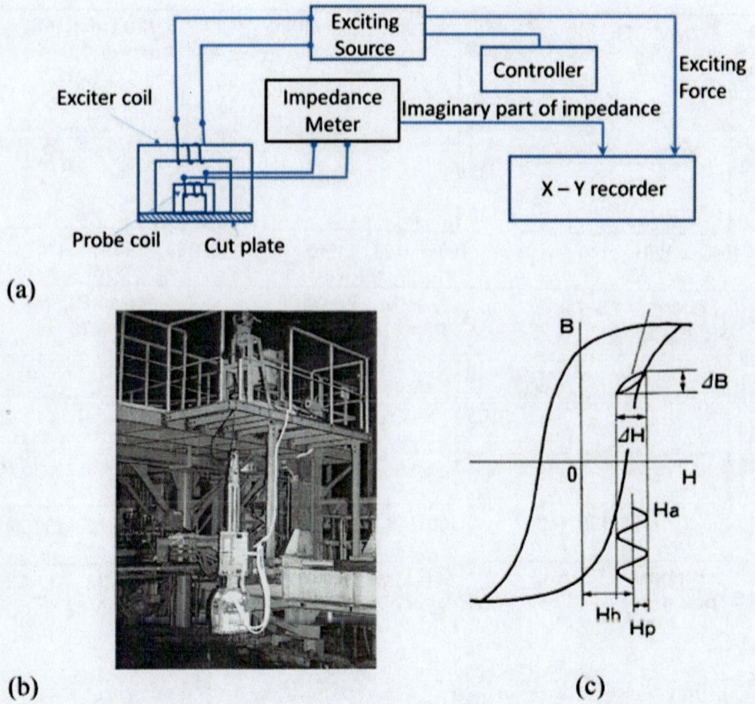

Fig. 3.37 **a** Control unit for measuring the so-called magnetic index (MI) in a production line; **b** Measuring device in the production line; and **c** Schematic hysteresis curve for measuring the incremental permeability (from [99], with permission from Springer Nature)

3.3.7 Applications of Micromagnetism for Hardness and Hardening Depth Measurements

For many applications, it is necessary that materials, here steels, simultaneously possess hardness as well as high strength and toughness. This combination of qualities can be achieved by hardening the surface while keeping the bulk "soft", retaining high strength and elasticity. Alternatively, only highly loaded volumes are hardened [102]. Induction hardened crank shafts or flame hardened turbine blades are examples of such optimizations.

There are a number of surface hardening procedures: case hardening, flame hardening, induction hardening, electron beam hardening, laser hardening, chrome plating, boriding, and nitriding. Some of the hardening procedures deliver shallow hardening depths like nitridation hardening (< 1 mm) and others deeper like induction hardening (> 1 mm). The hardness gradient between the surface and the bulk can be quite different due to the different hardening processes.

Hardening depth measurements by NDMC methods require calibrations with an appropriate standard such as Brinell, Vickers, Knoop, or Rockwell hardness [103].

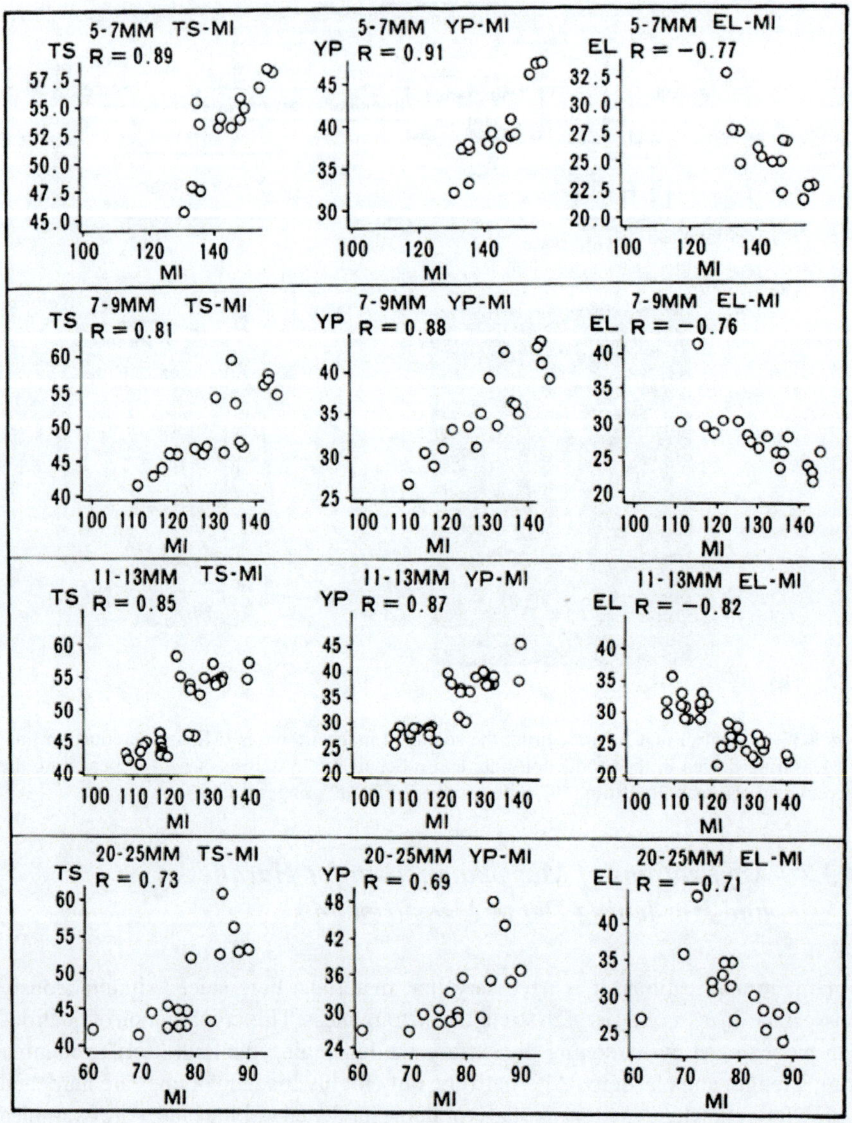

Fig. 3.38 Correlation between the mechanical properties and the index MI, which is the variation of the imaginary part of the coil impedance, for different metal plate thicknesses; TS: Tensile strength [kg/mm^2]; YP: Yield point [kg/mm^2]; EL: Elongation [%]; R: Correlation coefficient (from [99], with permission from Springer Nature)

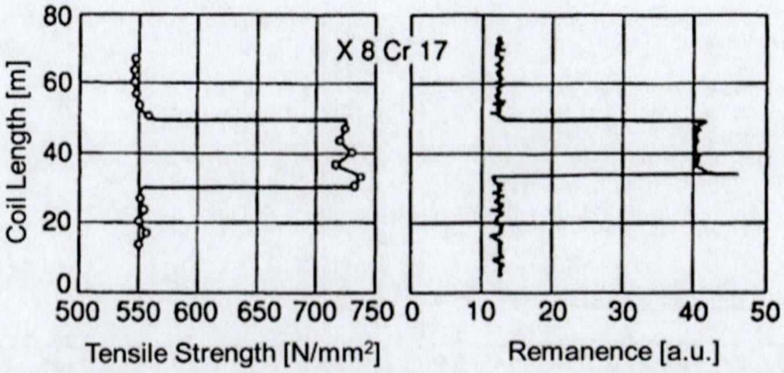

Fig. 3.39 Correlation between the magnetic remanence B and the tensile strength (in modified form from ([100], with permission from the Deutsche Gesellschaft für zerstörungsfreie Prüfung (DGZFP) and the author)

Because the depth sensitivity for hardness variations of the various NDMC methods differ, a generally valid standardization does not exist either for Barkhausen emissions, eddy current techniques, or the 3MA technique. There is no generally accepted procedure for defining hardness on solely NDMC measurement parameters. Based on the employed NDMC measurement technique, one may define therefore the hardness depth (DoH) as the depth at which the surface hardness value has dropped to a pre-defined value, for example 85% as shown in (3.40). The employed NDMC measurement technique must be calibrated against a destructive standard such as indentation hardness.

Magnetic Barkhausen emissions *M(H)* can be used to determine the DoH for depth ranges down to 4 mm as given by the skin depth (3.14). Often double peak Barkhausen emission profiles can be observed during magnetization cycles. One maximum, *A*, arises due to the contributions of the "soft" bulk material with a low coercivity H_c and the other maximum, *B,* arises due to the contributions of the hardened surface layer with a high coercivity H_c (3.41). The ratio *B/A* as a function of DoH depends on the magnetization frequency f_e and the analyzing frequency f_A or on the analyzing frequency bandwidth $<f_A>$. For a given material, a selection of optimal parameters

Fig. 3.40 Definition of hardness depth (DoH) for the employed NDMC technique

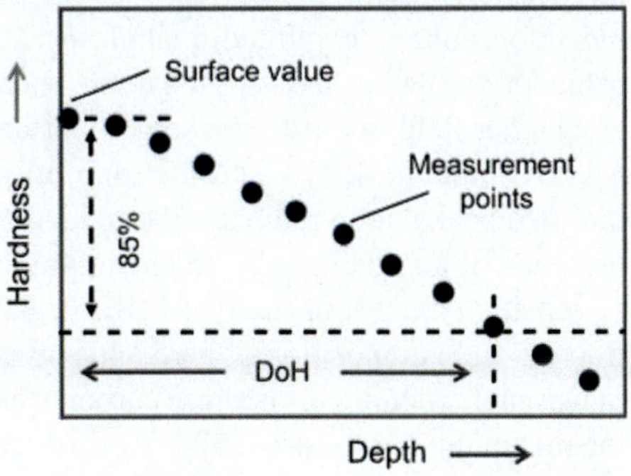

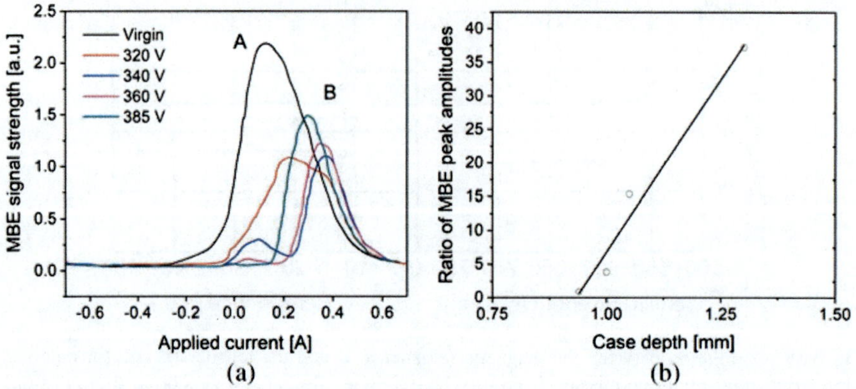

Fig. 3.41 **a** Variation of MBE signal strength (RMS voltage) as a function of current applied to the yoke for half a magnetization cycle for specimens subjected to given induction hardening voltages; **b** Relationship between the ratio of MBE peak of the green curve to the peak of the black curve versus case hardening depth showing a correlation coefficient of 0.98 (from [88] in modified form, with permission from Taylor and Francis)

f_e, f_A or $\langle f_A \rangle$ is necessary. With a given selection f_e, f_A or $\langle f_A \rangle$, a series of different types of steels were characterized in respect to their DoH [104].

Vaidyanathan et al. [88] studied the influence of the microstructure for different induction hardened carbon steel (Fe–0.42C–0.035P–0.035S–0.15Si–0.5Mn) shafts on the hysteresis loop and the magnetic Barkhausen emissions. In the induction hardening process, the surface of a component is heated by means of an alternating magnetic field and the ensuing eddy currents to a temperature beyond the upper critical (Ac_3) temperature i.e. in the austenitic range, followed by immediate quenching. This leads to hardening of the outer case (near-surface region), while the core of the component remains unaffected. The systematic changes in the MBE profiles for different voltages, i.e. the strength of the magnetic field applied during induction heating, indicated the microstructural variations within the case. A drastic drop in the MBE peak signal strength is observed in the surface hardened samples as compared to that in the virgin ferritic steel sample, due to the formation of the hard martensitic phase in the near surface region. The sample hardened at the maximum induction voltage exhibited a single peak (peak of the green curve in Fig. 3.41a) behavior similar to the virgin sample (peak of the black curve in Fig. 3.41a), but at a higher field, indicating a single martensitic structure in the interrogation thickness. A single peak MBE profile for a fully martensitic structure gradually changed into two peaks upon reducing the induction hardening voltage, which indicates the presence of an additional soft ferrite phase within the interrogation thickness. The systematic changes in the two MBE peak heights can be correlated to the volume fraction and composition of hard and soft phases within the case (Fig. 3.41b). In the induction hardening of medium carbon steel, as studied by Vaidyanathan et al. [88], the maximum hardness (~800 HV) is observed near the surface up to about 0.5 mm depth due to its martensitic structure. The hardness starts to decrease beyond this depth and reaches the base hardness of about 350 HV at about 1 mm depth.

In the case of laser hardening of X210Cr12 steel with 2.1% C and 12% Cr as studied by Kern et al. [105], a lower hardness was observed near the surface. This was attributed to a large amount of retained austenite (~80%) caused by the reduction in the martensitic transformation temperature due to high carbon content and up-hill diffusion of carbon to the surface [106]. The hardness values peaked to about 750 HV at a depth of about 1 mm, followed by a continuous decrease to the base hardness at a depth of about 4 mm (Fig. 3.42a). Furthermore, Kern et al. [105] analyzed the laser hardening process using three different techniques, namely, MBE, incremental permeability, and X-ray Mössbauer spectroscopy in backscattering technique in order to investigate the phase transformations occurring at various depths from the steel surface. The transformations were induced by different laser power levels in order to produce different hardness depth profiles. Similar to the study of Vaidyanathan et al. [88], two peaks corresponding to soft ferritic bulk and hard martensitic structures were observed in the MBE signals (Fig. 3.42b) obtained in all the three laser surface treated specimens. With increasing laser power, the amplitudes of both peaks decreased due to the formation of an increasing amount of nonmagnetic retained austenite phase near the surface, which covered the martensitic and ferritic phases below it. The depth profiles determined by means of both Mössbauer and magnetic techniques agreed with the metallographic and Vickers hardness data. Finally, the determination of the retained austenite fraction by Mössbauer spectroscopy was used to calibrate the micromagnetic measuring quantities.

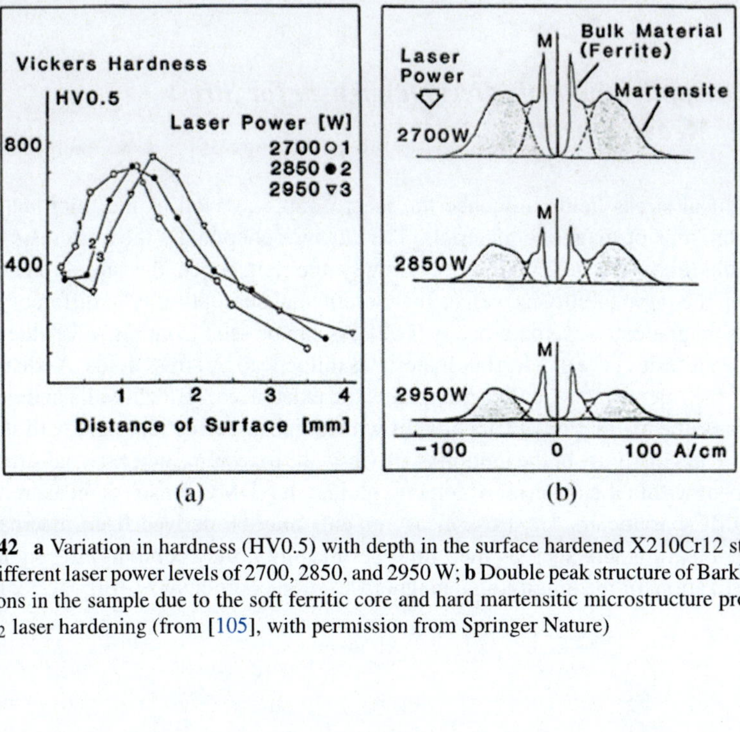

Fig. 3.42 **a** Variation in hardness (HV0.5) with depth in the surface hardened X210Cr12 steel for three different laser power levels of 2700, 2850, and 2950 W; **b** Double peak structure of Barkhausen emissions in the sample due to the soft ferritic core and hard martensitic microstructure produced by CO_2 laser hardening (from [105], with permission from Springer Nature)

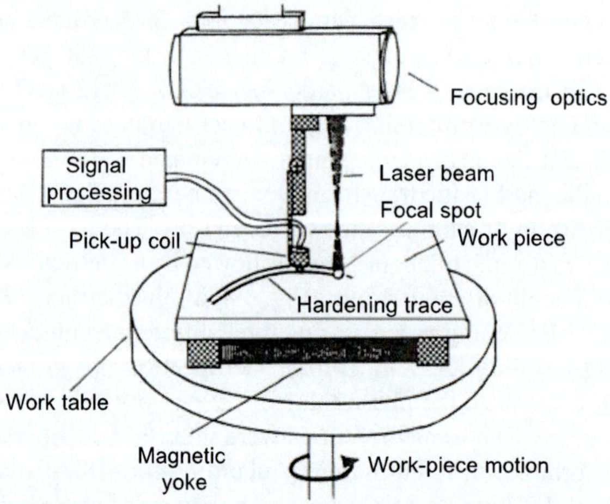

Fig. 3.43 On-line laser hardening process control by Barkhausen emission profiles ([107, 108], with permission from the Deutsche Gesellschaft für zerstörungsfreie Prüfung (DGZFP) and the authors)

A fast and reliable determination of the hardness depth by a micromagnetic technique is desired for in-line monitoring of laser-hardening processes with feedback in order to control the process parameter [107, 108]. Here, it is exploited that the hardening process by laser irradiation influences the Barkhausen noise in a specific way relative to the noise of the bulk material. Typical double peaks appear in the total profile of the Barkhausen emissions: one related to the bulk material and the other to the hardened layer. By measuring the noise profile immediately after the process zone, allows one to device a feed-back for hardening depth control (Fig. 3.43).

3.3.8 Applications of Micromagnetism for Stress Measurements

Mechanical stress fields influence the macroscopic as well as the micromagnetic characteristics of magnetic materials. The changes encompass the hysteresis curve, the initial (anhysteretic) curve, the coercivity, the remanence, the saturation magnetization, the permeability as well as the incremental susceptibility at different fields, and the magnetostrictive parameters [109]. As can be seen from Fig. 3.28, the shape of the hysteresis curve of a ferrous material is influenced by stress fields. As discussed earlier, the micromagnetic theory explains the existence of Bloch-walls and domain structures, the interaction of Bloch-walls with lattice imperfections, and with stresses (3.55). The sensitivity of the motion of Bloch-walls to residual and external stresses is the sensor within the material which is exploited in NDMC for stress measurements.

NDMC parameters for stress measurements may be derived from macroscopic magnetic parameters such as the initial permeability, the coercivity, the saturation magnetization, or the longitudinal magnetostriction coefficient λ_L (Fig. 3.28). In this case ferromagnetic materials are regarded as a bulk material with volumes encompassing many domains, which contributes to the measured magnetic quantities. As a result, absolute quantitative results can be obtained if test specimens or test components have a defined geometry and are excited by a constant magnetic field strength over their whole cross-section. Such tests are mainly performed in a laboratory setting. However, a relative quantitative measurement can still be performed on a test component in the field.

Which parameters can be derived from micromagnetic properties and which can serve as indicators of stresses in field applications? The micromagnetic theory predicts that 180° domain walls, 90° domain walls and domain rotations are contributing at different magnetic field strengths to the hysteresis curve, see Fig. 3.26. The wall motions and domain rotations depend on the stress level within the domain walls, the domains, and on the microstructure. Not only the stress level plays a role but also the spatial extent of the stress field and their gradients relative to the grain and domain sizes of the microstructure, i.e. whether macro- or micro-stresses are present in the material. In addition, precipitates and dislocations hinder the mobility of domain walls. Therefore, it is natural that those parameters are used, which depend on the microstructure and can be measured locally, such as the differential permeability recorded over the tangential magnetic field strength $\boldsymbol{H}$ besides measurements of the coercive field $\boldsymbol{H}_c$.

A difficulty for getting unambiguous stress values stems from the complicated behavior of the magnetostrictive coefficient λ_L as a function of the magnetic field strength. For example, polycrystalline iron exhibits for $\boldsymbol{H}$-field strengths smaller than approximately 3 kA/m, a positive longitudinal magnetostrictive coefficient λ_L, but at larger field strengths the coefficient λ_L becomes negative.

The calibration of the measured micromagnetic parameters versus the wanted quantities can be performed using tensile or bending tests, X-ray or neutron diffraction methods or any other method. Because of the complicated interdependencies of the measuring quantities in respect to the wanted quantities like stress levels or hardness, unambiguous values can only be achieved by using multiparameter approaches. Appropriate functions for quantitative ND-tests are usually obtained by multiple regression, by neural network analysis, and other mathematical tools, see following sections. In this way, several wanted NDMC parameters can be obtained, see Fig. 3.44.

Anglada-Rivera et al. [82] studied the influence of applied tensile stress and grain size on MBE and on the hysteresis loops in 1005 commercial steel. They found that the peak amplitude of the Barkhausen voltage increases with applied stress, reaching a maximum value at a certain stress value σ_c, and then decreases again at higher tensile stress. In addition, the amplitude of the MBE increased with decreasing grain size. This behavior was explained by the fact that the size of the magnetic domains is proportional to the square root of the grain diameter, and thus for microstructures with fine grains the number of domain walls that can move becomes larger. Furthermore, the effect of grain size on the magnetic hysteresis loop in the stressed materials showed that the maximum induction B_{max} decreased with increasing grain size as well as the slope of the hysteresis curve, and hence also the differential permeability

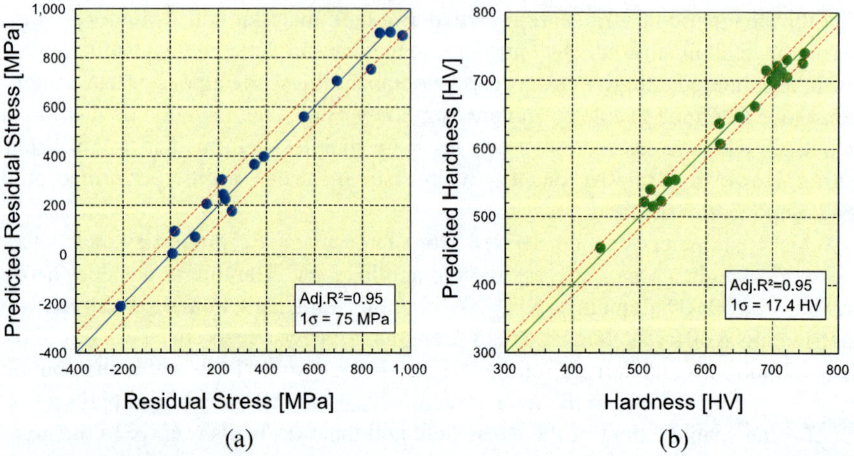

Fig. 3.44 **a** Predicted residual stress value after calibration of the micromagnetic measurement parameters with the hole-drilling method; **b** Predicted surface hardness value after calibration with Vickers hardness measurements (in modified form from [110]), and with permission and courtesy from Fraunhofer IZFP

$\Delta\mu_r$. Because the magnitude of d$\boldsymbol{B}$/d$\boldsymbol{H}$ determines the Barkhausen activity, it should decrease itself. Hence, the behavior of MBE and the hysteresis loop for samples with different grain size are closely related.

Altpeter et al. [111] examined the influence of IInd and IIIrd kind micro-residual stresses in steels due to Cu precipitates on the Barkhausen amplitude as a function of external stress and compared these data to the same measurement close to the material's Curie temperature (phase transition of the cementite). The shift of the maximum Barkhausen emission amplitude $M_{max}(\sigma)$ as a function of stress can be used as a measure for micro-residual stresses originating from the precipitates, which allows their quantitative determination without a reference method.

Sorsa et al. [112] extracted from Barkhausen emission signals features which allow one to predict the residual stress and the hardness of case-hardened steel samples (18CrNiMo7-6(EN10084). A data based approach for building a prediction model, consisted of feature generation, feature selection and model identification and validation steps. Features were selected with a simple forward-selection algorithm. Multivariable linear regression models were used in the predictions. Throughout the selection and identification procedures, a cross-validation was used in order to guarantee that the results were realistic and could be used for future predictions. The obtained prediction models were validated with an external validation data set. The accuracy of the method showed that the proposed scheme could be applied to predict material properties.

Kypris et al. [113] divided a ferromagnetic material into layers from the surface into the depth of the material. For each layer, a non-linear integral equation was established that describes the attenuation of the magnetic Barkhausen emissions in that layer due to eddy current damping as a function of frequency and distance from the surface. The Barkhausen emission spectrum measured at the surface by an induction coil was expressed as the sum of the individual layer spectra. In the case that stress is present, it influences the amplitude of the Barkhausen emissions. The evaluation of the tensile stress was based on the theory of ferromagnetic hysteresis using a linear relationship between stress and the reciprocal peak voltage envelope amplitude V^{-1} of the MBE [114]:

$$\frac{1}{V(\sigma_1)} - \frac{1}{V(\sigma_2)} = \frac{3b'(\sigma_1 - \sigma_2)}{\mu_0} \tag{3.62}$$

Here, σ_1 and σ_2 denote different stress states, such that $V(\sigma_1)$ and $V(\sigma_2)$ are the Barkhausen voltages corresponding to the stresses σ_1 and σ_2, and b' is a modified magnetostriction constant. The technique should also be applicable to compressive stresses. The quality of the measured data fit was also influenced by the sensitivity of the ferromagnetic material to strain, as well as by the sensor-specimen coupling. The proposed method was extended to evaluate stress gradients perpendicular to the surface as a function of depth, for example induced by shot-peening [115].

Ding et al. [116] presented experimental studies that the skewness, a measure of the asymmetry of the Barkhausen amplitude distribution as a function of applied magnetic field, is affected by compressive and tensile stresses in soft steels, see Fig. 3.45. With the appropriate parameters to measure the skewness, this can be exploited for stress measurements.

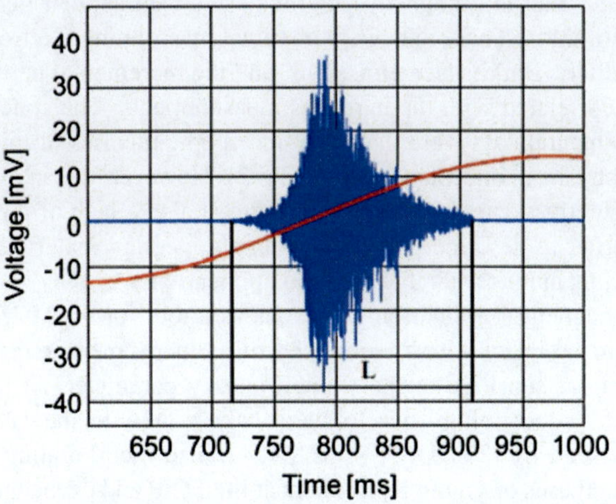

Fig. 3.45 Example of a skewed, i.e. an asymmetric magnetic Barkhausen emission signal in rel. amplitudes. The red curve is proportional to the exciting magnetic field strength. The temporal length of the Barkhausen emission signal is *L*. The MBN signals were amplified by 30 dB, bandpass filtered (2–40 kHz), and sampled with a rate of 200 kHz and with 14-bit in an analog-to-digital converter. The signal shape is used to determine the stress state of the material (from [116], with permission from Elsevier)

Eventually, the results obtained in non-destructive materials characterization using eddy current and MBE measurement techniques with various experimental set-ups and prototypes developed at Fraunhofer IZFP were integrated into the so-called 3MA technique, which will be discussed in the following section. The exchange with other groups, who also developed engineering instruments [117, 118], supported this endeavor. Their experience and that of other NDE groups was considered as well, in particular within international exchange programs.

3.3.9 3MA Technique

In two early studies on the steels 22 NiMoCr 3–7 and 15 MnMoNiV 5–3, relations between their microstructures, coercivities, hardness, and residual stress were established, already with the first version of the 3MA analyzer [119, 120]. These studies were further expanded in a PhD thesis guided by Prof. Höller, founding director of Fraunhofer IZFP, and submitted in 1987 [121] to the science faculty of the Saarland University. The influence of precipitates on the mechanical hardness and in parallel on the micromagnetic measuring parameters was studied in iron-copper alloys. These alloys were considered to be polycrystalline model substances allowing one to study the behavior of domain walls in engineering ferromagnetic materials. Great care was taken to control the influence of other lattice distortions, which might influence the mechanical and micromagnetic parameters as well. Thermal treatment was used to grow magnetic and non-magnetic precipitates as well as their density and their volumetric distribution. These data were obtained by scanning electron microscopy. Then, the coercivity, Barkhausen emissions, and the incremental permeability were measured and correlated with the hardness measurements. The conclusions drawn from the experimental data were that the microscopic theories of micromagnetism supported the physical concepts (see Sect. 3.3.2). However, the analytical relations were not sufficiently accurate. In addition, there is also a lack of input parameters (see 3.63) in order to describe the situation in engineering materials as a base for reliable NDMC techniques suitable for field applications.

To exploit magnetic Barkhausen emissions as a tool for NDMC, an integrative approach had to be taken which comprised of as many measurement parameters as possible that are sensitive to the wanted quality parameter. At the Fraunhofer IZFP this led to a technology development called 3MA in the 1980s [122] and was further pursued by [31, 83, 109, 123, 124], also incorporating the results of master and PhD theses of young researchers at the IZFP and exchange visitors, also from the IGCAR. These efforts also led to the norm "Hardness testing of metallic materials" (issued by VDI/VDE-Gesellschaft Mess- und Automatisierungstechnik (GMA), Beuth Verlag, Berlin, (2012) pp. 1–69), which describes the procedures to measure hardness by a combination of eddy current and Barkhausen emission measurements.

The acronym "3MA" stood originally for a combination of three, then four micromagnetic measurement procedures or methods, namely Barkhausen Noise (BN),

Harmonic Analysis of the tangential magnetic field strength (HA), multi-frequency Eddy Current analysis (EC), and incremental permeability (IP). The acronym 3MA stands now for micromagnetic, multi-parameter, microstructure and stress analysis, a methodology which has been validated in many applications.

In the 3MA technology [31, 124], the ferromagnetic material is magnetized by a yoke having a coil wound around its central part (Figs. 3.46 and 3.47). The coil is driven by a slowly varying AC current at the frequency f_e. The yoke injects the magnetic field generated by the coil into the material or the component being examined until saturation within the skin depth $\delta(f_e)$ (3.14). For steel with $\mu_r = 200$ and $\sigma = 5$ Ms/m, $\delta \approx 0.5$ mm for $f_e \approx 1$ kHz, $\delta \approx 2.3$ mm for $f_e \approx 50$ Hz, and $\delta \approx 71$ mm for $f_e = 50$ mHz. The following signals are measured during the magnetization process of the material:

1. *Magnetic Barkhausen Emissions*: The signals of the domain-wall motions are detected by pick-up coils placed on the component. The sudden domain-wall motions generate local eddy currents and hence stray-fields at frequencies much higher than f_e. Selecting a frequency range Δf_a allows one to analyze the signals emanating from different depths due to the skin effect (3.14). Therefore, different microstructural states, caused for example by a hardened surface and a softer material core, may become distinguishable by multiple maxima in the *BN(H)* signal. After rectification, the signal is recorded and analyzed based on experience using measurement parameters, which characterize the shape and amplitude of the noise distribution, see Figs. 3.45 and 3.46c.
2. *Incremental Permeability IP*: In addition to the yoke coil, a second small coil placed between the poles of the yoke is excited at a frequency f_Δ, (Fig. 3.46a). This coil generates small hysteresis loops (Fig. 3.46c). The signal detected with the receiver coil yields the incremental permeability μ_Δ, which is defined as $\mu_\Delta = (1/\mu_0) d\boldsymbol{B}/d\boldsymbol{H}$ (Fig. 3.7). The incremental permeability may be measured at 25, 50, and 75% of the saturation magnetization or at another useful point of the

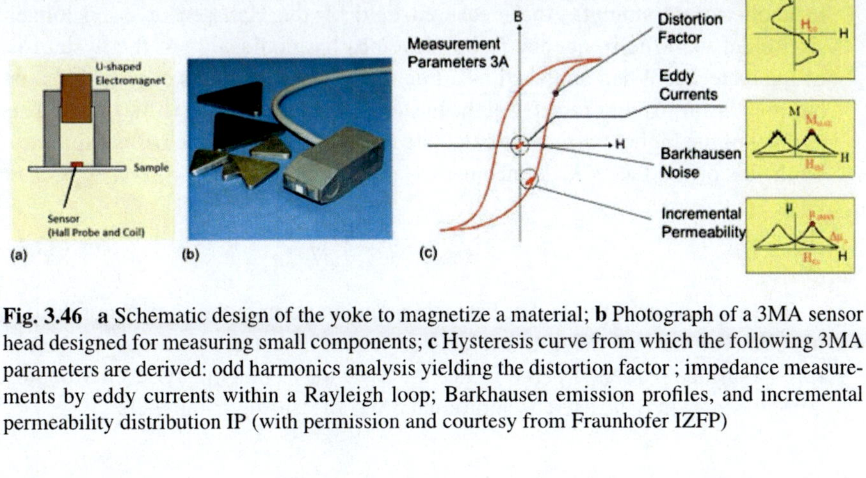

Fig. 3.46 **a** Schematic design of the yoke to magnetize a material; **b** Photograph of a 3MA sensor head designed for measuring small components; **c** Hysteresis curve from which the following 3MA parameters are derived: odd harmonics analysis yielding the distortion factor ; impedance measurements by eddy currents within a Rayleigh loop; Barkhausen emission profiles, and incremental permeability distribution IP (with permission and courtesy from Fraunhofer IZFP)

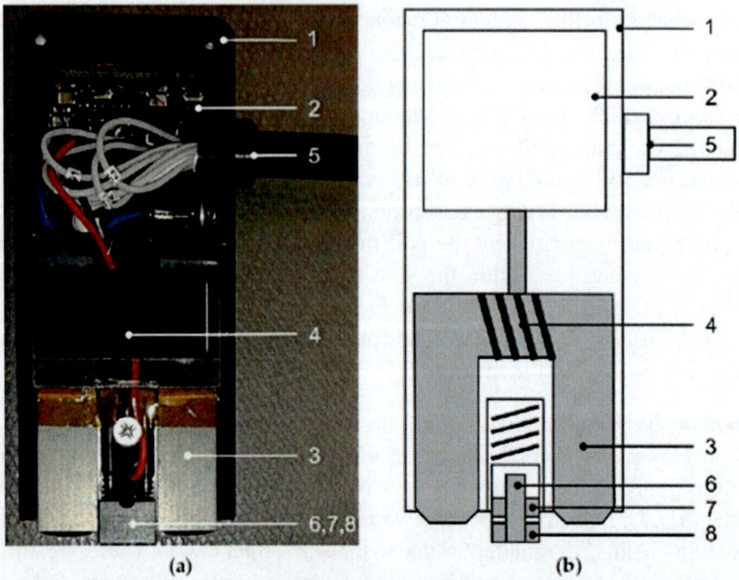

Fig. 3.47 **a** Photograph and **b** schematic design of a 3MA sensor probe with its components: **1**-housing, **2**-preamplifier, **3**-magnetic yoke, **4**-magnetization coil, **5**-connection cable, **6**-Hall sensor, **7**-transmitter coils and **8**-receiver coils for measuring the EC and IP parameters (from [124], with permission under Creative Commons Attribution License (https://creativecommons.org/licenses/by/4.0/))

hysteresis curve. This provides further parameters for the 3MA technique, which are correlated to material parameters.

3. *Harmonic analyses of the tangential field (HA)*: The magnetic induction $B(H)$ within the material follows the hysteresis curve. Because the tangential field is continuous, one can monitor its amplitude in the material with a Hall probe, see Figs. 3.46a and 3.47a and b. When the hysteresis curve is measured as a function of applied field strength, the measured field by the Hall sensor is no longer sinusoidal with the frequency f_e but develops harmonics due to the hysteresis of the material. When analyzed as a Fourier series, it contains only odd higher harmonics due to the symmetry of the hysteresis curve in respect to the origin. The amount of the higher harmonics calculated from their amplitudes A_i is expressed by the distortion factor K defined as:

$$K = \sqrt{\left(A_3^2 + A_5^2 + A_7^2\right)/A_1^2} \tag{3.63}$$

The evaluation of the higher harmonic content permits the analysis of deeper material volumes allowing to evaluate structural and stress gradients as occurring in surface hardened parts. Either the K-factor or individual odd harmonics amplitudes A_i may be used as additional 3MA measuring parameters.

4. *Eddy Current Parameters (EC)*: A small AC current at a frequency f_{HF}, typically between 10 kHz and 1 MHz, is fed into a transmitter coil with a diameter of a few mm. This coil results in a low level of induction of a few μT and thus excites a reversible Rayleigh loop (Eq. 3.37) [17]. The AC magnetic field induces eddy currents in the test component according to Faraday's law. A receiver coil is used to detect the magnetic flux as an induced voltage, and the real and imaginary parts of the coil's impedance are recorded.

The present 3MA system allows the operator to apply up to four different frequencies f_{HF} in a single measuring cycle, resulting in 16 3MA measuring parameters. They depend on the conductivity σ and on the permeability $\mu(H)$ of the material, which in turn are influenced by internal stresses, hardness, yield stress, and on a number of other material characteristics.

The numerous measuring parameters are fed into a multiple regression algorithm in order to obtain the material properties to be measured. This algorithm must be calibrated, and the proper selection of the calibration specimen is the most important task when applying the 3MA technique. Each calibration specimen should represent a relevant material state. These states should represent a standard distribution, thus allowing to take into account the natural scatter of the material properties. Furthermore, disturbing effects must be considered as well. Examples of disturbing effects are temperature variations and/or residual stresses. In this case, the calibration must be performed with test samples whose temperature and stress level are varied in a controlled way. Also surface roughness may be a cause for disturbing effects. In this case, the calibration samples should exhibit the same surface roughness parameters as the components to be tested non-destructively. The number of test specimens which represent a material's defect or microstructure states, multiplied by the number of the measuring positions, should be appreciably larger than the number of unknown parameters which are to be measured. Finally, in order to render the calibration more reliable, the results of reference methods like hardness indenter tests, tensile yield tests, or X-ray diffraction data for obtaining the residual stress level have to be taken into account as well.

Eventually, the measured 3MA parameters and the target quantities are stored in a common database, and then the correlations between both are established. The 3MA software offers different methods for calculation of the calibration functions, including regression analysis and pattern recognition algorithms. In the case of regression analysis, the calibration function can be written as:

$$Y = a_0 + a_1X_1 + a_2X_2 + a_3X_3 + ... + a_nX_n \tag{3.64}$$

where Y is the target quantity. The X_n (n = 1, 2, 3,..) are the 3MA measuring parameters and a_n are the coefficients that have to be determined by least-squares methods. Several calibration functions for different target quantities can be determined in parallel. Besides the 3MA measuring parameters, also parameters that are measured by external sensors, such as temperature, sensor lift-off, sheet thickness, strip velocity, and mechanical stress levels can be included in the calibration functions. For certain materials and components, large calibration databases were accumulated, which allow the development of universally applicable calibration functions. In addition to laboratory calibrations, simulation results can also be used to complement the database. These additional data minimize the risk of calibration errors. Finally, machine-learning algorithms are applied for determining the calibration functions. Recently, the calibration of the 3MA techniques was further underpinned by FEM modeling [125–127].

Theiner et al. [110] presented applications of the 3MA technique for measuring hardness and case depth using the then industrial-grade equipment developed by the Fraunhofer IZFP (Fig. 3.48a). At first, destructive hardness tests were performed according to usual standards. Then, the following 3MA parameters were measured: (1) Eddy currents (EC); (2) Barkhausen noise (BN); (3) Time signal of tangential magnetic field-strength (HA), and (4) incremental permeability (IP). Figure 3.48b shows a manually handled 3MA probe on induction hardened input shafts in order to obtain the 3MA measuring quantities. In Fig. 3.48c the case depth measured with the 3MA technique is shown in comparison to conventional tests, which were obtained by metallography. The mean standard error of the non-destructively measured values was comparable to that of conventional tests. In both applications, ten 3MA parameters were used as input variables in order to evaluate the hardness and the case depth values. In another application, the 3MA-sensor was integrated in a CO_2 laser-hardening machine where guide bars were produced. Hardness and case depth were measured during the hardening process, approximately 5 cm behind the laser spot. For calibration, eight 3MA parameters were used. When measuring the case hardening depth, scale, retained austenite and stresses varying on a large scale are the dominant disturbing factors.

Ewen et al., [128] demonstrated the implementation of the 3MA-technique into the production line of a forge master's production. Here, the tensile strength and the yield strength were the target parameters. High accuracy of the 3MA calibration was sought in order to evaluate the mechanical properties at rim segments of shafts. Calibrations were performed for each steel grade and with standard deviations of about 5 MPa for the tensile strength (R_{TS}) value for the steel grades 22 CrMoNiWV 8–8, 27 NiCrMoV 11–5 and 27 NiCrMoV 15–6, see Fig. 3.49. The evaluation took 15 min for one segment of a shaft. This implied cost savings because the 3MA technology replaced the cutting of test segments for testing them destructively and thus shortening the internal quality procedures. In this application, the 3MA measuring parameters comprised the analysis of magnetic Barkhausen emissions, incremental permeability, analysis of the harmonics in the magnetic tangential field strength, and eddy current impedance. Approximately 40 measuring quantities span a data space for different material states that were separated by means of pattern recognition or multivariate regression analysis. A set of calibration data was obtained on samples of well-defined, quantitatively known target properties such as Vickers hardness, yield strength, and residual stress.

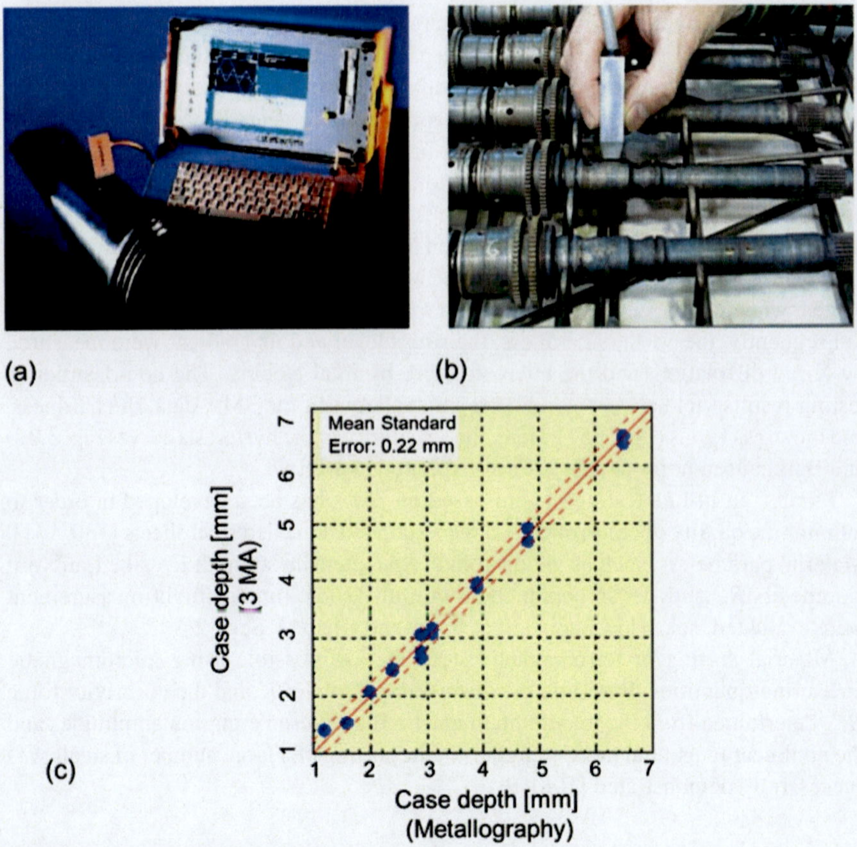

Fig. 3.48 **a** 3MA testing unit; **b** 3MA probe and measured shafts; **c** Correlation between case depth determined by the 3MA technique and the conventionally determined values for the input shafts (in modified form from [110], and with permission and courtesy from Fraunhofer IZFP)

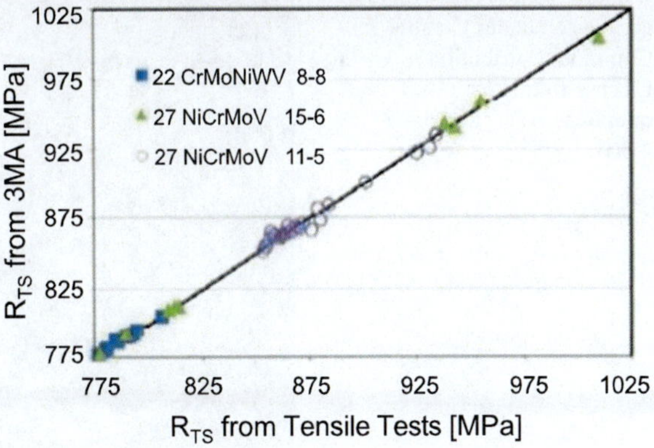

Fig. 3.49 Calibration of 3MA data by data obtained from destructive tests for production shafts made of 22 CrMoNiWV 8–8, 27 NiCrMoV 15–6, and 27 NiCrMoV 11–5 (in modified form from [128], open access at NDT.net; www.ndt.net/?id=16684; with permission)

In order to provide quality information regarding the microstructure of a grinded component, the nital etching method is often used. Here, the 3 MA technique has been used for detecting grinding burns in several applications [124, 129]. To this end, the relationships between the load-bearing capacity of the gearwheel teeth, their hardness, and the presence of residual stress in the tooth-flanks, which are influenced by the grinding process, were examined to check the tooth-flank load capacity, and material deterioration caused by grinding. To this end, a specially adapted and miniaturized 3MA probe was developed and then integrated into a multi-axis manipulation system (see Fig. 3.50). For calibration, the 3MA measuring parameters were recorded on gear wheels in the undamaged state and with different degrees of grinding damage. Subsequently, the Vickers hardness, the stress level and its gradient were measured by X-ray diffraction, and the microstructure by nital etching. The non-destructive testing results yielded very good correlations between the 3MA data, the hardness, and the stress levels (Fig. 3.51). Here, the sensitivity of the hysteresis curve (Fig. 3.28) and Barkhausen noise to stresses was exploited as well.

Finally, an in-line testing system based on 3MA has been developed in order to monitor the quality of cold-rolled and recrystallized annealed steel sheets [130, 131]. Material parameters, such as yield strength $Rp_{0.2}$, tensile strength R_{TS}, the Lankford parameters R_m and ΔR in conjunction with ultrasonic time-of-flight measurement were evaluated, see also Chap. 1, Sect. 1.15, and Chap. 2, Sect. 2.4.12.

Material sorting of ferromagnetic steels is also possible using micromagnetic measuring quantities. Based on the distortion parameter K and the coercivity force H_{cm} determined from the maximum magnetic Barkhausen emission amplitudes and the nonlinear tangential field strength H_{nl}, the sorting of a large number of steels was successfully demonstrated [108].

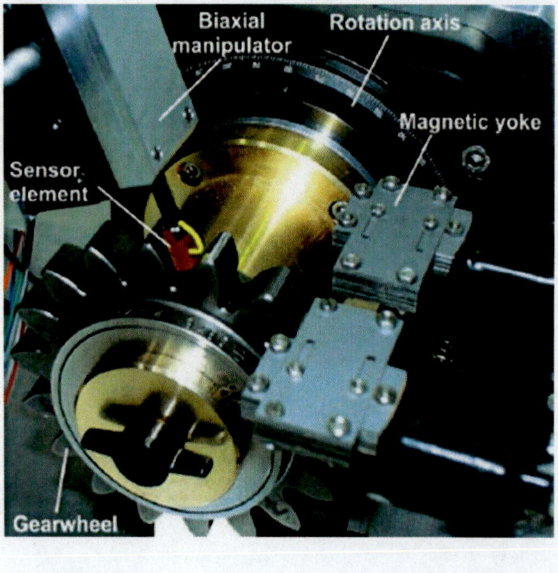

Fig. 3.50 3MA gearwheel inspection system in order to scan tooth flanks of a gear-wheel in meander-like fashion (from [124], with permission under Creative Commons Attribution License (https://creativecommons.org/licenses/by/4.0/))

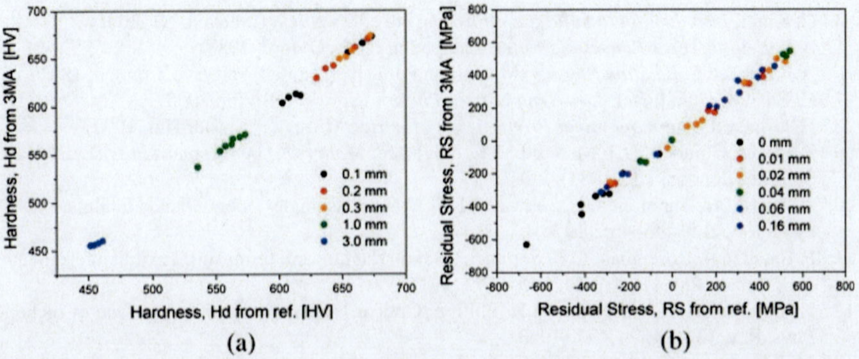

Fig. 3.51 Hardness and stress data obtained by the 3MA technique versus calibration tests; **a** Hardness at depths of 0.1, 0.2, 0.3, and 3.0 mm; **b** Residual stress at depths of 0, 0.01, 0.02, 0.04, and 0.06 mm (in modified form from [124], with permission under Creative Commons Attribution License (https://creativecommons.org/licenses/by/4.0/)

3.3.10 Commercial Instruments

There are several commercial or semi-commercial non-destructive systems based on eddy current and Barkhausen noise emission measurements on the market, for example the 3MA system which is commercialized by Fraunhofer IZFP. Similarly, an instrument has been developed called "Magnescope" [117, 118]. Stresstech (www.stresstech.com) produces probes, control units and readily automated solutions for Barkhausen emission inspections for a number of applications, for example the detection of grinding burns. The Introscan system is a Barkhausen emission measurement system for qualitative stress assessment (http://iaph.bas-net.by/VDev/Introscan/introscane.html) [132]. QASS manufactures a commercial Barkhausen emission inspection device, the QASS μ-Magnetic, mainly for hardness inspection utilizing known hardness references (Start—QASS). It is relatively easy to assemble the corresponding hardware to magnetize a material, record magnetic fields with suitable sensors, digitize the data, and write a software for data evaluation. Therefore, there are most likely many more commercial or semi-commercial instruments on the market.

References

1. F. Förster, *The Origin of Nondestructive Determination of Characteristic Material Parameters Using Electromagnetic Methods*, ed. by P. Höller, V. Hauk, G. Dobmann, C.O. Ruud, R. E. Green. Nondestructive Characterization of Materials (Springer, 1989)
2. Y.K. Fedosenko, V.G. Gerasimov, A.D. Pokrovskiy, Y.Y. Ostanin, Eddy Current Testing, in *Nondestructive Testing*. ed. by V.V. Klyuev (Publishing House Spektr, Moscow, Russia, 2010), pp.331–632
3. N.W. Ashcroft, N.D. Mermin, *Solid State Physics* (Saunders College Publishing, Philadelphia, PA, 1976), p.19105

4. C. Kittel, *Introduction to Solid State Physics*, 8th edn. (Wiley, Hoboken, NJ, 2005)
5. A.J. Walton, *Three Phases of Matter* (Clarendon Press, Oxford, 1983)
6. A.J. Dekker, *Solid State Physics* (Macmillan Publisher Limited, Palgrave, London, 1981)
7. C. Enss, S. Hunklinger, *Low-Temperature Physics* (Springer, Berlin, 2005)
8. G. Grimvall, *Thermophysical Properties of Materials* (Elsevier, Amsterdam, 1999)
9. P.V.E. McClintock, D.J. Meredith, J.K. Wigmore, *Matter at Low Temperatures* (Blackie & Son Ltd, London, UK, 1984)
10. A. Bansil, H. Ehrenreich, L. Schwartz, R.E. Watson, Complex energy bands in alpha-brass. Phys. Rev. B **9**, 445–465 (1974)
11. F. Batallan, I. Rosenman, C.B. Sommers, Band-structure and fermi surface of hcp ferromagnetic cobalt. Phys. Rev. B **11**, 545–557 (1975)
12. H. Ehrenreich, D.J. Olechna, H.R. Philipp, Optical properties and fermi surface of nickel. Phys. Rev. **131**, 2469–2477 (1963)
13. M.F. Crommie, C.P. Lutz, D.M. Eigler, Imaging standing waves in a 2-dimensional electron gas. Nat. **363**, 524–527 (1993)
14. V. Nalladega, S. Sathish, K.V. Jata, M.P. Blodgett, Development of eddy current microscopy for high resolution electrical conductivity imaging using atomic force microscopy. Rev. Sci. Instrum. **79** (2008)
15. J.D. Jackson, *Classical Electrodynamics* (Wiley, New York, 1999)
16. P.A. Tipler, G. Mosca, *Physics for Scientist and Engineers* (W.H. Freeman and Company, New York, 2007)
17. D. Jiles, *Introduction to Magnetism and Magnetic Materials* (CRC Press, Boca Raton, 2015)
18. V. Natarajan, *Modern Atomic Physics* (CRC Press, Boca Raton, 2015)
19. G. Bertotti, *Hysteresis in Magnetism* (Academic Press, San Diego, CA, 1998)
20. R.M. Bozorth, *Ferromagnetism* (Wiley&Sons and IEEE Press, Hoboken, N.J., 2003)
21. C. Kittel, Physical theory of ferromagnetic domains. Rev. Mod. Phys. **21**, 541–583 (1949)
22. J. Abuthahir, A. Kumar, Magnetic force microscopy studies in bulk polycrystalline iron. J. Magn. Magn. Mater. **448**, 107–116 (2018)
23. L. Batista, U. Rabe, S. Hirsekorn, Determination of the easy axes of small ferromagnetic precipitates in a bulk material by combined magnetic force microscopy and electron backscatter diffraction techniques. Ultramicroscopy **146**, 17–26 (2014)
24. M. Kronmüller, M. Fähnle, *Micromagnetism and the Microstructure of Ferromagnetic Solids* (Cambridge University Press, Cambridge, UK, 2003)
25. R.E. Beissner, Theory of Eddy Current Characterization of Magnetic Conductors, in *Nondestructive Characterization of Materials*. ed. by P. Höller, V. Hauk, G. Dobmann, C. Ruud, R.E. Green (Springer, Saarbrücken, Germany, 1989), pp. 541–548
26. H.A. Crostack, W. Bischoff, J. Nehring, *Nondestructive Testing of Forged Components Using CS-Pulsed Eddy-Current Technique.*, ed. by P. Höller, V. Hauk, G. Dobmann, C.O. Ruud, R.E. Green. Nondestructive Characterization of Materials (Springer, 1989), pp. 574–581
27. H. Heptner, H. Stroppe, *Magnetische und Magnetinduktive Werkstoffprüfung* (VEB Deutscher Verlag für Grundstoffindustrie, Leipzig, Germany, 1972)
28. J. Blitz, *Electrical and Magnetic Methods of Nondestructive Testing* (Springer, Netherlands, 1997)
29. N. Bowler, *Eddy-Current Nondestructive Evaluation* (Springer, New York, NY, 2019), p.10013
30. J. Garcia-Martin, J. Gomez-Gil, E. Vazquez-Sanchez, Non-destructive techniques based on eddy current testing. Sens. **11**, 2525–2565 (2011)
31. I. Altpeter, R. Becker, G. Dobmann, R. Kern, W. Theiner, A. Yashan, Robust solutions of inverse problems in electromagnetic non-destructive evaluation. Inverse Prob. **18**, 1907–1921 (2002)
32. R. Becker, *Über die Prüfung auf Fehler in metallischen Werkstoffen und Bauteilen mittels eines zerstörungsfreien Mehrfrequenz-Wirbelstrom-Prüfverfahrens*, Ph.D. Thesis, (Faculty of Mathematics and Natural Sciences, Saarland University, Saarbrücken, 1980), p. 109
33. H.L. Libby, *Introduction to Electromagnetic Nondestructive Test Methods* (Wiley, 1971)

34. A. Sophian, G.Y. Tian, M.B. Fan, Pulsed eddy current non-destructive testing and evaluation: a review. Chin. J. Mech. Eng. **30**, 500–514 (2017)
35. G.Y. Tian, Y.Z. He, I. Adewale, A. Simm, Research on spectral response of pulsed eddy current and NDE applications. Sens. Actuators A **189**, 313–320 (2013)
36. C.V. Dodd, W.E. Deeds, Analytical solutions to eddy-current probe-coil problems. J. Appl. Phys. **39**, 2829–2838 (1968)
37. D. Mackintosh, in *Electromagnetic Testing*, ed. by L. Mordfin. Handbook of Reference Data for Nondestructive Testing, (ASTM International, 2002), pp. 90–95
38. J. Fava, M. Ruch, Design, construction and characterisation of ECT sensors with rectangular planar coils. Insight **46**, 268–274 (2004)
39. T. Uchimoto, T. Takagi, S. Konoplyuk, T. Abe, H. Huang, M. Kurosawa, Eddy current evaluation of cast irons for material characterization. J. Magn. Magn. Mater. **258**, 493–496 (2003)
40. S. Konoplyuk, T. Abe, T. Uchimoto, T. Takagi, M. Kurosawa, Characterization of ductile cast iron by eddy current method. NDT E Int. **38**, 623–626 (2005)
41. G. Dobmann, C. König, U. Hoffmann, G. Schneibel, *Development and Qualification of the Eddy-Current Testing Techniques "EC" and "EC+" in Combination with Leeb-Hardness Measurements for Detection and Verification of Hardness Spots on Heavy Steel Plates*. 46th National Conference on Nondestructive Testing (KKBN) (Badania Nieniszczące i Diagnostyka, SIMP Publishing Agenda, Starachowice, Poland, 2017)
42. K.V. Rajkumar et al., Characterization of aging behaviour in M250 grade maraging steel using eddy current non-destructive methodology. Mater. Sci. Eng. A **464**, 233–240 (2007)
43. M.P. Blodgett, P.B. Nagy, Eddy current assessment of near-surface residual stress in shot-peened nickel-base superalloys. J. Nondestr. Eval. **23**, 107–123 (2004)
44. M. Disqué, R. Becker, Multifrequency eddy current inspection of layered aluminium aircraft structures to detect and quantify hidden corrosion. Mater. Corros. Werkstoffe Korrosion **49**, 796–798 (1998)
45. P. Neumeier, Erweiterte Einsatzmöglichkeiten des Magnet-Induktiv-Verfahrens durch modernste Digitaltechnik, in *Jahrestagung der Deutschen Gesellschaft für Zerstörungsfreie Prüfung*, ed. by D. Berlin. DGZFP Berlin, vol. 10 (Lindau, Germany, 1987), pp.193–202
46. P. Neumeier, Rechnergesteuertes Prüfgerät zur zerstörungsfreien Ermittlung von Materialeigenschaften. Qualität und Zuverlässigkeit. **32**, 174–176 (1987)
47. S.C. Chan et al., Nonlinear eddy current technique for characterizing case hardening profiles. IEEE Trans. Magn. **46**, 1821–1824 (2010)
48. S.J. Xie et al., A numerical method for simulation of nonlinear eddy current testing signals based on transient A(r) formulation. Mater. Trans. **54**, 964–968 (2013)
49. K. Grotz, B. Lutz, *Simultaneous Electromagnetic Determination of Various Material Characteristics*, ed. by P. Höller, V. Hauk, G. Dobmann, C.O. Ruud, R.E. Green. Nondestructive Characterization of Materials. (Springer, Saarbrücken, 1989), pp. 565–573
50. R. Becker, C. Rodner, K. Betzold, *Werkstoffsortierung und Bestimmung der Rand- und Einhärtetiefe; Sigma, mu und Impedanzspektren als Prüfgrößen*, ed. by DGZFP. DACH-Jahrestagung, vol. 10 (DGZFP, Berlin, Germany, Lindau, Germany, 1987), pp. 683–693
51. W. Karpen, R. Becker, W. Arnold, Characterization of electric and magnetic material properties with eddy-current measurements. Nondestr. Test. Eval. **15**, 93–107 (1998)
52. T. Sakamoto, in *Multiparameter Eddy Current Technique for Material Sorting of Steel Bars*, ed. by J. Boogaard, G.M. van Dijk. 12th World Conference on Non-Destructive Testing, Vol. II (Elsevier, Amsterdam, The Netherlands, 1989), pp. 1369–1374
53. J. Nehring, *Wirbelstromprüfung - Ein Baustein im Rahmen moderner Qualitätssicherungssyteme*, ed. by DGZFP. Kolloquium Qualitätssicherung durch Werkstoffprüfung, vol. 26 (Deutsche Gesellschaft für zerstörungsfreie Prüfung, Zwickau, Germany, 1991), pp. 249–262
54. N. Mack, J. Nehring, J. Sand, B. Fischer, in *Gerätetechnische Weiterentwicklungen als Reaktion auf den Wandel in der Prüfphilosophie am Beispiel moderner CS-Wirbelstromprüfsysteme*, ed. by DGZFP. Frühjahrstagung der Deutschen Gesellschaft für Zerstörungsfreie Prüfung,

vol. 32 (Deutsche Gesellschaft für zerstörungsfreie Prüfung, Garmisch-Partenkirchen, 1993), pp. 883–898
55. K. Grotz, M. Junger, *Verbesserte Möglichkeiten der automatischen Wirbelstromprüfung durch Rechnertechnik*, ed. by DGZFP. Kolloquium Qualitätssicherung durch Werkstoffprüfung, vol. 26. (Deutsche Gesellschaft für zerstörungsfreie Prüfung, Berlin, Zwickau, 1991), pp. 290–294
56. H. Heuer et al., Review on quality assurance along the CFRP value chain - Non-destructive testing of fabrics, preforms and CFRP by HF radio wave techniques. Compos. B Eng. **77**, 494–501 (2015)
57. G. Mook, R. Lange, O. Koeser, Non-destructive characterisation of carbon-fibre-reinforced plastics by means of eddy-currents. Compos. Sci. Technol. **61**, 865–873 (2001)
58. G. Bardl et al., Automated detection of yarn orientation in 3D-draped carbon fiber fabrics and preforms from eddy current data. Compos. B Eng. **96**, 312–324 (2016)
59. J.A. Oberteuffer, Magnetic separation: review of principles, devices, and applications. IEEE Trans. Magn. **10**, 223–238 (1974)
60. J.D. Ray, J.R. Nagel, D. Cohrs, R.K. Rajamani, Forces on particles in time-varying magnetic fields. Kona Powder Part. J. **35**, 251–257 (2018)
61. Y.R. Smith, J.R. Nagel, R.K. Rajamani, Eddy current separation for recovery of non-ferrous metallic particles: a comprehensive review. Miner. Eng. **133**, 149–159 (2019)
62. N. Menad, S. Guignot, J.A. van Houwelingen, New characterisation method of electrical and electronic equipment wastes (WEEE). Waste Manage. **33**, 706–713 (2013)
63. J.J. Ruan, Z.M. Xu, A new model of repulsive force in eddy current separation for recovering waste toner cartridges. J. Hazard. Mater. **192**, 307–313 (2011)
64. M.D. O'Toole, N. Karimian, A.J. Peyton, Classification of nonferous metals using magnetic induction spectroscopy. IEEE Trans. Industr. Inf. **14**, 3477–3485 (2018)
65. R. Becker, W. Döring, *Ferromagnetismus* (Springer, Berlin, 1939)
66. W. Döring, Mikromagnetismus, in *Ferromagnetismus*. ed. by S. Flügge (Springer, Heidelberg, 1966), pp. 341–406
67. M. Kersten, *Grundlagen der Theorie der ferromagnetischen Hysterese und der Koerzitivkraft* (S. Hirzel, Leipzig, 1944)
68. C. Kittel, Domain theory and the dependence of the coercive forces of fine ferromagnetic powders on particle size. Phys. Rev. **73**, 810–811 (1948)
69. E. Kneller, Theorie der Magnetisierungskurve kleiner Kristalle, in *Ferromagnetismus*. ed. by S. Flügge (Springer, Berlin, 1966), pp.438–544
70. T.L. Manh, J.A.P. Benitez, J.H.E. Hernández, J.M.H. López (eds.), *Barkhausen Noise for Nondestructive Testing and Materials Characterization in Low-Carbon Steels* (Woodhead Publishing-Elsevier, 2020)
71. H. Barkhausen, Two phenomena revealed with the help of new amplifiers. Phys. Z. **29** (1919)
72. A. Nowick, B.S. Berry, *Anelastic Relaxation of Solids* (Academic Press, Ney York, 1972)
73. R. Truell, C. Elbaum, B.B. Chick, *Ultrasonic Methods in Solid State Physics* (Elsevier, New York, 1969)
74. M. Paul, B. Haberer, W. Arnold, Materials characterization at high-temperatures using laser ultrasound. Mater. Sci. Eng. A **168**, 87–92 (1993)
75. M. Kersten, Zur Theorie der Koerzitivkraft. Z. Angew. Phys. **124**, 714–741 (1948)
76. J.M. Makar, B.K. Tanner, The in situ measurement of the effect of plastic deformation on the magnetic properties of steel: part I—hysteresis loops and magnetostriction. J. Magn. Magn. Mater. **184**, 193–208 (1998)
77. W. Reimers, Residual stress analysis in single crystallites, in *Structural and Residual Stress Analysis by Nondestructive Methods*. ed. by V. Hauk (Elsevier, Amsterdam, 1997), pp. 461–494
78. L. Néel, Base d'une nouvelle théorie générale du champ coercitif. Ann. Univ. Grenoble **22**, 299–343 (1946)
79. D.C. Jiles, D.L. Atherton, Theory of ferromagnetic hysteresis. J. Magn. Magn. Mater. **61**, 48–60 (1986)

80. L. Batista, U. Rabe, S. Hirsekorn, Magnetic micro- and nanostructures of unalloyed steels: domain wall interactions with cementite precipitates observed by MFM. NDT E Int. **57**, 58–68 (2013)
81. A. Kumar, K. Laha, T. Jayakumar, K.B.S. Rao, B. Raj, Comprehensive microstructural characterization in modified 9Cr-1Mo ferritic steel by ultrasonic measurements. Metall. Mater. Trans. A **33**, 1617–1626 (2002)
82. J. Anglada-Rivera, L.R. Padovese, J. Capo-Sanchez, Magnetic barkhausen noise and hysteresis loop in commercial carbon steel: influence of applied tensile stress and grain size. J. Magn. Magn. Mater. **231**, 299–306 (2001)
83. C. Boller et al., Electromagnetism as a means for understanding materials mechanics phenomena in magnetic materials. Materialwiss Werkstofftech **42**, 269–278 (2011)
84. C. Gatelier-Rothea, J. Chicois, R. Fougeres, P. Fleischmann, Characterization of pure iron and (130 PPM) carbon-iron binary alloy by barkhausen noise measurements: study of the influence of stress and microstructure. Acta Mater. **46**, 4873–4882 (1998)
85. A. Ktena et al., Barkhausen noise as a microstructure characterization tool. Phys. B **435**, 109–112 (2014)
86. V. Moorthy, S. Vaidyanathan, T. Jayakumar, B. Raj, Microstructural characterization of quenched and tempered 0.2% carbon steel using magnetic Barkhausen noise analysis. J. Magn. Magn. Mater. **171**, 179–189 (1997)
87. R. Ranjan, D.C. Jiles, O. Buck, R.B. Thompson, Grain-size measurement using magnetic and acoustic Barkhausen noise. J. Appl. Phys. **61**, 3199–3201 (1987)
88. S. Vaidyanathan, V. Moorthy, T. Jayakumar, B. Raj, Evaluation of induction hardened case depth through microstructural characterisation using magnetic barkhausen emission technique. Mater. Sci. Technol. **16**, 202–208 (2000)
89. S.S.M. Tavares, J.M. Pardal, L.F. Noris, M.G. Diniz, Microstructural characterization and non-destructive testing and of welded joints of duplex stainless steel in flexible pipes. J. Mater. Res. Technol. **15**, 3399–3408 (2021)
90. D.C. Jiles, Dynamics of domain magnetization and the Barkhausen effect. Czech J. Phys. **50**, 893–924 (2000)
91. V. Moorthy, S. Vaidyanathan, T. Jayakumar, B. Raj, On the influence of tempered microstructures on magnetic Barkhausen emission in ferritic steels. Philos. Mag. A **77**, 1499–1514 (1998)
92. V. Moorthy, S. Vaidyanathan, B. Raj, T. Jayakumar, B.P. Kashyap, Insight into the microstructural characterization of ferritic steels using micromagnetic parameters. Metall. Mater. Trans. A **31**, 1053–1065 (2000)
93. M.J. Sablik, Modeling the effect of grain size and dislocation density on hysteretic magnetic properties in steels. J. Appl. Phys. **89**, 5610–5613 (2001)
94. H. Sakamoto, M. Okada, M. Homma, Theoretical analysis of Barkhausen noise in carbon steels. IEEE Trans. Magn. **23**, 2236–2238 (1987)
95. S. Segalini, M. Putignani, P. Kleine, B. Backelandt, in *Online Detection of Coarse Grains by Barkhausen Noise Methods*, ed. by J. Boogaard, G.M. van Dijk. 12th World Conference on Non-Destructive Testing, Vol. I (Elsevier, Amsterdam, The Netherlands, 1989), pp. 342–347
96. S. Santa-aho, A. Laitinen, A. Sorsa, M. Vippola, Barkhausen noise probes and modelling: a review. J. Nondestruct. Eval. **38** (2019)
97. A. Mitra, P.K. Srivastava, P.K. De, D.K. Bhattacharya, D.C. Jiles, Ferromagnetic properties of deformation-induced martensite transformation in AISI 304 stainless steel. Metall. Mater. Trans. A **35A**, 599–605 (2004)
98. A. Mitra, Z.J. Chen, F. Laabs, D.C. Jiles, Micromagnetic Barkhausen emissions in 2.25 wt% Cr—1wt% Mo steel subjected to creep. Philos. Mag. A **75**, 847–859 (1997)
99. I. Komine, K. Nishifuji, Nondestructive Measurement of Mechanical Properties of Steel Plates, in *Nondestructive Characterization of Materials*. ed. by P. Höller, V. Hauk, G. Dobmann, C. Ruud, R.E. Green (Springer, Saarbrücken, Germany, 1989), pp.557–564
100. K. Berner, *Stand der zerstörungsfreien Mess- und Prüftechnik zur on-line Qualitätssicherung von Flachprodukten im Hüttenwerk*, ed. by DGZFP. Kolloquium für Qualitätssicherung durch

Werkstoffprüfung, vol. 26 (Deutsche Gesellschaft für zerstörungsfreie Prüfung, Zwickau, Germany, 1991), pp. 70–79

101. J.F. Bussière, On-line measurement of the microstructure and mechanical properties of steel. Mater. Eval. **44**, 560–567 (1986)
102. J.F. Shackelford, *Introduction to Material Science for Engineers* (Pearson Education, Upper Saddle River, NJ, USA, 2005)
103. S. Kalpakjian, R.S. Schmid, S. Werner, *Werkstofftechnik* (*Manufacturing Processes for Engineering Materials*) (Pearson Education Germany, Munich, 2011)
104. I. Altpeter, W.A. Theiner, *Bestimmung der Einsatz- und Randhärtetiefe mit mikromagnetischen Verfahren*, ed. by GZFP. DACH-Jahrestagung, vol. 10 (Deutsche Gesellschaft für zerstörungsfreie Prüfung, Lindau, Germany, 1987), pp. 836–843
105. R. Kern, W.A. Theiner, P. Schaaf, U. Gonser, *Comparative Micromagnetic and Mössbauer Spectroscopic Depth Profile Analysis of Laserhardened Steel* $X_{210}Cr_{12}$, ed. by P. Höller, V. Hauk, G. Dobmann, C. Ruud, R.E. Green. Nondestructive Characterization of Materials (Springer, 1989) pp. 598–605
106. S.A. Fedosov, Laser beam hardening of carbon and low alloyed steels: discussion of increased quantity of retained austenite. J. Mater. Sci. **34**, 4259–4264 (1999)
107. G. Dobmann, M. Kröning, *Integrierte automatische zerstörungsfreie Prüfung als Werkzeug der Qualitätssicherung in der Prozess- und Qualitätssteuerung*, ed. by DGZFP. Kolloquium Qualitätssicherung durch Werkstoffprüfung, vol. 26 (Deutsche Gesellschaft für zerstörungsfreie Prüfung, Zwickau, Germany, 1991), pp. 108–118
108. T.D. Koblé, A.K. Sengupta, G. Dobmann, M. Kröning, *Steel Grading by Micromagnetiques Techniques*, ed. by T. Kishi, T. Saito, C. Ruud, R. Green. 5th International Symposium on Nondestructive Characterization of Materials (Iketani Science and Technology Foundation, 1991), pp. 615–621
109. W. Theiner, Micromagnetiques Techniques, in *Structural and Residual Stress Analysis by Nondestructive Methods*. ed. by V. Hauk (Elsevier, Amsterdam, 1997), pp. 564–589
110. W. Theiner, R. Kern, M. Stroh, in *Process-Integrated Nondestructive Testing of Ground and Case Hardened Parts*, ed. by AEND. 8th European Conference on Nondestructive Testing. Spanish Society for NDT (AEND), Barcelona, Spain, pp. Code MC-4.1.
111. I. Altpeter, G. Dobmann, M. Kröning, M. Rabung, S. Szielasko, Micro-magnetic evaluation of micro residual stresses of the IInd and IIIrd order. NDT E Int. **42**, 283–290 (2009)
112. A. Sorsa, K. Leiviska, S. Santa-aho, T. Lepisto, Quantitative prediction of residual stress and hardness in case-hardened steel based on the Barkhausen noise measurement. NDT E Int. **46**, 100–106 (2012)
113. O. Kypris, I.C. Nlebedim, D.C. Jiles, Barkhausen spectroscopy: non-destructive characterization of magnetic materials as a function of depth. J. Appl. Phys. **115** (2014)
114. O. Kypris, I.C. Nlebedim, D.C. Jiles, A new method for obtaining stress-depth calibration profiles for non-destructive evaluation using a frequency-dependent model of Barkhausen emissions. IEEE Trans. Magn. **49**, 3893–3896 (2013)
115. O. Kypris, I.C. Nlebedim, D.C. Jiles, Measuring stress variation with depth using Barkhausen signals. J. Magn. Magn. Mater. **407**, 377–395 (2016)
116. S. Ding, G.Y. Tian, G. Dobmann, P. Wang, Analysis of domain wall dynamics based on skewness of magnetic Barkhausen noise for applied stress determination. J. Magn. Magn. Mater. **421**, 225–229 (2017)
117. M.K. Devine, D.C. Jiles, A.R. Eichmann, D.A. Kaminski, S. Hardwick, Magnescope: applications in nondestructive evaluation. J. Appl. Phys. **73**, 5617–5619 (1993)
118. A.R. Eichmann, M.K. Devine, D.C. Jiles, D.A. Kaminski, New procedures for in-situ measurement of the magnetic properties of materials - application of the magnescope. IEEE Trans. Magn. **28**, 2462–2464 (1992)
119. P. Deimel, D. Kuppler, K. Herz, W.A. Theiner, Bloch wall arrangement and Barkhausen noise in steels 22 NiMoCr 3 7 and 15 MnMoNiV 5 3. J. Magn. Magn. Mater. **36**, 277–289 (1983)
120. W.A. Theiner, P. Deimel, Nondestructive testing of welds with the 3MA analyzer. Nucl. Eng. Des. **102**, 257–264 (1987)

121. O. Ayère, *Analyse von Mikrogefügezuständen ausscheidungshärtender Modelllegierungen mit transmissionselektronenmikroskopischen Verfahren und mikromagnetischen zerstörungsfreien Messgrößen*, Ph.D. Thesis, (Faculty of Mathematics and Natural Sciences, Saarland University, Saarbrücken, Germany, 1987)
122. P. Höller, in *Nondestructive Analysis of Structure and Stresses by Ultrasonic and Micromagnetic Methods*, ed. by J.F. Bussière, J.P. Monchalin, C.O. Ruud, R.E. Green. Nondestructive Characterization of Materials II (Springer, 1987) pp. 211–212
123. G. Dobmann, M. Kröning, W. Theiner, H. Willems, U. Fiedler, Non-destructive characterization of materials (ultrasonic and micromagnetic techniques) for strength and toughness prediction and the detection of early creep damage. Nucl. Eng. Des. **157**, 137–158 (1995)
124. B. Wolter, Y. Gabi, C. Conrad, Nondestructive testing with 3MA - an overview of principles and applications. Appl. Sci. Basel **9**, 29 (2019)
125. Y. Gabi, K. Jacob, B. Wolter, C. Conrad, B. Strass, J. Grimm, Analysis of incremental and differential permeability in NDT via 3D-simulation and experiment. J. Magn. Magn. Mater. **505**, 166695-1–166695-6 (2020)
126. Y. Gabi, O. Martins, B. Wolter, B. Strass, Combination of electromagnetic measurements and FEM simulations for nondestructive determination of mechanical hardness. AIP Adv. **8**, 047502-1–047502-8 (2018)
127. Y. Gabi, B. Wolter, A. Gerbershagen, M. Ewen, P. Braun, O. Martins, FEM Simulations of incremental permeability signals of a multi-layer steel with consideration of the hysteretic behavior of each layer. IEEE Trans. Magn. **50**, 7026504-1–7026504-4 (2014)
128. M. Ewen et al., *Nondestructive Determination of Mechanical Properties of Open-Die Forgings and Potentials for Full Implementation in Production Process Chain*, ed. by Z. Prevorovsky. 11th European Conference on Nondestructive Testing (ECNDT), Brno (2014)
129. B. Wolter et al., in *Detection and Quantification of Grinding Damage by Using EC and 3MA Techniques*, ed. by G. Donzella. 4th International Conference on Barkhausen Noise and Micromagnetic Testing (Stresstech Oy, Brescia, Italy, 2003), pp. 159–170
130. M. Borsutzki, G. Dobmann, W. Theiner, *Online ND-Characterization and Mechanical Property Determination of Cold Rolled Steel Strips*, ed. by B. Brusey, J.P. Bussière, M. Dubois, A. Moreau. Annual Conference of Metallurgists of CIM - Gateway to the 21th Century (Canadian Institute of Mining, Metallurgy and Petroleum, Quebec, 1999), pp. 77–84
131. M. Borsutzki, J. Kroos, W. Reimche, E. Schneider, Magnetische und akustische Verfahren zur Materialcharakterisierung von Stahlblechen. Stahl Eisen **120**, 115–121 (2000)
132. V. Vengrinovich, V. Tsukerman, in *Stress and Texture Measurement Using Barkhausen Noise and Angular Scanning of Driving Magnetic Field*, ed. by J.P. Bussière. 16th World Conference on Non-destructive Testing (WCNDT), Montreal, Canada (2004)
133. J.M.D. Coey, Louis Neel: Retrospective (invited). J. Appl. Phys. 93, 8224–8229 (2003)

Index

W. Arnold et al., *Non-destructive Materials Characterization and Evaluation*, Springer Series in Materials Science 329,
https://doi.org/10.1007/978-3-662-66489-6